CBS PROBLEMS and SOLUTIONS *Series*

Problems and Solutions in CONTROL SYSTEMS

✳ This book is suitable for examinations of all Indian universities.

✳ Solutions of **315 problems** are given in most simplified manner.

✳ All types of problems are included and the solutions given in detail.

✳ To explain the basics lucidly, computerised figures have been liberally added in the solutions. This volume contains more than **380 line diagrams**.

CBS Problems and Solutions *Series*

📖 Engineering Thermodynamics
📖 Strength of Materials
📖 Integrated Electronics
📖 Communication Systems
📖 Signals and Systems
📖 Power Systems
📖 Electrical Machines and Transformers
📖 Network Analysis
📖 Power Electronics
📖 Control Systems
📖 Analog Systems

plus many ... more new topics

CBS PROBLEMS and SOLUTIONS *Series*

Problems and Solutions in
CONTROL SYSTEMS

S. K. PRASAD
Engineer

C B S

CBS Publishers & Distributors Pvt. Ltd.

New Delhi • Bengaluru • Chennai • Kochi • Kolkata • Mumbai
Hyderabad • Uttarakhand • Nagpur • Patna • Pune • Jharkhand

ISBN: 81-239-1201-3

First Edition: 2005
Reprint: 2009, 2013, 2019

Published by **Satish Kumar Jain** and produced by **Varun Jain** for
CBS Publishers & Distributors Pvt. Ltd.,
4819/XI Prahlad Street, 24 Ansari Road, Daryaganj, New Delhi - 110002
delhi@cbspd.com, cbspubs@airtelmail.in • www.cbspd.com
Ph.: 23289259, 23266861, 23266867 • Fax: 011-23243014

Corporate Office: 204 FIE, Industrial Area, Patparganj, Delhi - 110 092
Ph: 49344934 • Fax: 011-49344935
E-mail: publishing@cbspd.com • publicity@cbspd.com

Branches:
- *Bengaluru:* 2975, 17th Cross, K.R. Road, Bansankari 2nd Stage, Bengaluru - 70 • Ph: +91-80-26771678/79 • Fax: +91-80-26771680
  E-mail: cbsbng@gmail.com, bangalore@cbspd.com
- *Chennai:* No. 7, Subbaraya Street, Shenoy Nagar, Chennai - 600030
  Ph: +91-44-26681266, 26680620 • Fax: +91-44-42032115
  E-mail: chennai@cbspd.com
- *Kochi:* Ashana House, 39/1904, A.M. Thomas Road, Valanjambalam, Ernakulum, Kochi • Ph: +91-484-4059061-65
  Fax: +91-484-4059065 • E-mail: cochin@cbspd.com
- *Kolkata:* 6-B, Ground Floor, Rameshwar Shaw Road, Kolkata - 700014
  Ph: +91-33-22891126/7/8 • E-mail: kolkata@cbspd.com
- *Mumbai:* 83-C, Dr. E. Moses Road, Worli, Mumbai - 400018
  Ph: +91-9833017933, 022-24902340/41 • E-mail: mumbai@cbspd.com

Representatives:

- Hyderabad: 0-9885175004
- Patna: 0-9334159340
- Jharkhand: 0-9811541605
- Nagpur: 0-9021734563
- Pune: 0-9623451994
- Uttarakhand: 0-9716462459

Printed at:
J.S. Offset Printers, Delhi (India)

Preface

Control Systems is one of the essential subjects for the students of electronics, electronics and telecommunication, electrical engineering and all the related fields. The author feels it a great pleasure in presenting this very complex subject through typical problems and their solutions.

This book covers all important and major topics taught at the undergraduate level in an engineering college. All the 310 problems have been discussed at length, and more than 380 detailed line diagrams given extensively to help the reader understand the basics and grasp the logic. This book is suitable for preparing for all university examinations and also to face the competitions. This feature is expected to help the reader in visualising the patterns of problems and the extent and level of difficulty in solving the questions likely to be encountered in the examinations, and thus tackling the problems with confidence and success.

Although every care has been taken to ensure accuracy, yet some errors might have crept in. The author will be grateful if these are brought to his notice so that these could be rectified in the subsequent printings and editions of the book. Readers are requested to send their suggestions for further improvement and revision to me through the publishers at the following e-mail address: cbspubs@del3.vsnl.net.in.

S. K. PRASAD

Contents

CHAPTER 1

Concepts of Block Diagram

Problem 1.1 Derive the transfer function of the ckt shown in Fig. P.1.1.

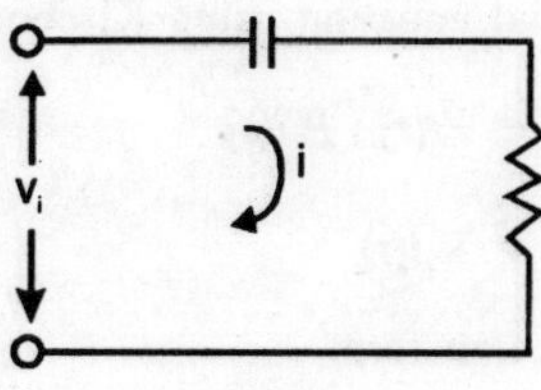

Fig. P. 1.1

Solution:

Applying Kirchoff's law, we get

$$V_i(t) = R_i(t) + \frac{1}{C} \int i(t)\, dt$$

Taking laplace transform, we have,

$$V_i(s) = RI(s) + \frac{1}{sC} I(s)$$

or $\qquad \dfrac{I(s)}{V_i(s)} = \left(\dfrac{sC}{1+sCR} \right) = (T(s))$

Problem 1.2. Find the transfer function of the electrical Network shown in Fig. P.1.2.

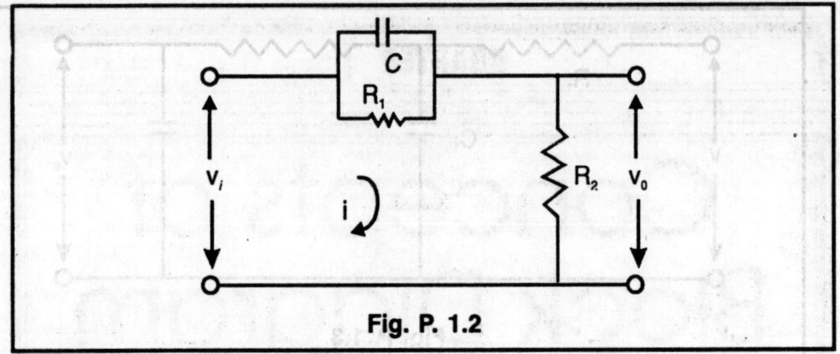

Fig. P. 1.2

Solution:

Equivalent impedance of the parallel combination of R_1 and C, is say Z

$$\therefore \qquad Z = \left(\frac{R_1}{1+sCR_1}\right)$$

Now writing differential equation using Kirchoff's law, we get

$$V_i(t) = Zi(t) + R_2 i(t)$$

and

$$V_0(t) = R_2 i(t)$$

Again, taking Laplace transform

$$V_i(s) = Z_1 I(s) + R_2 I(s) \tag{1}$$

and

$$V_0(s) = R_2 I(s) \quad \left(I(s) = \frac{V_0(s)}{R_2}\right) \tag{2}$$

From equations (1) and (2), we have

$$\text{or} \qquad V_i(s) = Z\frac{V_0(s)}{R_2} + V_0(s) = V_0(s) = \left(\frac{1}{R_2}Z + 1\right)$$

$$\frac{V_o(s)}{V_i(s)} = \frac{1}{\left(\dfrac{Z}{R_2}+1\right)} = \left(\frac{R_2(1+sCR_1)}{R_1+R_2+sCR_1R_2}\right)$$

Problem 1.3.   Find the transfer function of the electrical network shown in Fig. P.1.3.

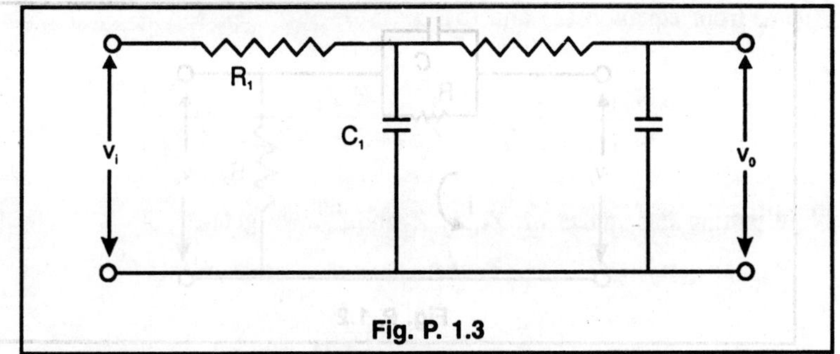

Fig. P. 1.3

Solution:

Redrawing the above circuit by assuming

$$R_1 = Z_1, \frac{1}{sC_1} = Z_i, R_2 = Z_3 \text{ and } \frac{1}{sC_2} = Z_4$$

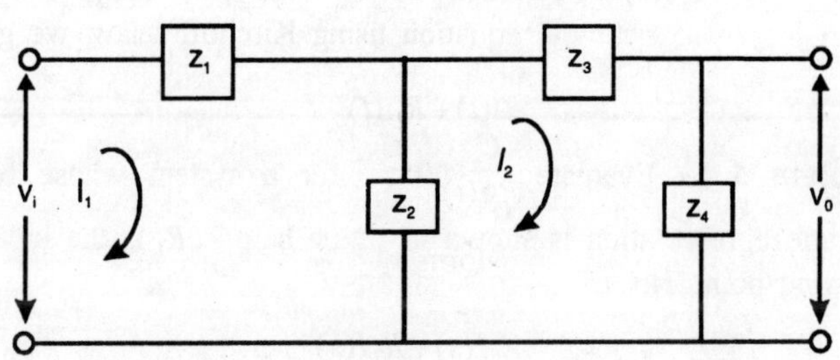

Fig. P. 1.3 (a)

By using Kirchoff's law, writing the differential equations

$$V_i(s) = (Z_1 + Z_2)I_1 - Z_2 I_2(s) \tag{1}$$

$$0 = Z_3 I_2(s) + Z_4 I_2(s) + Z_2 I_2(s) - Z_2 I_1(s) \tag{2}$$

and

$$V_0(s) = Z_4 I_2(s) \tag{3}$$

Now, from equation (2)

$$I_1(s) = \left[\frac{(Z_2 + Z_3 + Z_4)I_2(s)}{Z_2} \right] \tag{4}$$

Now from equations (1) and (4)

$$\frac{V_i(s)}{I_2(s)} = \left[\frac{(Z_1 + Z_2)(Z_3 + Z_4 + Z_2) - Z_2^2}{Z_2} \right] \tag{5}$$

Again, from equation (3) and (5)

$$\frac{V_0(s)}{V_i(s)} = \left[\frac{Z_2 Z_4}{Z_1 Z_2 + Z_1 Z_3 + Z_1 Z_4 + Z_2 Z_3 + Z_2 Z_4}\right]$$

Now putting the values of Z_1, Z_2, Z_3 and Z_4, we get

$$\frac{V_0(s)}{V_i(s)} = \left(\frac{\dfrac{1}{s^2 C_1 C_2}}{\dfrac{R_1}{sC_1} + R_1 R_2 + \dfrac{R_1}{sC_2} + \dfrac{R_2}{sC_1} + \dfrac{1}{s^2 C_1 C_2}}\right)$$

$$\frac{V_0(s)}{V_i(s)} = \frac{1}{1 + s(R_1 C_1 + R_2 C_2 + R_1 C_2) + s^2 R_1 R_2 C_1 C_2}$$

Problem 1.4. Evaluate $\dfrac{C}{R_1}$ and $\dfrac{C}{R_2}$ for a system whose block diagram representation is shown in figure below: R_1 is the input to summing point No. 1.

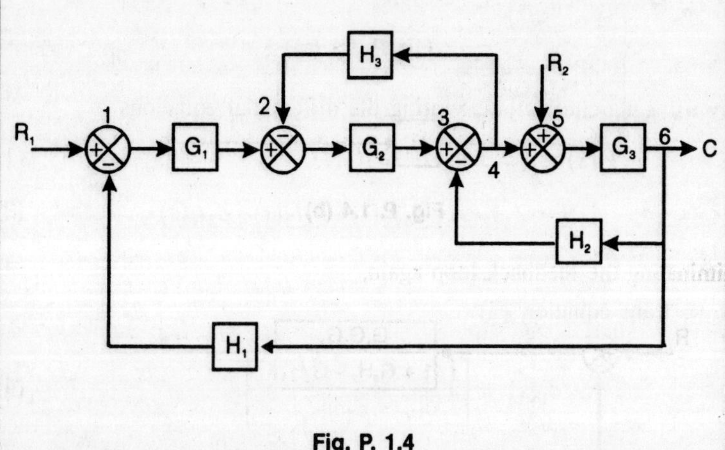

Fig. P. 1.4

Solution:

For finding C/R_1;

Assume, $R_2 = 0$, hence the summing pt. 5 can be removed. Shifting take off point 4 beyond block G_3.

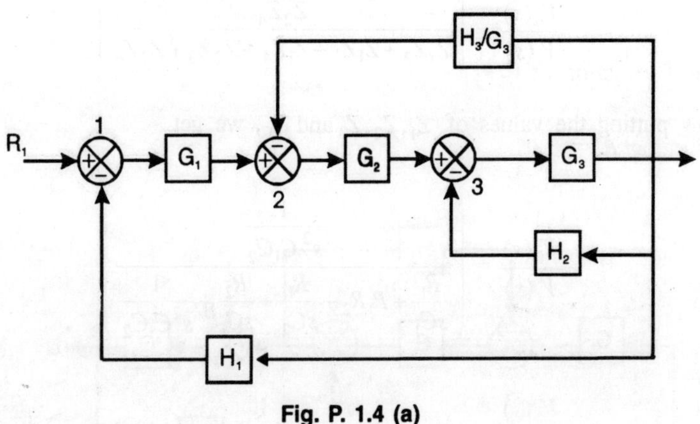

Fig. P. 1.4 (a)

Eliminating the internal feedback loop,

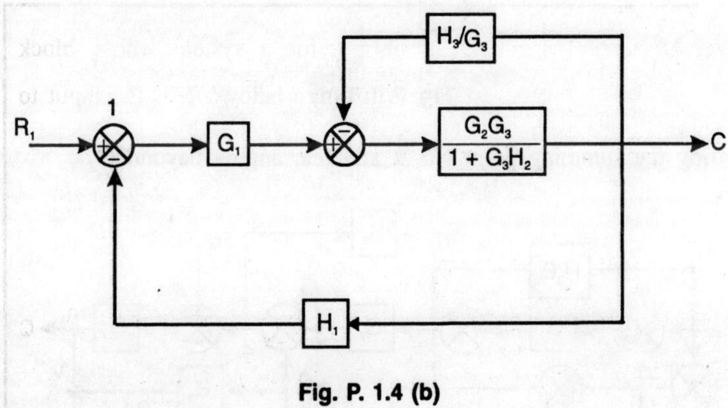

Fig. P. 1.4 (b)

Eliminating the feedback loop again,

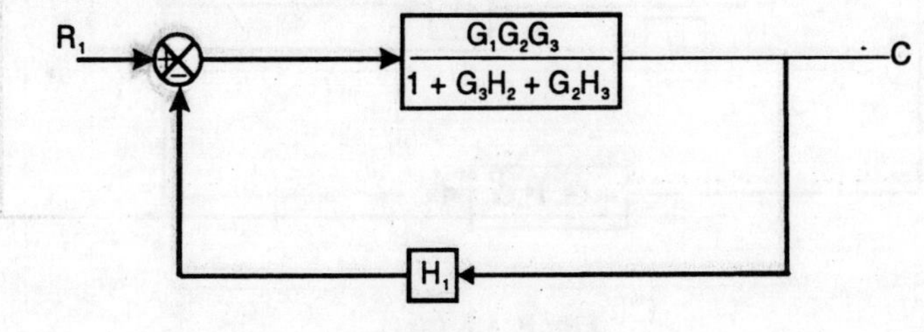

Fig. P. 1.4 (c)

$$\therefore \qquad \frac{C}{R_1} = \left(\frac{G_1 G_2 G_3}{1 + G_3 H_2 + H_3 G_2 + G_1 G_2 G_3 H_1} \right) \textbf{ Ans.}$$

Now Evaluation of C/R_2

Assuming $R_1 = 0$

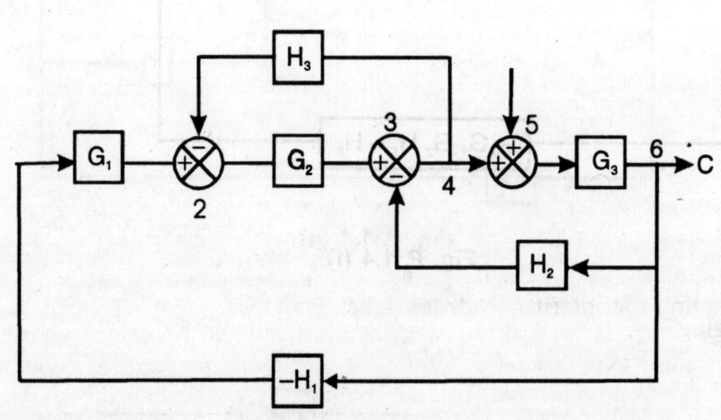

Fig. P. 1.4 (d)

Shifting the summing point No. 2 and rearranging beyond G_2

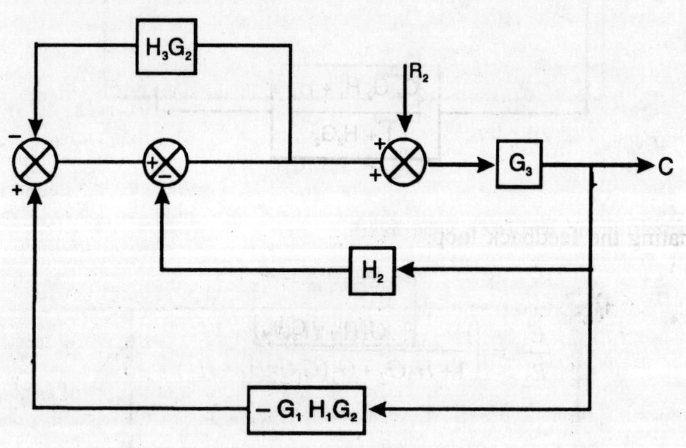

Fig. P. 1.4 (e)

Now rearranging and eliminating the feedback loop

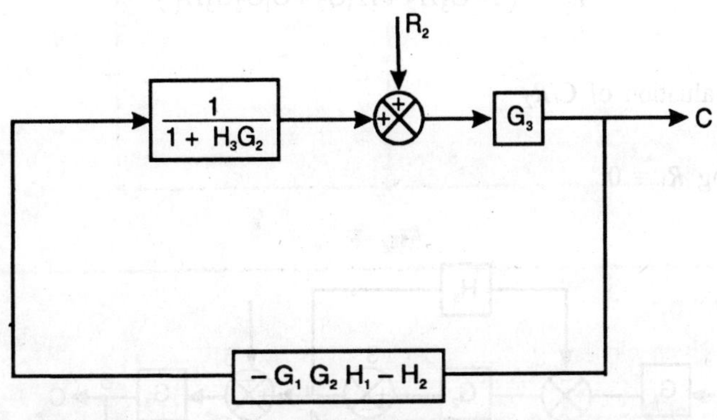

Fig. P. 1.4 (f)

Rearranging

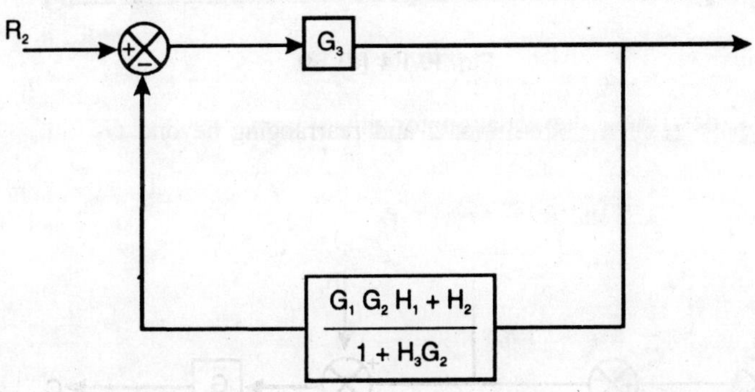

Eliminating the feedback loop,

$$\frac{C}{R_2} = \left(\frac{G_3(1+H_3G_2)}{1+H_3G_2 + G_3(G_1G_2H_1 + H_2)} \right) \quad \textbf{Ans.}$$

Problem 1.5. Find the transfer function of the ckt as shown in Fig.P.1.5

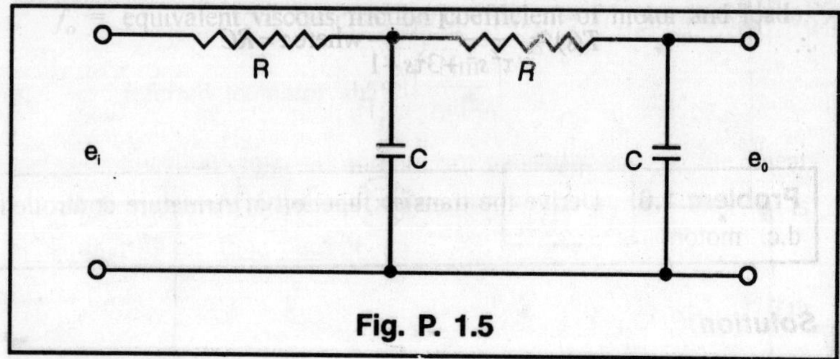

Fig. P. 1.5

Solution:

The given ckt is,

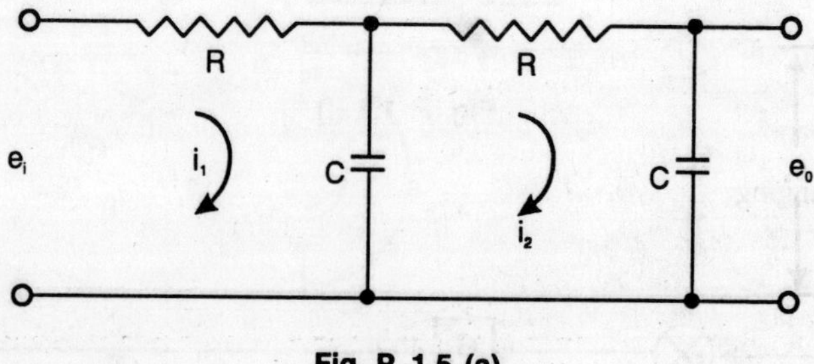

Fig. P. 1.5 (a)

Now, describing the equation for this system,

$$\frac{1}{C}\int_{-\infty}^{t}(i_1-i_2)\,dt + Ri_1 = e_i \tag{1}$$

and

$$\frac{1}{C}\int_{-\infty}^{t}(i_2-i_1)\,dt + Ri_2 = \frac{1}{C}\int_{-\infty}^{t}i_2\,dt = -e_0 \tag{2}$$

Taking laplace transform,

$$\frac{1}{sC}[I_1(s)-I_2(s)] + RI_1(s) = E_i(s)$$

and

$$\frac{1}{sC}[I_2(s)-I_1(s)] + RI_2(s) = -1/sC\,I_2(s) = -E_0(s)$$

Now, eliminating $I_1(s)\,\&\,I_2(s)$ from these two equations,

$$\frac{E_o(s)}{E_i(s)} = T(s) = \frac{1}{(RC)^2 s^2 + 3RCs + 1}$$

$$\therefore \quad T(s) = \dfrac{1}{\tau^2 s^2 + 3\tau s + 1} \quad \text{where } \tau = RC$$

Problem 1.6. Derive the transfer function of Armature controlled d.c. motor.

Solution:

Armature Control

Consider the armature-controlled d.c. motor shown in Fig. P.1.6.

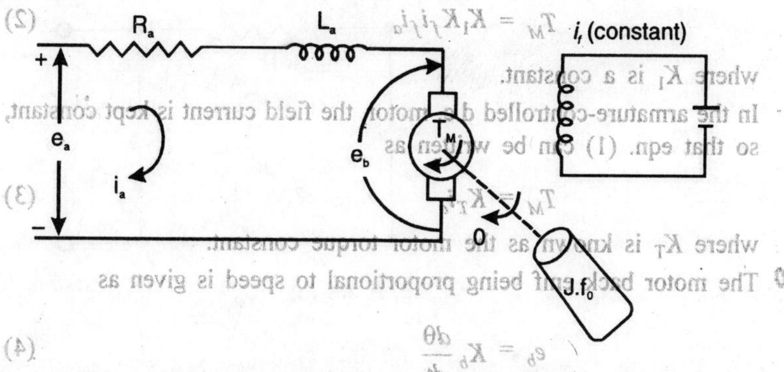

Fig. P. 1.6 Armature-controlled d.c. motor

In this system,

R_a = resistance of armature (Ω)

L_a = inductance of armature winding (H).

i_a = armature current (A).

i_f = field current (A).

e_a = applied armature voltage (V).

e_b = back emf (volts).

T_M = torque developed by motor (Nm).

θ = angular displacement or motor-shaft (rad).

J = equivalent moment of inertia of motor and load referred to motor shaft (kg-m^2)

$$f_o = \text{equivalent viscous friction coefficient of motor and load}$$

$$\text{referred to motor shaft} \left(\frac{\text{Nm}}{\text{rad/s}} \right)$$

In servo applications, the d.c. motors are generally used in the linear range of the magnetization curve. Therefore, the air gap flux ϕ is proportional of the field current, i.e.,

$$\phi = K_f i_f \tag{1}$$

where K_f is a constant.

The torque T_M developed by the motor is proportional to the product of the armature current and air gap flux, i.e.,

$$T_M = K_1 K_f i_f i_a \tag{2}$$

where K_1 is a constant.

In the armature-controlled d.c. motor, the field current is kept constant, so that eqn. (1) can be written as

$$T_M = K_T i_a \tag{3}$$

where K_T is known as the motor torque constant.

The motor back emf being proportional to speed is given as

$$e_b = K_b \frac{d\theta}{dt} \tag{4}$$

where K_b is the back emf constant.

The differential equation of the armature circuit is

$$L_a \frac{di_a}{dt} + R_a i_a + e_b = e_a \tag{5}$$

The torque equation is $\quad j\dfrac{d^2\theta}{dt^2} + f_0 \dfrac{d\theta}{dt} = T_M = K_T i_a \tag{6}$

Taking the Laplace transforms of eqns. (3) to (5), assuming zero initial conditions, we get,

$$E_b(s) = K_b s\theta(s) \tag{7}$$

$$(L_a s + R_a) I_a(s) = E_a(s) - E_b(s) \tag{8}$$

$$(Js^2 + f_0 s)\theta(s) = T_M(s) = K_T I_a(s) \tag{9}$$

From eqns. (6) to (8), the transfer function of the system is obtained as

$$G(s) = \frac{\theta(s)}{E_a(s)} = \frac{K_T}{s[(R_a + sL_a)(Js + f_o) + K_T K_b]}$$

Problem 1.7. Derive the transfer function of field controlled d.c. motor.

Solution:

Field-control

A field-controlled d.c. motor is shown in Fig. P.1.7.

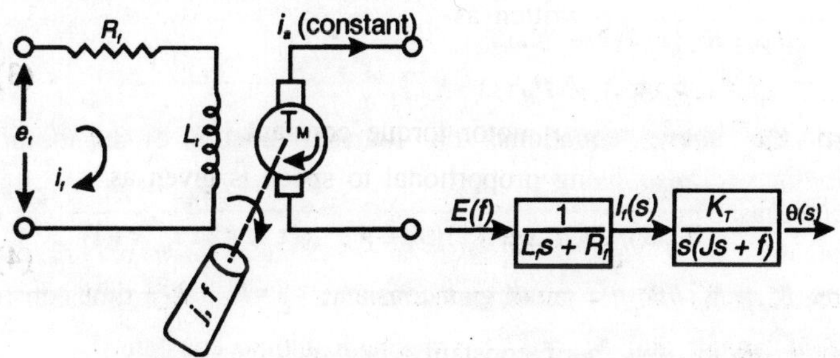

Fig. P.1.7 (a) Field-controlled d.c. motor
(b) Block diagram of field-controlled motor

In this system,

R_f = field winding resistance (Ω).

L_f = field winding inductance (H).

e = field control voltage (V).

i_f = field current (A).

T_M = torque development by motor (Nm).

J = equivalent moment of inertia of motor and load referred to motor shaft (kg-m^2).

$$f = \text{equivalent viscous friction coefficient of motor and load}$$

referred to motor shaft $\dfrac{\text{Nm}}{\text{rad/s}}$.

θ = angular displacement of motor shaft (rad.)

In the field-controlled motor, the armature current is fed from a constant current source. Therefore,

$$T_M = K_1 K_f i_f i_a = K_T' i_f$$

where K_T' is a constant.

The equation for the field circuit is $L_f \dfrac{di_f}{dt} + R_f i_f = e_f$ (1)

The torque equation is $J \dfrac{d^2\theta}{dt^2} + f \dfrac{d\theta}{dt} = T_M = K_T' i_f$ (2)

Taking the Laplace transform of Previous Problem of eqns. (3) and (4) of armature controlled, assuming zero initial conditions, we get

$$\left(L_j s + R_f\right) I_f(s) = E_f(s) \tag{3}$$

$$(Js^2 + fs)\,\theta(s) = T_M(s) = K_T' I_f(s) \tag{4}$$

From the above equations, the transfer function of the motor is

obtained as $\dfrac{\theta(s)}{E_f(s)} = \dfrac{K_T'}{s(L_f s + R_f)(Js + f)} = \dfrac{K_m}{s(\tau_f s + 1)(\tau_{me} s + 1)}$ (6)

where $K_m = K_T' / R_f f$ = motor gain constant; $\tau_f = L_f / R_f$ = time constant of field circuit; and $\tau_{me} = J / f$; mechanical time constant.

Problem 1.8. The figure below shows the block diagram of a two system. Find the total output *C(s)*.

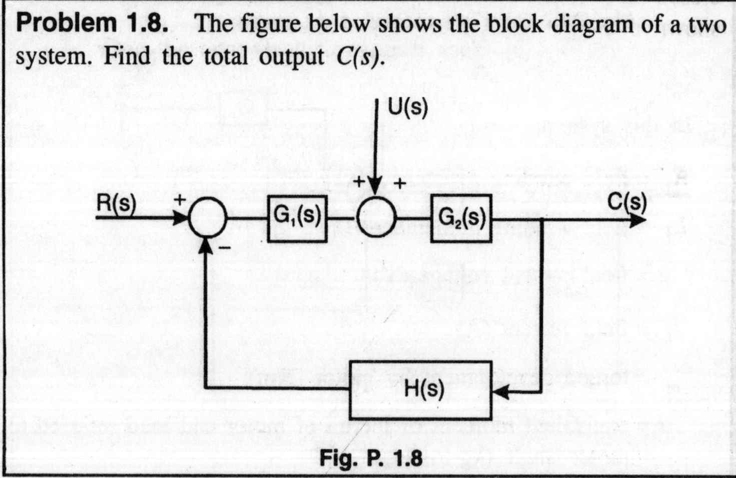

Fig. P. 1.8

Solution:

The output net is determined by one input taken only at a time.

Output due to *R(s)* acting alone,

$$C_R(s) = \frac{G_1(s).G_2(s)}{1+G_1(s).G_2(s).H(s)}R(s)$$

Now the output due to *U(s)* acting alone,

$$C_U(s) = \frac{G_2(s)}{1+G_1(s).G_2(s).H(s)}U(s)$$

Hence the response due to simultaneous application of *R(s)* & *U(s)* will be,

$$C(s) = C_R(s)+C_U(s)$$

$$= \frac{G_2(s)}{1+G_1(s).G_2(s).H(s)}\left[G_1(s).R(s)+U(s)\right]$$

Problem 1.9.   Using block diagram reduction technique, find the closed loop Transfer function of the system whose block diagram is shown in Fig. P. 1.19.

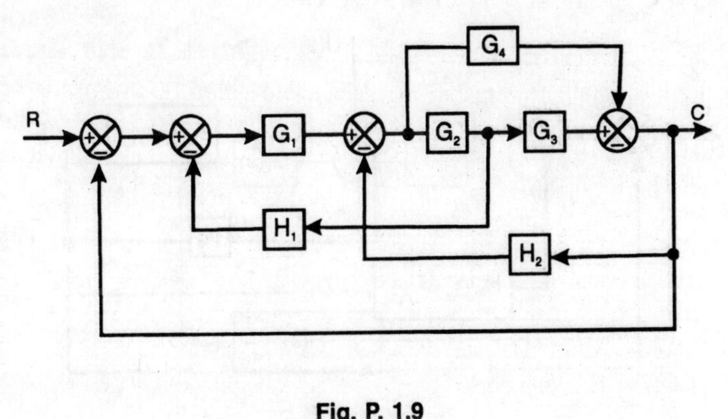

Fig. P. 1.9

Solution:

The given block diagram is reduced in the following steps:

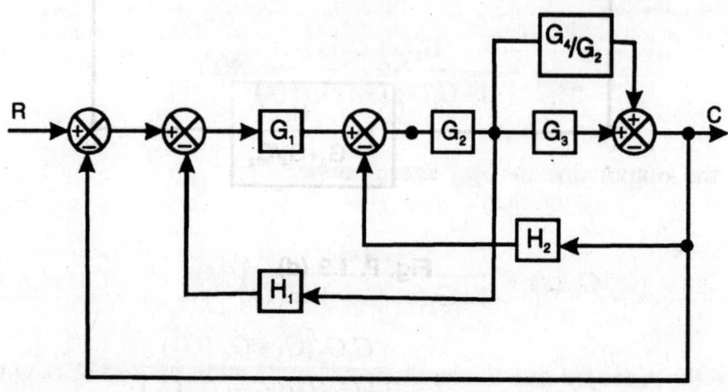

Fig. P. 1.9 (a)

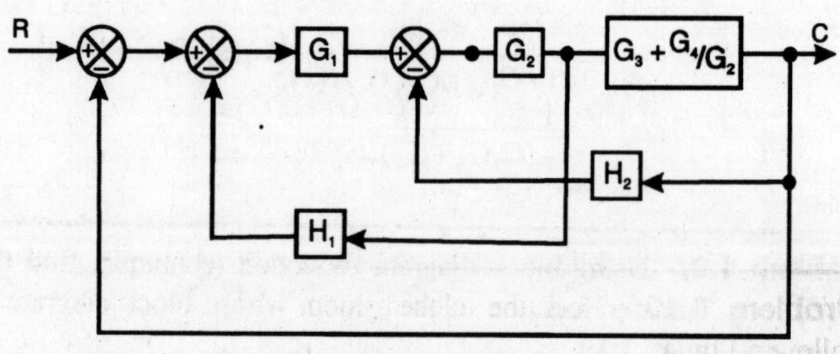

Fig. P. 1.9 (b)

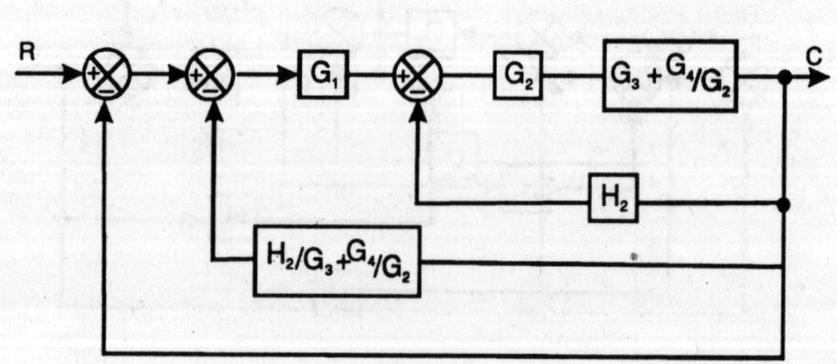

Fig. P. 1.9 (c)

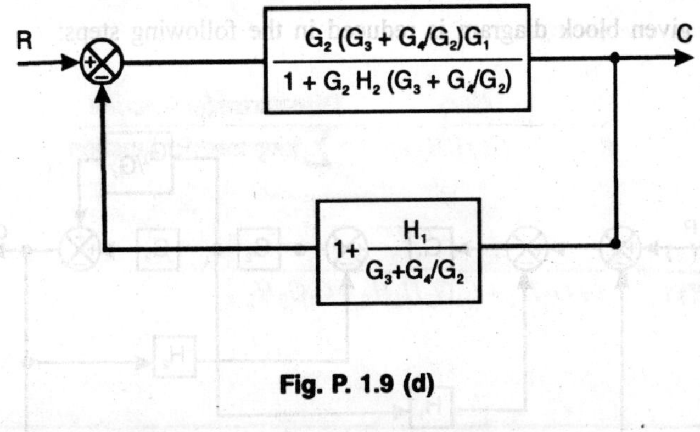

Fig. P. 1.9 (d)

$$\therefore \quad T(s) = \frac{C(s)}{R(s)} = \cfrac{\cfrac{G_1 G_2 (G_3 + G_4/G_2)}{1 + G_2 H_2 (G_3 + G_4/G_2)}}{1 + \cfrac{G_1 G_2 (G_3 + G_4/G_2)}{1 + G_2 H_2 (G_3 + G_4/G_2)} \times \cfrac{G_3 + \dfrac{G_4}{G_2} + H_1}{\left(G_3 + \dfrac{G_4}{G_2}\right)}}$$

$$\therefore \quad T(s) = \frac{G_1(G_2 G_3 + G_4)}{1 + (G_2 G_3 + G_4)(G_1 + H_1) + G_1 G_2 . H_1}$$

Problem 1.10. Find the closed loop transfer function of the following figure

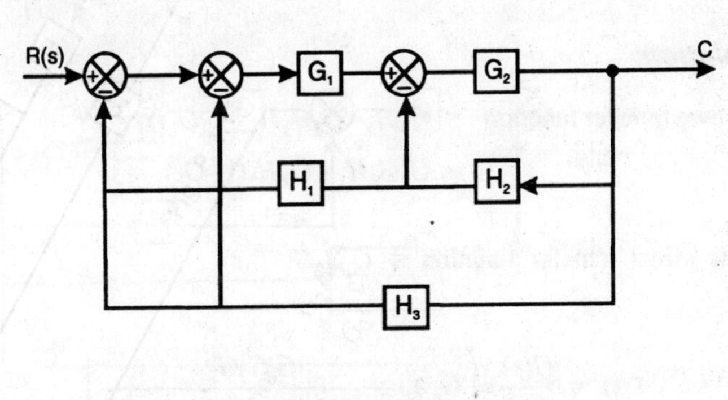

Fig. P. 1.10

Solution:

Applying the formula,

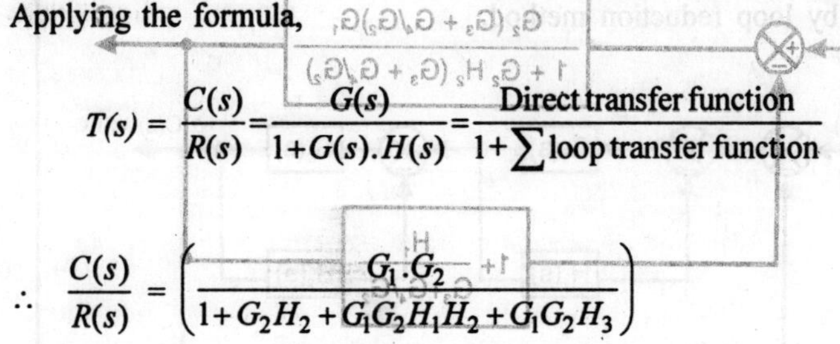

$$T(s) = \frac{C(s)}{R(s)} = \frac{G(s)}{1+G(s).H(s)} = \frac{\text{Direct transfer function}}{1+\sum \text{loop transfer function}}$$

$$\therefore \quad \frac{C(s)}{R(s)} = \left(\frac{G_1 . G_2}{1+G_2 H_2 + G_1 G_2 H_1 H_2 + G_1 G_2 H_3} \right)$$

Problem 1.11. Find the transfer function (overall) of the figure below:

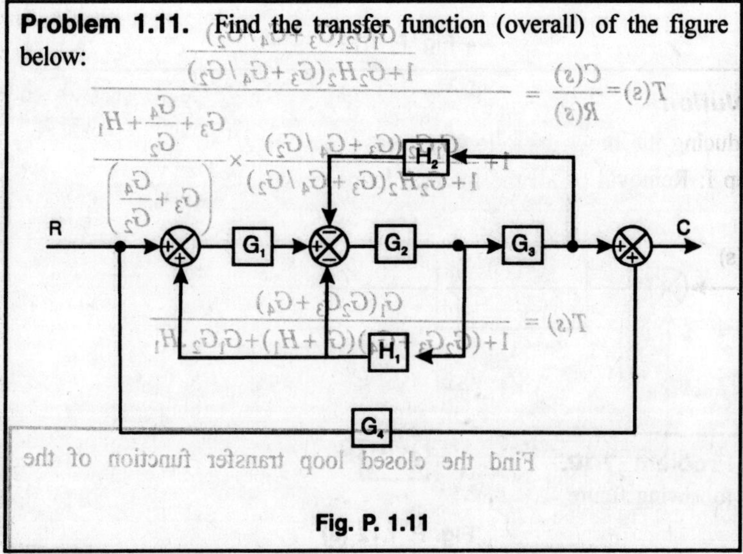

Fig. P. 1.11

Solution:

$$\sum \text{loop transfer function} = G_2^2 H_1 + G_2 G_3 H_1 - G_1 G_2 H_1$$

$$= G_2 G_3 H_1 + G_2 H_1 (1-G_1)$$

Now direct transfer function is $G_1 G_2 G_3$

$$\therefore \quad T(s) = \frac{C(s)}{R(s)} = \left(G_4 + \frac{G_1 G_2 G_3}{1+G_2 G_3 H_2 + G_2 H_1 (1-G_1)} \right)$$

Problem 1.12. Find the overall transfer function of the given block diagram by loop reduction method.

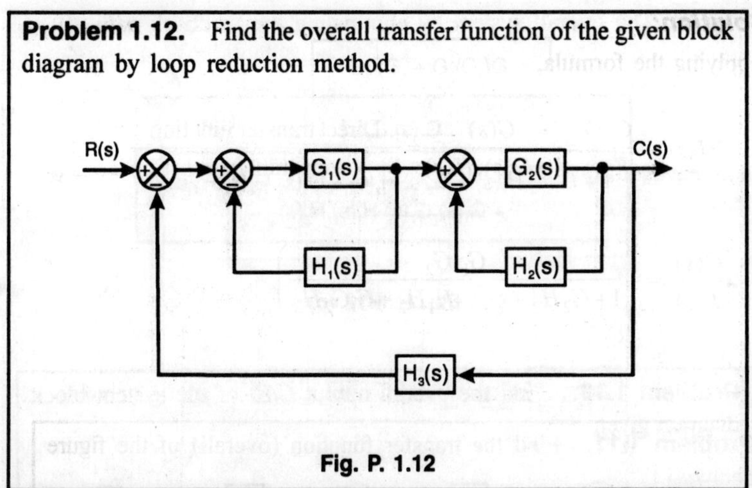

Fig. P. 1.12

Solution:

Reducing the block in following steps:

Step I: Removal of all the internal loops.

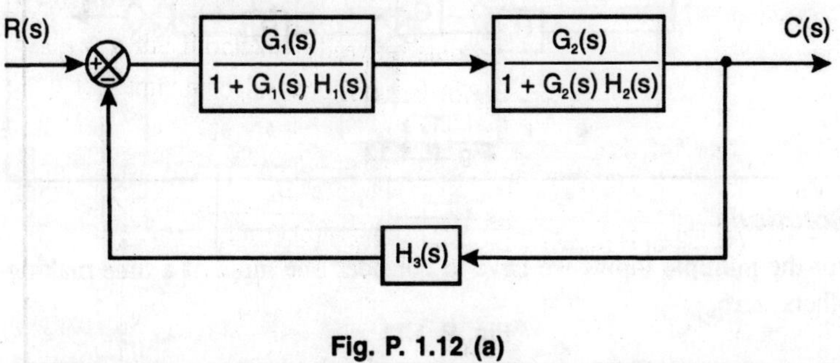

Fig. P. 1.12 (a)

Step II: Cascading the two forward blocks.

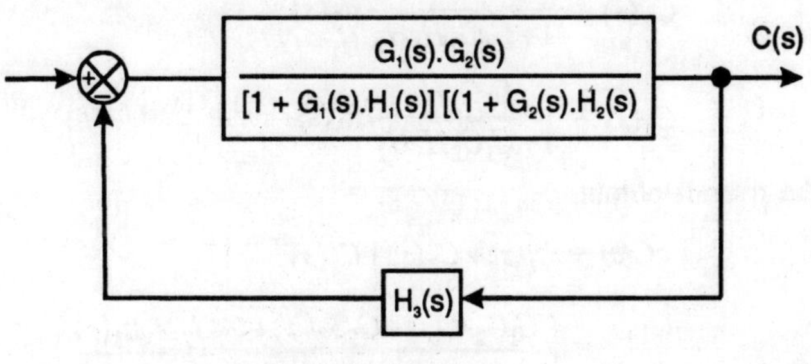

Fig. P. 1.12 (b)

Step III: The overall system by eliminating the feedback path.

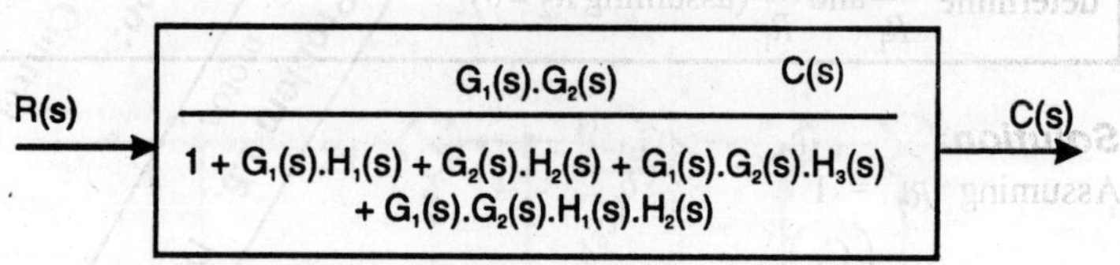

$$\text{R(s)} \longrightarrow \boxed{\dfrac{G_1(s).G_2(s)}{1 + G_1(s).H_1(s) + G_2(s).H_2(s) + G_1(s).G_2(s).H_3(s) + G_1(s).G_2(s).H_1(s).H_2(s)}} \overset{C(s)}{\longrightarrow} C(s)$$

Fig. P. 1.12 (c)

Problem 1.13. Find the overall output $C(S)$ of the system block diagram given

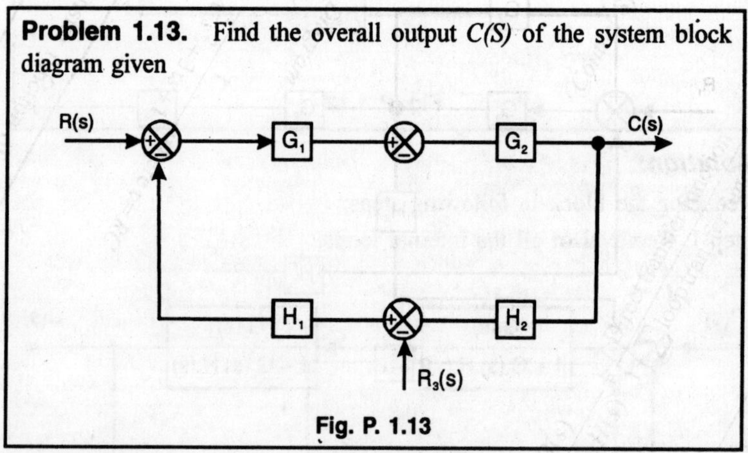

Fig. P. 1.13

Solution:

For the multiple inputs we have to consider one input at a time making others zero.

$$C_1(s) = \frac{G_1 G_2}{1 + G_1 G_2 H_1 H_2} R_1(s)$$

$$C_2(s) = \frac{G_2}{1 + G_1 G_2 H_1 H_2} R_2(s)$$

$$C_3(s) = \frac{G_1 G_2 H_1}{1 + G_1 G_2 H_1 H_2} R_3(s)$$

Now the overall output,

$$C(s) = C_1(s) + C_2(s) + C_3(s)$$

$$= \left(\frac{G_1 G_2 R_1(s) + G_2 R_2(s) + G_1 G_2 H_1 R_3(s)}{1 + G_1 G_2 H_1 H_2} \right)$$

> **Problem 1.14.** From the block diagrams shown in figure below, determine $\dfrac{C_1}{R_1}$ and $\dfrac{C_2}{R_1}$ (assuming $R_2 = 0$).

Solution:

Assuming $R_2 = 1$

Evaluation of $\left(\dfrac{C_1}{R_1}\right)$, Assuming $C_2 = 0$ and rearranging,

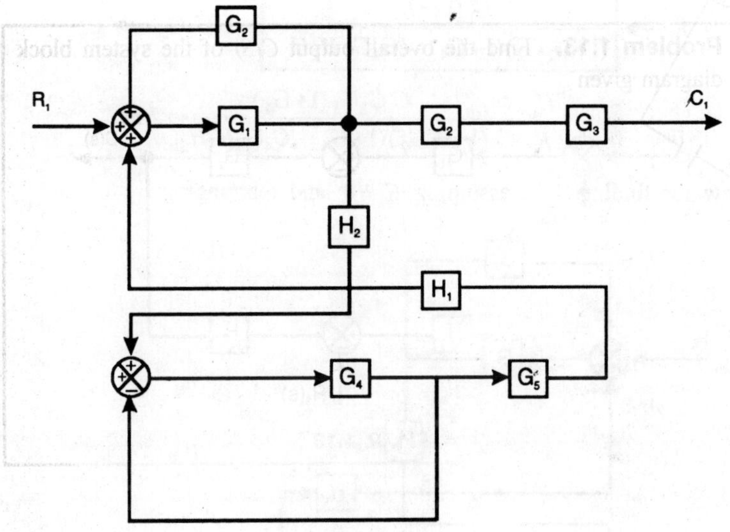

Fig. P. 1.14

Again,

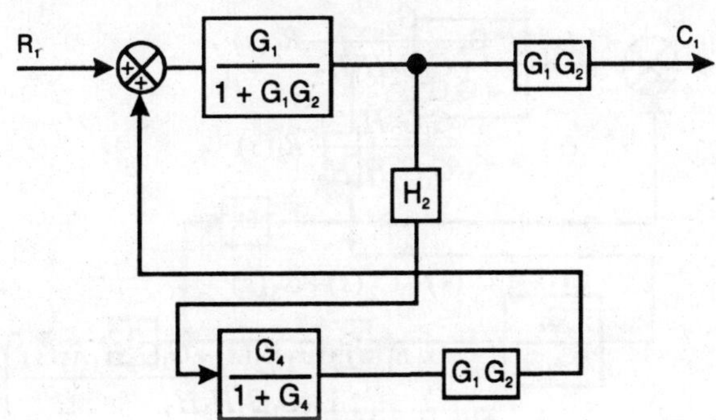

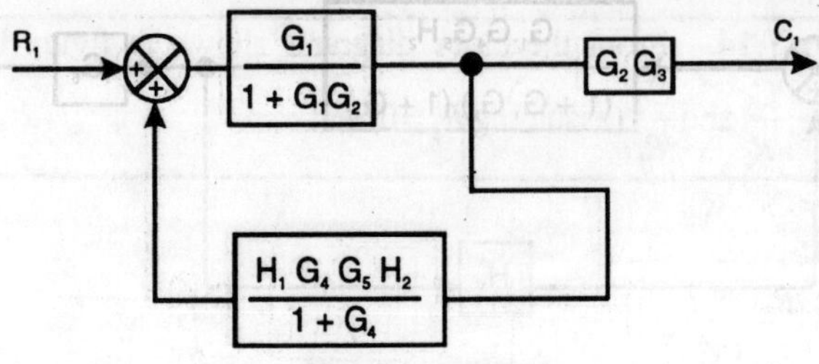

Fig. P. 1.14 (a)

Now eliminating feedback loop;

$$R_1 \longrightarrow \boxed{\dfrac{G_1(1+G_4)}{(1+G_1G_2)(1+G_4)-(G_4G_5H_1H_2)}} \longrightarrow G_2G_3 \longrightarrow C_1$$

$$\therefore \quad \frac{C_1}{R_1} = \frac{G_1G_2G_3(1+G_4)}{(1+G_1G_2)(1+G_4)-G_4G_5H_1H_2} \quad \textbf{Ans.}$$

Now for finding $\dfrac{C_2}{R_1}$, assuming $C_1 = 0$ and rearranging:

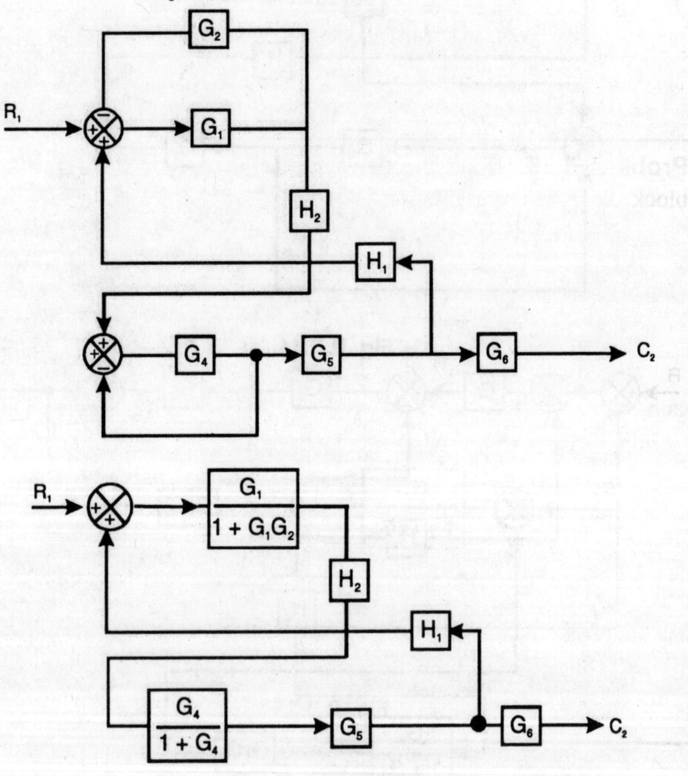

Fig. P. 1.14 (b), (c)

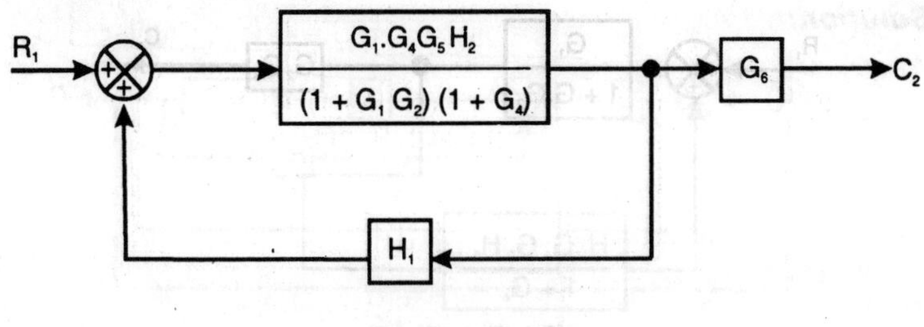

Fig. P. 1.14 (d)

$$\frac{G_1 G_4 G_5 H_2}{(1+G_1 G_2)(1+G_4)-G_1 G_4 G_5 H_1 H_2} \longrightarrow G_6 \longrightarrow C_2$$

$$\therefore \quad \frac{C_2}{R_1} = \frac{G_1 G_4 G_5 G_6 H_2}{(1+G_1 G_2)(1+G_4)-G_1 G_4 G_5 H_1 H_2} \quad \textbf{Ans.}$$

Problem 1.15. Find the transfer function for the system whose block diagram representation is shown below:

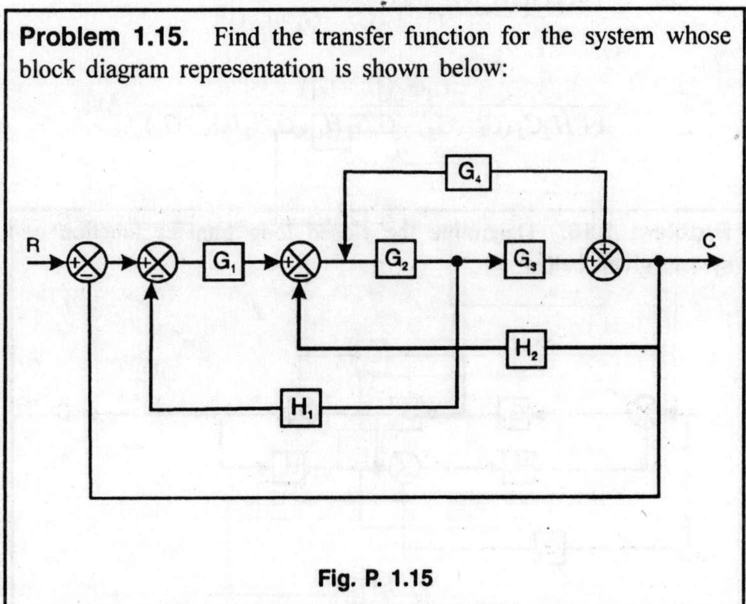

Fig. P. 1.15

Solution:

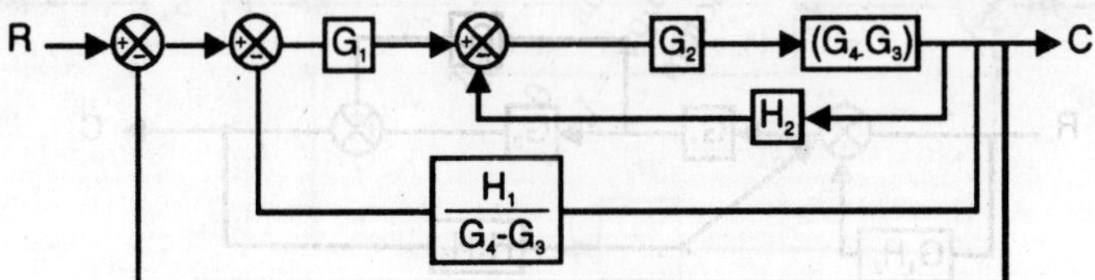

Fig. P. 1.15 (a)

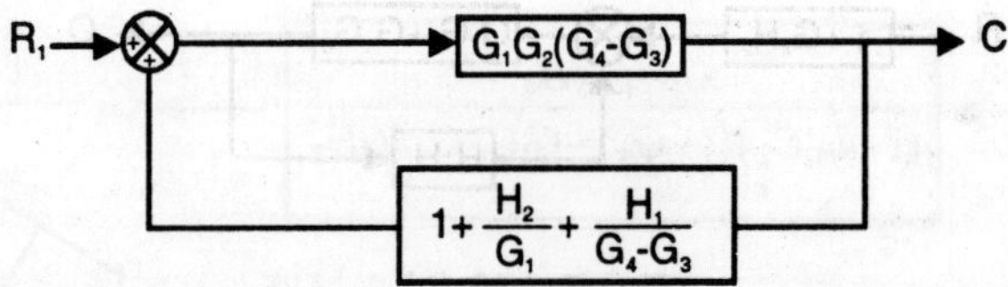

Fig. P. 1.15 (b)

$$\therefore \quad \frac{C}{R} = \frac{G_1 G_2\,(G_4 - G_3)}{1 + G_1 G_2\,(G_4 - G_3)\left(1 + \dfrac{H_2}{G_1} + \dfrac{H_1}{G_4 - G_3}\right)}$$

$$= \frac{G_1 G_2\,(G_4 - G_3)}{1 + H_2 G_2\,(G_4 - G_3) + G_1 G_2 H_1 + G_1 G_2\,(G_4 - G_3)} \quad \textbf{Ans.}$$

Problem 1.16. Determine the closed loop transfer function of system given below:

Fig. P. 1.16

Solution:

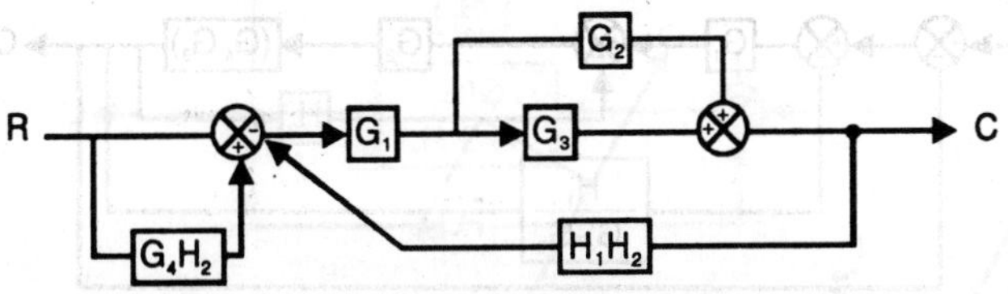

Fig. P. 1.16 (a)

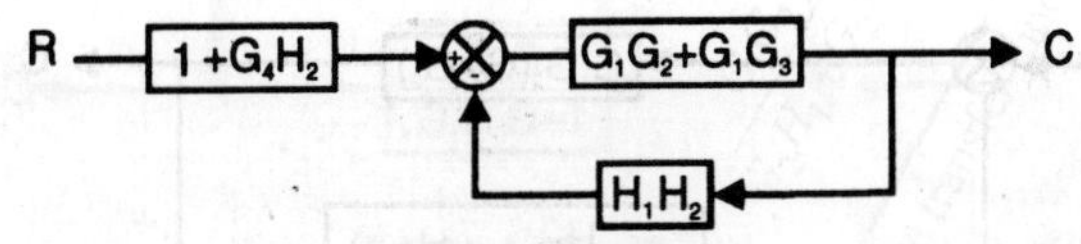

Fig. P. 1.16 (b)

Replacing the feedback loop,

$$\longrightarrow \boxed{1+G_4H_2} \longrightarrow \boxed{\dfrac{G_1G_2G_1G_3}{1+G_1G_2H_1H_2+G_1G_3H_1H_2}} \longrightarrow C$$

$$\therefore \quad \frac{C}{R} = \left(\frac{(1+G_4H_2)(G_1G_2+G_1G_3)}{1+G_1H_1H_2(G_2+G_3)}\right)$$

$$\Rightarrow \quad \frac{C}{R} = \left(\frac{G_1(G_2+G_3)(1+G_4H_2)}{1+G_1H_1H_2(G_2+G_3)}\right) \textbf{ Ans.}$$

Problem 1.17. Find the closed loop T.F. of the system given in the block diagram as below shown. (AMIE)

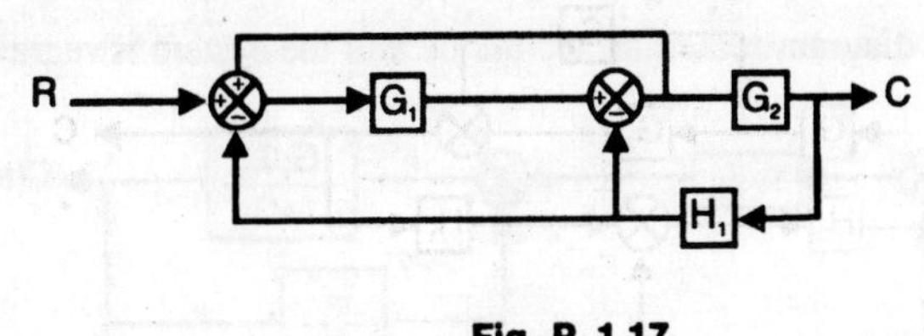

Fig. P. 1.17

Solution:

Rearranging and reducing the feedback loop;

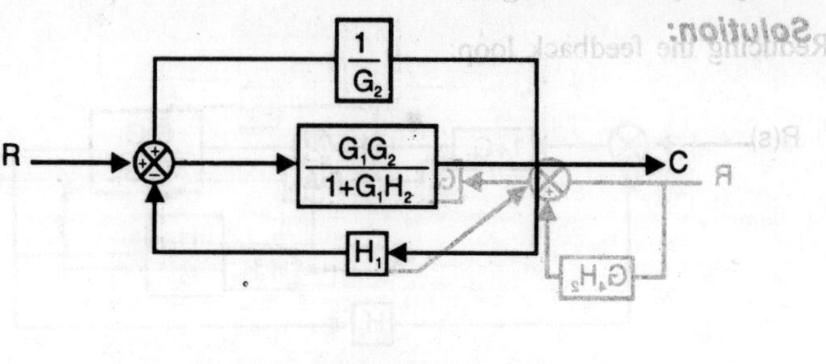

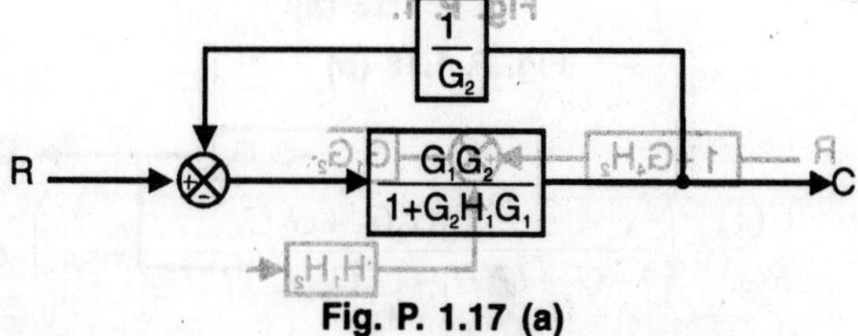

Fig. P. 1.17 (a)

$$\therefore \quad \frac{C}{R} = \left(\frac{G_1 G_2}{1 + G_1 G_2 H_1 + G_2 H_1 - G_1} \right) \textbf{ Ans.}$$

Problem 1.18. Determine the transfer function using block diagram reduction technique. (AMIE)

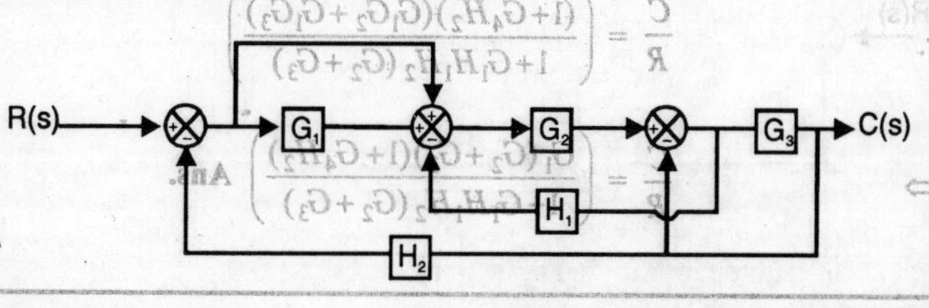

Fig. P. 1.18

Solution:

Applying block diagram reduction technique and the system is redrawn:

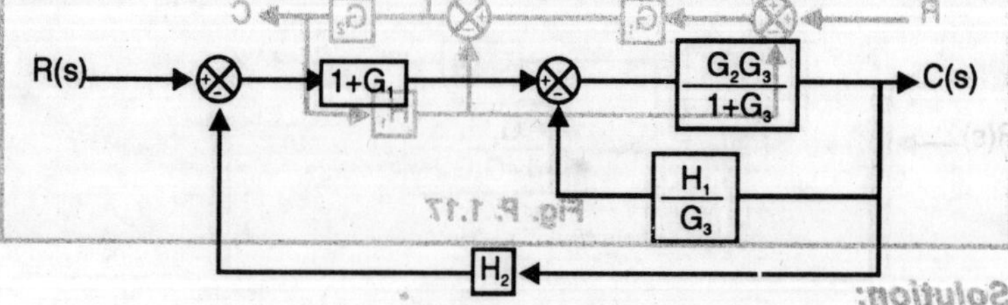

Fig. P. 1.18 (a)

Reducing the feedback loop:

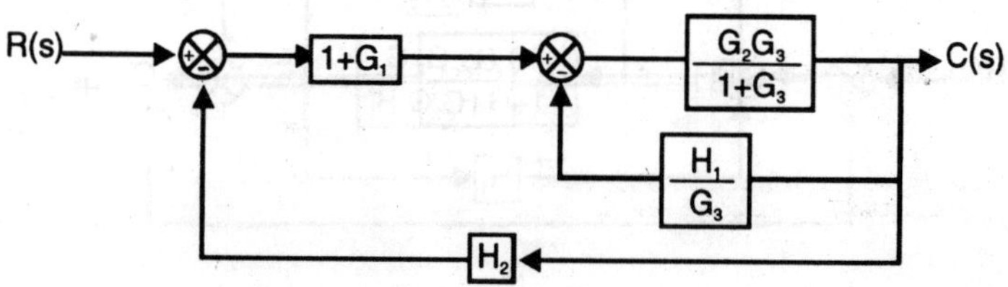

Fig. P. 1.18 (b)

$$\therefore \qquad \frac{C(s)}{R(s)} = \left(\frac{G_1 G_2 G_3 + G_2 G_3}{1 + G_3 + G_2 H_1 + G_2 G_3 H_2 + G_1 G_2 G_3 H_2} \right) \textbf{ Ans.}$$

Problem 1.19. Find the transfer function using block diagram reduction technique of the system below:

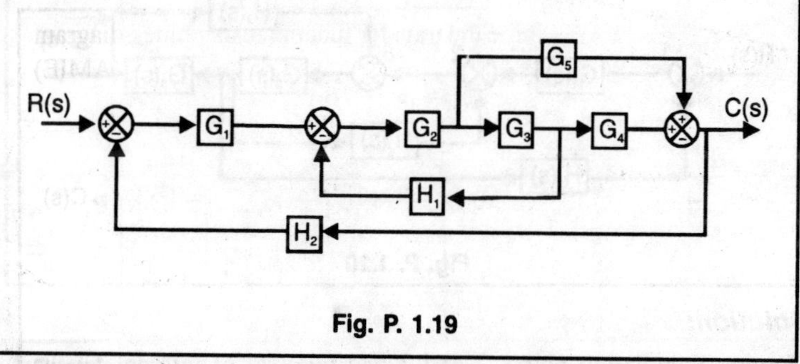

Fig. P. 1.19

Solution:

Reducing the system block diagram

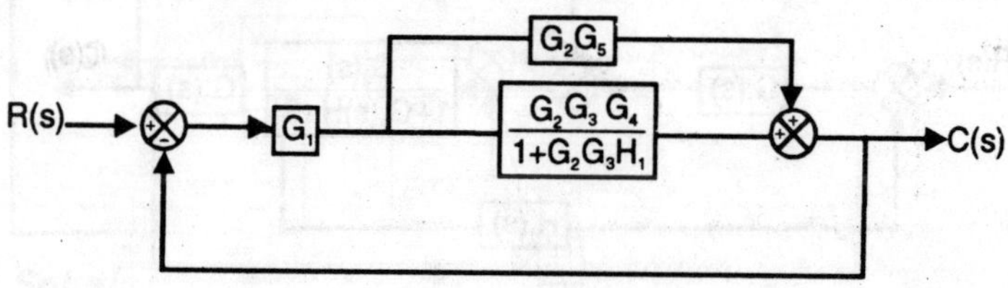

Fig. P. 1.19 (a)

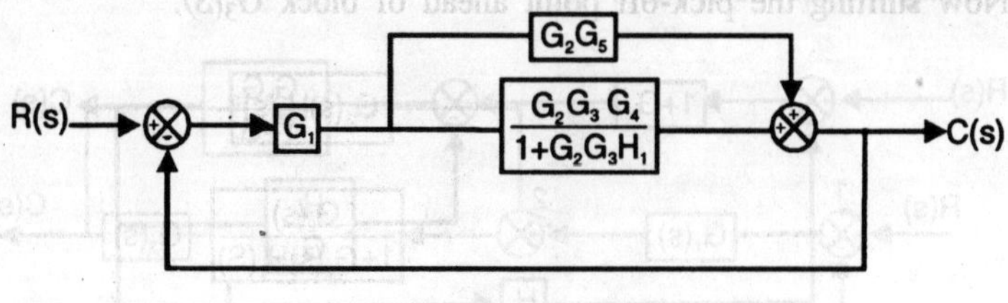

Fig. P. 1.19 (b)

$$\therefore \quad \frac{C(s)}{R(s)} = \left(\frac{G_1G_2G_3G_4 + G_1G_2G_3G_5H_1}{1 + G_2G_3H_1 + G_1G_2G_3G_4H_2 + G_1G_2G_3G_5H_1H_2} \right) \textbf{ Ans.}$$

Problem 1.20. Reduce the number of blocks into an equivalent one.

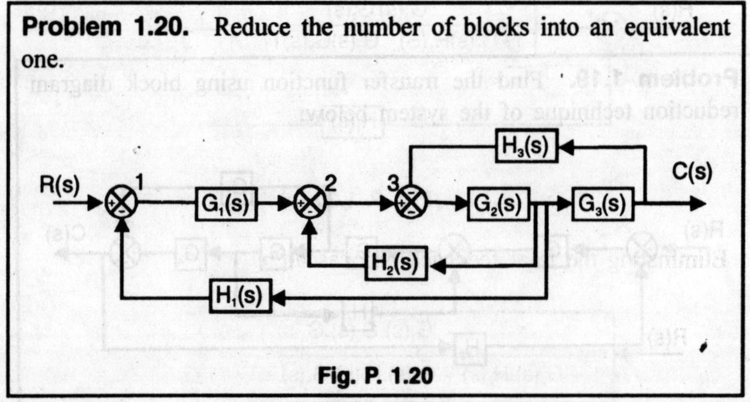

Fig. P. 1.20

Solution:

Merging Summation points 2 and 3 and removal of internal feedback loops.

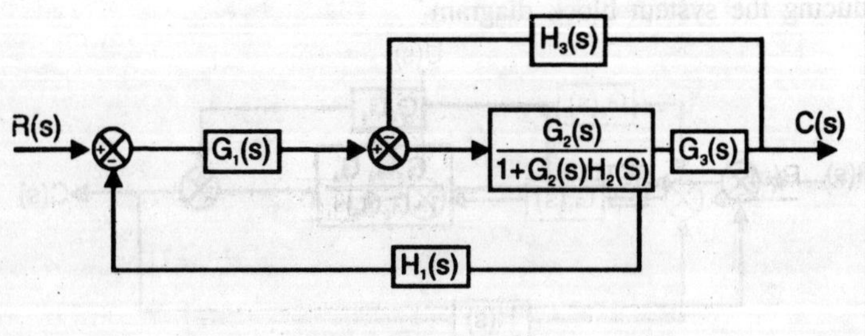

Fig. P. 1.20 (a)

Now shifting the pick-off point ahead of block $G_3(S)$.

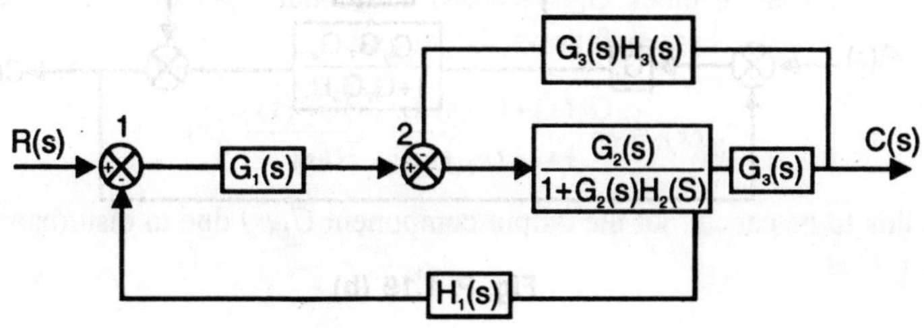

Fig. P. 1.20 (b)

Eliminating of internal feedback loop and cascading of block $G_1(s)$; then,

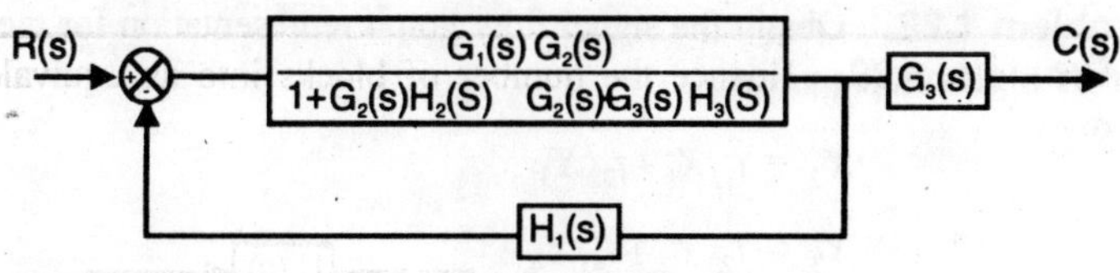

Fig. P. 1.20 (c)

Eliminating the final feedback and cascading with $G_3(s)$;

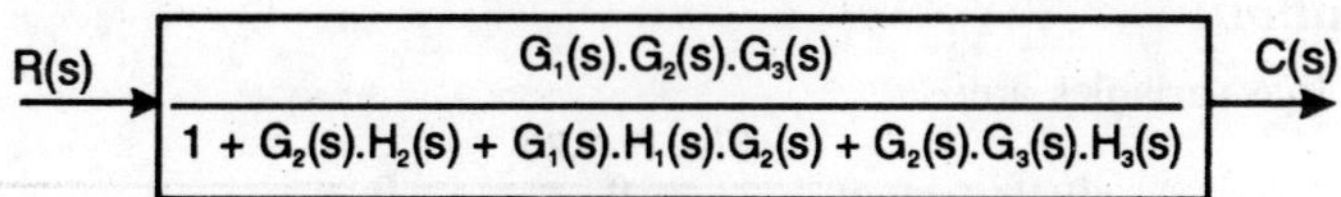

Problem 1.21. Consider the following diagram, $U(s)$ is a disturbance input. Find the value of compensating block $G_C(s)$ to kept the error introduced by $U(s)$. Within acceptable limits.

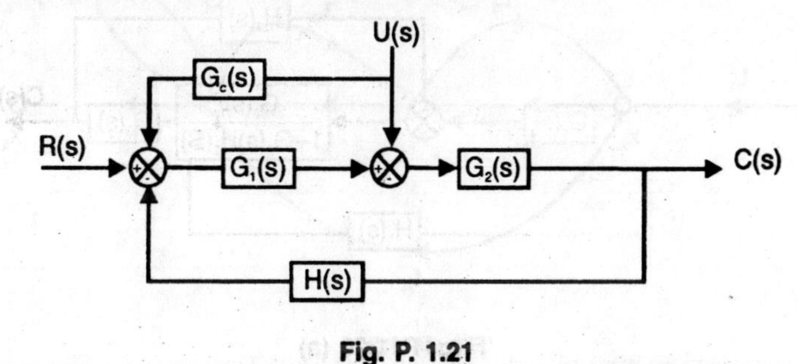

Fig. P. 1.21

Solution:

The compensating block $G_C(s)$ causes additional input of $G_C(s)$. $U(s)$ along with $U(s)$. It then follows the output,

$$C_U(s) = \frac{G_2(s)+G_1(s).G_2(s)G_C(s)}{1+G_1(s).G_2(s).H(s)}U(s)$$

For this to be cancel out the output component $C_U(s)$ due to disturbance input $U(s)$,

$$G_2(s)+G_1(s).G_2(s).G_C(s) = 0$$

$$\therefore \qquad G_C(s) = -\frac{1}{G_1(s)}$$

Problem 1.22.   Obtain the signal flow graph representation for the set of algebraic equations

$$X_2 = t_{12}X_1+t_{32}X_3$$

$$X_3 = t_{23}X_2+t_{43}X_4$$

$$X_4 = t_{24}X_2+t_{34}X_3+t_{44}X_4$$

$$X_5 = t_{45}X_4$$

Solution:

The five variables are:

$$\begin{array}{ccccc} 0 & 0 & 0 & 0 & 0 \\ X_1 & X_2 & X_3 & X_4 & X_5 \end{array}$$

Connecting all the nodes by the corresponding gains:

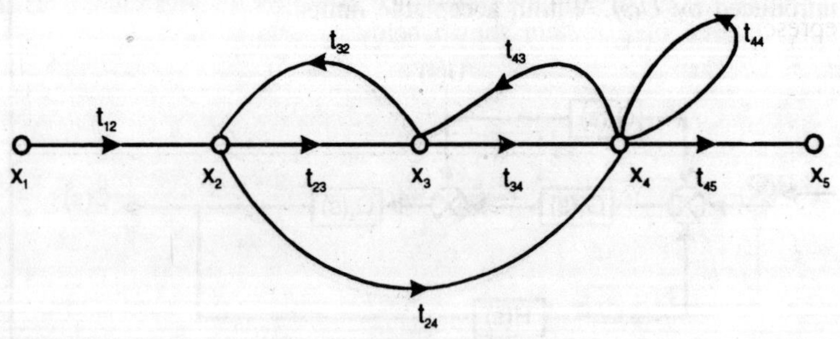

Fig. P. 1.22

Here the input variable is at X_1 and the output variable is X_5.

$$t_{pq} = \text{gains between two nodes.}$$

Problem 1.23. Draw the signal flow graph representation for the feedback control system shown in figure below.

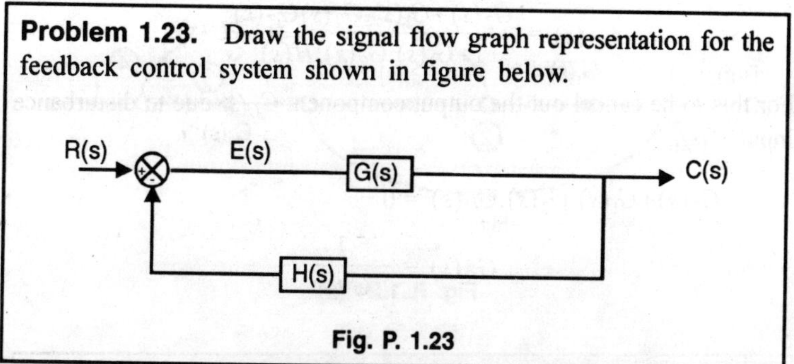

Fig. P. 1.23

Solution:

The input *R(s)* and output *C(s)* are represented by 2 nodes and connected the system by unit gain branches:

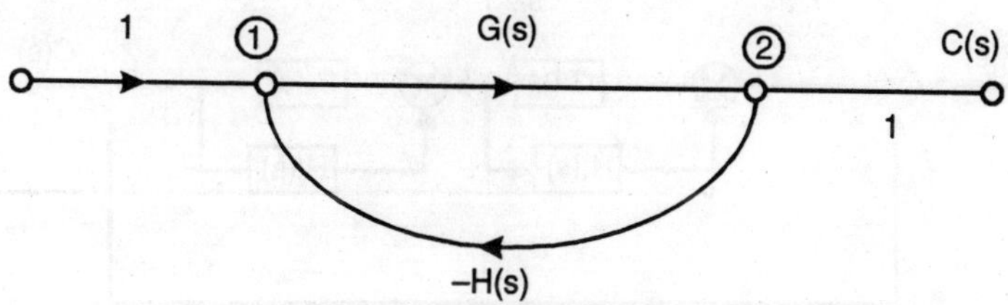

Fig. P. 1.23 (a)

The pick-off point and the summing point by 2 and 1.

Problem 1.24. Obtain the signal flow graph for the block diagram representation of a system shown below.

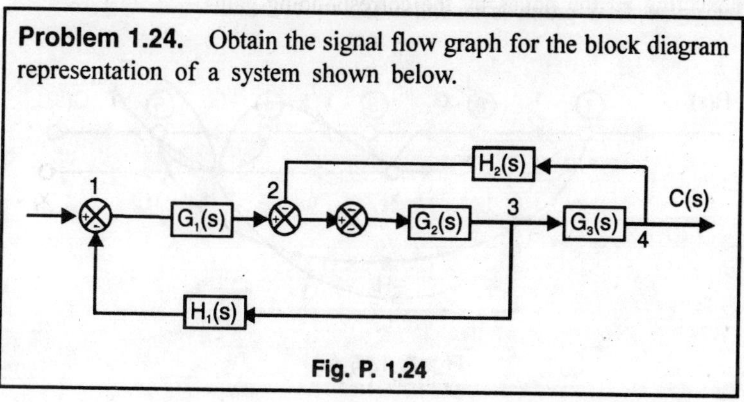

Fig. P. 1.24

Solution:

The summation point 1 & 2 and 3 & 4 are the required nodes of the signal flow graph. The signal flow graph is as below:

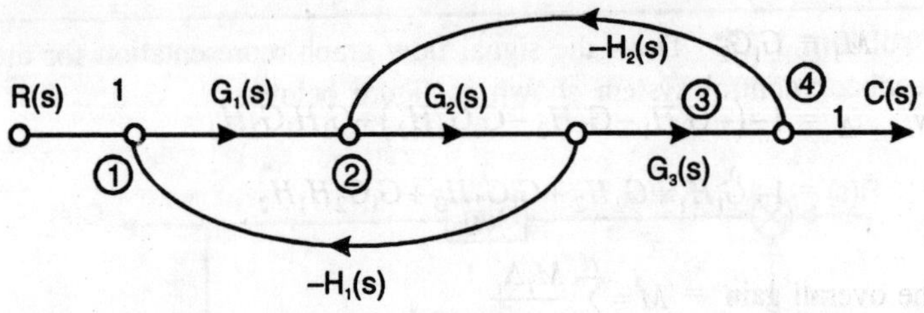

Fig. P. 1.24 (a)

Problem 1.25. Consider the system shown below. Obtain the signal flow graph. Using Mason's gain formula, determine the one overall gain.   (I.E.S.-88)

Fig. P. 1.25

Solution:

The signal flow graph is as follows:

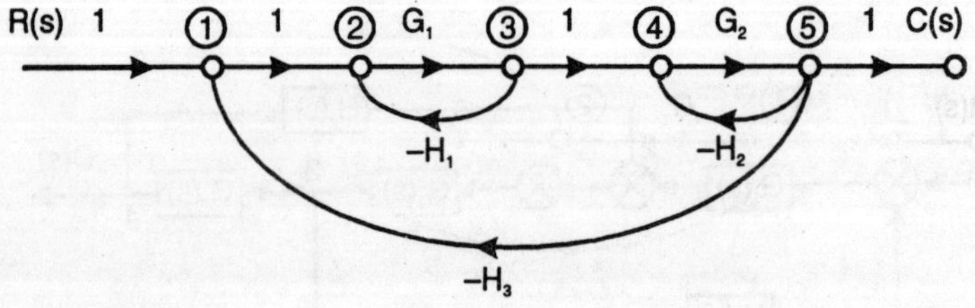

Fig. P. 1.25 (a)

No. of forward path $= N = 1$

$$P_{11} = -G_1H_1 \qquad\qquad P_{12} = -G_2H_2$$

$$P_{13} = -G_1G_2H_3 \qquad\qquad P_{21} = G_1H_1G_2H_2$$

$$M_1 = G_1G_2$$

Now, $\Delta = 1-(-G_1H_1-G_2H_2-G_1G_2H_3)+G_1H_1G_2H_2$

$$= 1+G_1H_1+G_2H_2+G_1G_2H_3+G_1G_2H_1H_2$$

$\therefore$ The overall gain $= M = \displaystyle\sum_{i=1}^{N}\frac{M_i\Delta_i}{\Delta}$

$$M = \left(\frac{G_1G_2}{1+G_1H_1+G_2H_2+G_1G_2H_3+G_1G_2H_1H_2}\right) \textbf{ Ans.}$$

Problem 1.26.   Obtain the signal flow graph representation for the block diagram shown below. Determine the overall gain. (I.E.S.-90)

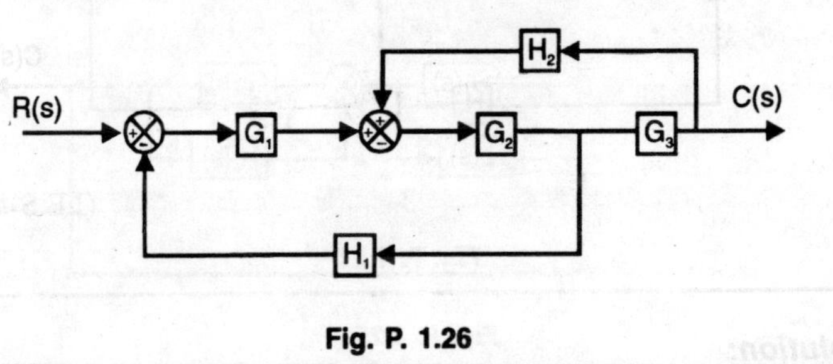

Fig. P. 1.26

Solution:

The signal flow graph is,

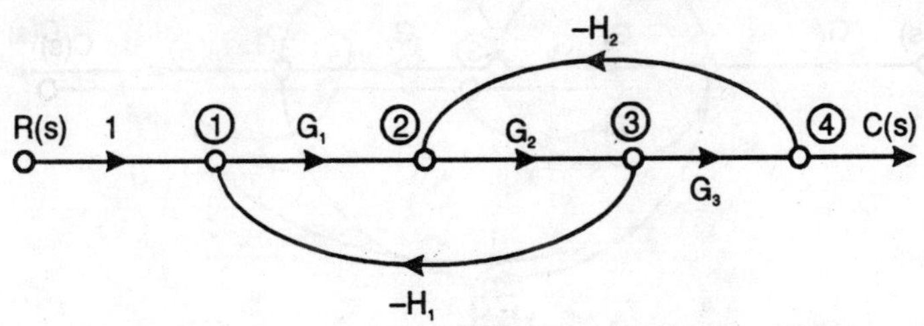

Fig. P. 1.26 (a)

$$P_{11} = -G_1 G_2 H_1$$

$$P_{12} = -G_2 G_3 H_2$$

$$M_1 = G_1 G_2 G_3$$

$\therefore$ The overall gain $= \dfrac{C(S)}{R(S)} = \left(\dfrac{G_1 G_2 G_3}{1 + G_1 G_2 H_1 + G_2 G_3 H_2} \right)$

Problem 1.27. Draw the signal flow graph for the system shown in fig. below and hence obtain the transfer function $\dfrac{C(s)}{R(s)}$ Using Mason's gain formula.

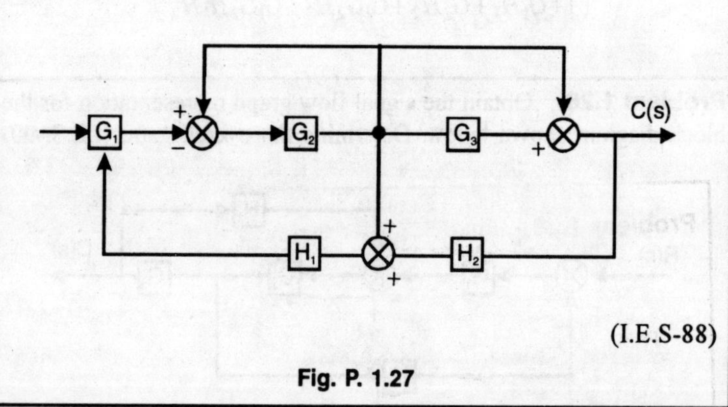

(I.E.S-88)

Fig. P. 1.27

Solution:

The signal flow graph for the system;

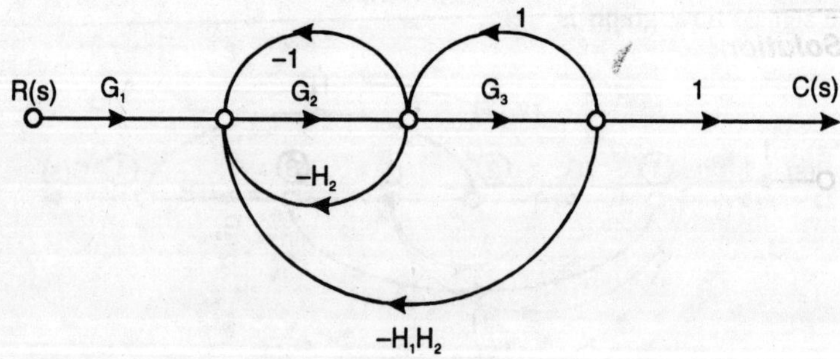

Fig. P. 1.27 (a)

Forward paths, $\quad M_1 = G_1 G_2 G_3$

$$M_2 = G_1 G_2$$

Loops

$-G_2$

$-G_2 H_2$

$-G_2 H_1 H_2$

$-G_2 G_3 H_1 H_2$

Now Applying Mason's gain formula,

$$T = \frac{C(s)}{R(s)} = \frac{M_1 \Delta_1 + M_2 \Delta_2}{\Delta}$$

Here $\Delta_1 = 1$ and $\Delta_2 = 1$
as both forward path touches all loops.

$$\Delta = 1 + G_2 + G_2 H_2 + G_2 H_1 H_2 + G_2 G_3 H_1 H_2$$

$$\therefore \quad T = \left(\frac{G_1 G_2 + G_1 G_2 G_3}{1 + G_2 + G_2 H_2 + G_2 H_1 H_2 + G_2 G_3 H_1 H_2} \right)$$

Problem 1.28. Find $\dfrac{C(s)}{R(s)}$ for the system given in figure below:

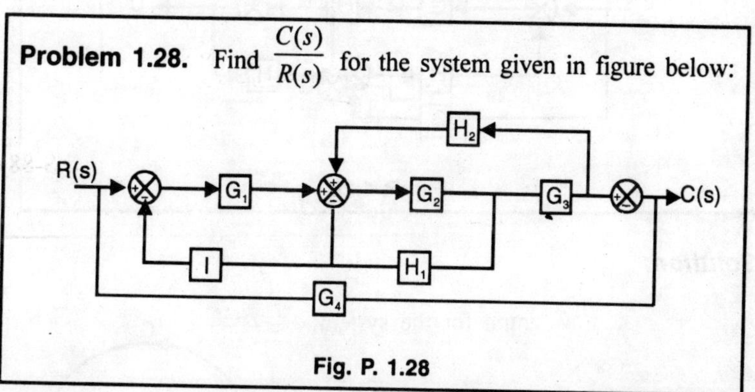

Fig. P. 1.28

Solution:

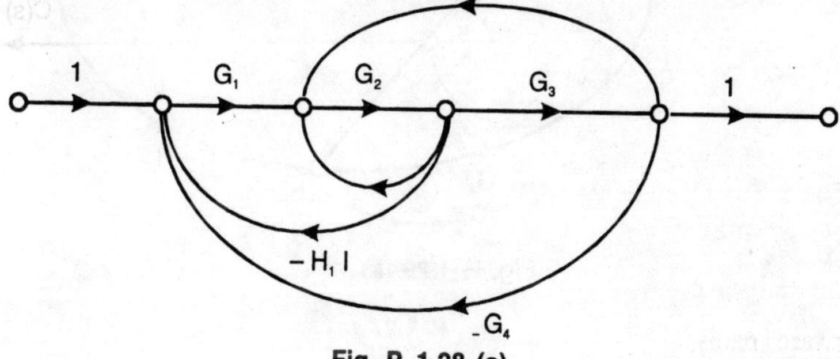

Fig. P. 1.28 (a)

Forward paths: Loops

$$M_1 = G_1 G_2 G_3 \qquad -G_1 G_2 H_1 I$$

$$-G_1 G_2 G_3 G_4$$

$$-G_2 H_1$$

$$-H_2 G_2 G_3$$

$$\Delta = 1 + G_1 G_2 H_1 I + G_1 G_2 G_3 G_4 + G_2 H_1 + H_2 G_2 G_3$$

$\therefore$ The transfer function,

$$T = \frac{C(s)}{R(s)} = \left(\frac{G_1 G_2 G_3}{1 + G_1 G_2 H_1 I + G_1 G_2 G_3 G_4 + G_2 H_1 + H_2 G_3 G_2} \right)$$

Problem 1.29. Draw a signal flow graph and evaluate the closed-loop transfer function of a system whose block diagram is given in Fig. P.1.29.

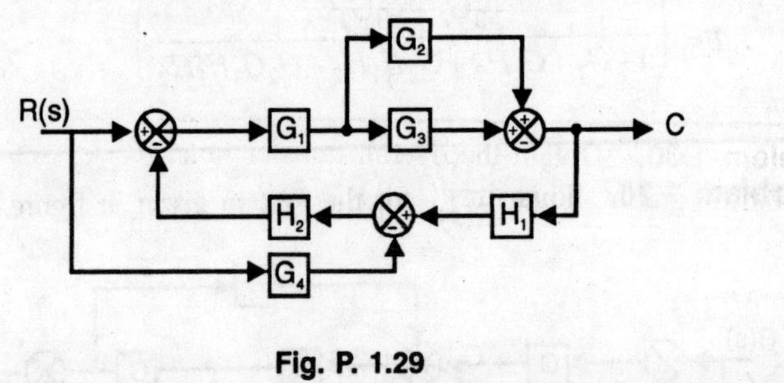

Fig. P. 1.29

Solution:

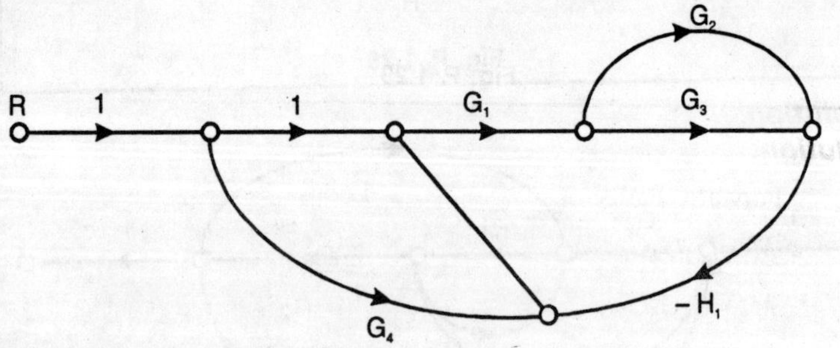

Fig. P. 1.29 (a)

Forward paths,

$$M_1 = G_1 G_3 \qquad\qquad \Delta_1 = 1$$

$$M_2 = G_1 G_2 \qquad\qquad \Delta_2 = 1$$

$$M_3 = G_1 G_3 G_4 H_2 \qquad\qquad \Delta_3 = 1$$

$$M_4 = G_1 G_2 G_4 H_2 \qquad\qquad \& \; \Delta_4 = 1$$

Loops

$$-G_1 G_3 H_1 H_2$$

$$-G_1 G_2 H_1 H_2$$

$$\therefore \qquad \Delta = 1 + G_1 G_2 H_1 H_2 + G_1 G_3 H_1 H_2$$

$$\therefore \qquad T = \frac{M_1 \Delta_1 + M_2 \Delta_2 + M_3 \Delta_3 + M_4 \Delta_4}{\Delta}$$

$$= \left(\frac{G_1 (G_3 + G_2)(1 + G_4 H_2)}{1 + G_1 H_1 H_2 (G_3 + G_2)} \right) \quad \textbf{Ans.}$$

Problem 1.30. Obtain the overall transfer function C/R from the signal flow graph shown in Fig. P.1.29.

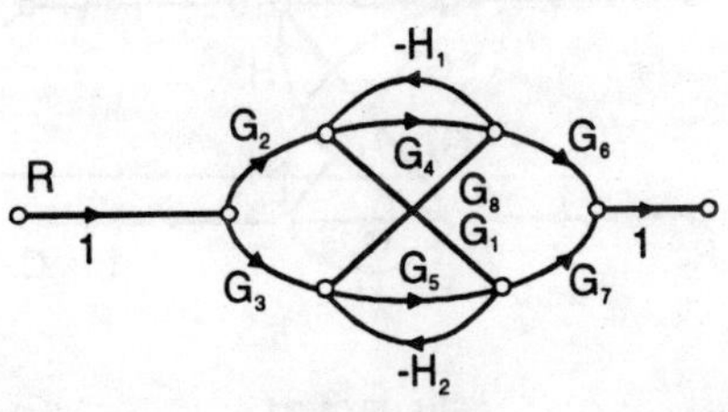

Fig. P. 1.29

Solution:

Forward Paths:

$$M_1 = G_2 G_4 G_6$$

$$M_2 = G_3 G_5 G_7$$

$$M_3 = G_2 G_1 G_7$$

$$M_4 = G_3 G_6 G_8$$

$$M_5 = -G_2 G_1 H_2 G_8 G_6$$

Loops:

$$-G_4 H_1$$

$$-G_5 H_2$$

$$+G_1 H_2 G_8 H_1$$

Non touching loops:

$$G_4 H_1 G_5 H_2$$

$$M_6 = -G_3 G_8 H_1 G_7 G_1$$

$$\Delta_1 = \left(1 + G_5 H_2\right)$$

$$\Delta_2 = \left(1 + G_4 H_1\right)$$

$$\Delta_3 = \Delta_4 = \Delta_5 = \Delta_6 = 1$$

and $$\Delta = 1 + G_4 H_1 + G_5 H_2 - G_1 H_2 G_8 H_1 + G_4 H_1 G_5 H_2$$

$$\therefore T = \left(\frac{\begin{array}{c} G_4 G_4 G_6 \left(1 + G_5 H_2\right) + G_3 G_5 G_7 \left(1 + G_4 H_1\right) + G_2 G_1 G_7 + G_3 G_8 G_6 \\ -G_2 G_6 G_8 G_1 H_2 - G_3 G_7 G_8 G_1 H_1 \end{array}}{1 + G_4 H_1 + G_5 H_2 + G_4 G_5 H_1 H_2 - G_1 G_8 H_1 H_2} \right) \quad \textbf{Ans.}$$

Problem 1.31.   Figure P. 1.31 gives the signal flow graph of a system with two inputs and two outputs. Find expressions for the outputs C_1 and C_2.   Also determine the condition that makes C_1 independent of R_2 and C_2 independent of R_1.

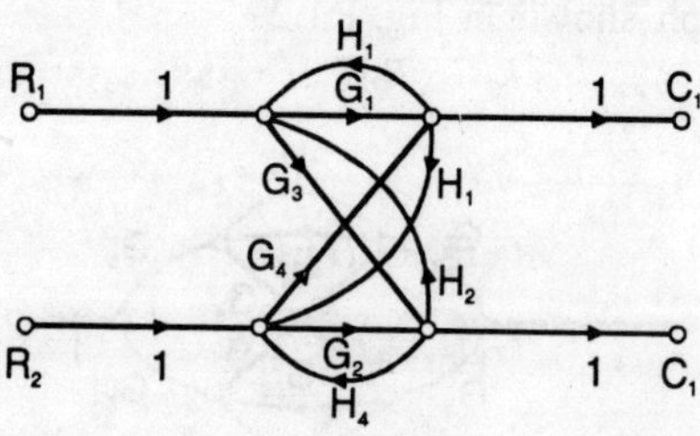

Fig. P. 1.31

Solution:

Expression for G

1. No. of forward path gains:

$$R_1 G_1 ; R_1 G_3 H_4 G_4$$

$$R_2 G_4 ; R_2 G_2 H_2 G_1$$

2. No of loop gains:

$$G_1 H_3 ; G_3 H_2 ; G_2 H_1 ; G_2 H_4$$

$$G_1 H_1 G_2 H_2 ; G_4 H_3 G_3 H_4$$

3. No. of non-touching loop gains:

$$G_1H_3; G_4H_4; G_3H_2; G_4H_1$$

$$\Delta_1 = \left(1-G_2H_4\right)$$

$$\Delta_2 = 1$$

$$\Delta_3 = \left(1-G_3H_2\right)$$

$$\Delta_4 = 1$$

The overall gain,

$$T = \frac{R_1G_1\left(1-G_2H_4\right)+R_1G_3H_4G_4+R_2G_4\left(1-G_3H_2\right)+R_2H_2G_1G_2}{1-\left(G_1H_3+G_2H_4+G_3H_2+G_4H_1+G_1G_2H_1H_2+G_3G_4H_3H_4\right)+\left(G_1H_3G_2H_4+G_3H_2G_4H_1\right)}$$

$$\therefore\ T = \left(\frac{R_1\left(G_1\left(1-G_2H_4\right)+G_3H_4G_4\right)+R_2\left(G_4\left(1-G_3H_2\right)+H_2G_1G_2\right)}{\Delta}\right)$$

Now expression for C_2

1. No. of forward paths:

$$R_2G_2; R_2G_4H_3G_3; R_1G_1H_1G_2$$

2. No. of loop gains:

$$G_1H_3; G_3H_2; G_4H_1; G_2H_4$$

$$G_4H_3G_3H_4; G_1H_1G_2H_2$$

3. No. of individual loop gains:

$$G_1H_3G_2H_4; G_3H_2G_4H_1$$

$$\Delta_1 = \left(1-G_1H_3\right); \Delta_2 = 1$$

$$\Delta_3 = \left(1-G_4H_1\right); \Delta_4 = 1$$

$$\therefore\ T = \left(\frac{R_2G_2\left(1-G_1H_3\right)+R_2G_4H_3G_3+R_1G_3\left(1-G_4H_1\right)+R_1G_1H_1G_2}{\Delta}\right)$$

$$= \left(\frac{R_2\left(G_2\left(1-G_1H_3\right)+G_4H_3G_3\right)+R_1\left(G_3\left(1-G_4H_1\right)+G_1H_1G_2\right)}{\Delta}\right)$$

Ans.

> **Problem 1.32.** For the system represented by the following equations, find the transfer function $X(S)U(S)$ by signal flow graph technique.
>
> $$x = x_1 + \beta_3 u$$
> $$\dot{x}_1 = -a_1 x_1 + x_2 + \beta_2 u$$
> $$\dot{x}_2 = -a_2 x_1 + \beta_1 u$$

Solution:

Consider x = output and u = input, writing the given equation in laplace transform,
$$x = x_1 + \beta_3 U \tag{1}$$

$$S x_1 = -a_1 x_1 + x_2 + \beta_2 U$$

$$x_1 = \frac{x_2}{S+a_1} + \left(\frac{\beta_2}{S+a_1}\right) U \tag{2}$$

and
$$S x_2 = -a_2 x_1 + \beta_1 U$$

$$\therefore \qquad x_2 = \frac{-a_2}{S} x_1 + \frac{\beta_1}{S} U \tag{3}$$

The signal flow graph is

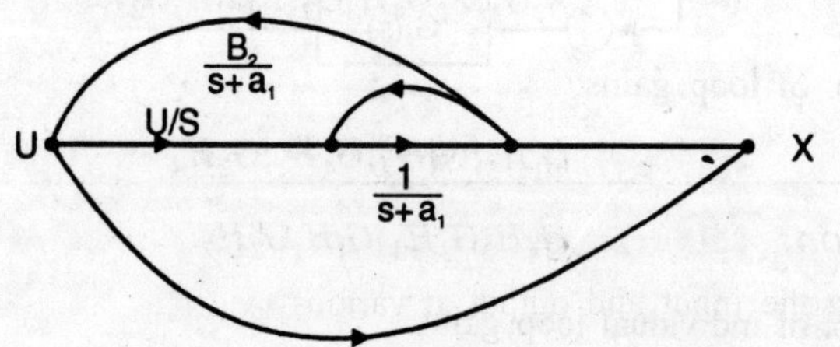

Fig. P. 1.32

Forward paths:

$$M_1 = \frac{\beta_1}{S(S+a_1)}$$

$$M_2 = \left(\frac{\beta_2}{S+a_1}\right), M_3 = \beta_3$$

Loops

$$\frac{a_2}{S(S+a_1)}; \quad \Delta = 1 + \frac{a_2}{S(S+a_1)}$$

$$\Delta_1 = \Delta_2 = 1, \Delta_3 = \left(1 + \frac{a_2}{s(s+a_1)}\right) = \frac{s^2 + a_1 s + a_2}{s(s+a_1)}$$

$$T = \frac{M_1 \Delta_1 + M_2 \Delta_2 + M_3 \Delta_3}{\Delta}$$

$$= \left(\frac{\beta_1 + \beta_2 S + \beta_3 (s^2 + a_1 s + a_2)}{s^2 + a_1 s + a_2}\right) \textbf{ Ans.}$$

Problem 1.33. Determine the Transfer Function $\dfrac{C(s)}{R(s)}$ of the system shown in Fig. P.1.33. (I.I.S.-85)

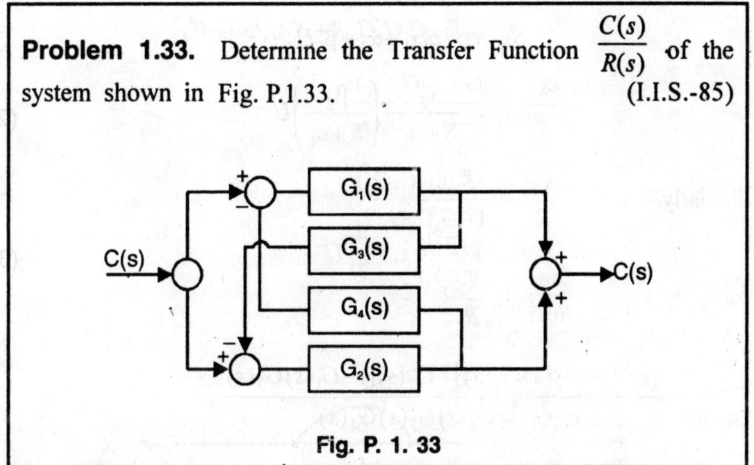

Fig. P. 1. 33

Solution:

Marking the input and output at various stages as follows:

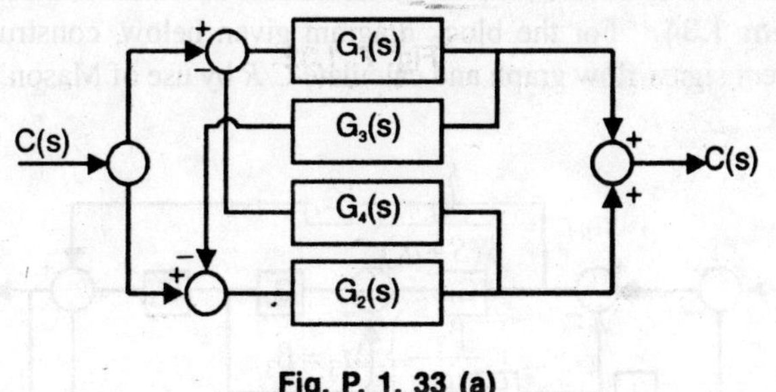

Fig. P. 1. 33 (a)

Dropping the *s* within parenthesis,

$$X_1 = E_1 G_1 = G_1 [R - G_2 G_4 (R - X_1 G_3)]$$

$$E_1 = R - B_1 = R - G_2 G_4 (R - X_1 G_3)$$

$$B_1 = X_2 G_4 = G_2 G_4 (R - X_1 G_3)$$

$$X_2 = E_2 G_2 = G_2 (R - X_1 G_3)$$

$$E_2 = R - B_2 = R - X_1 G_3$$

$$B_2 = X_1 G_3$$

Expanding equation (1),

$$X_1 = G_1 \left[R - G_2 G_4 R + X_1 G_2 G_3 G_4 \right]$$

$$= G_1 R - G_1 G_2 G_4 R + X_1 G_1 G_2 G_3 G_4$$

$$\frac{X_1}{R} = \frac{G_1 (1 - G_2 G_4)}{1 - G_1 G_2 G_3 G_4}$$

Similarly,

$$\frac{X_2}{R} = \frac{G_2 (1 - G_1 G_3)}{1 - G_1 G_2 G_3 G_4}$$

$$\therefore \quad \frac{C(s)}{R(s)} = \frac{X_1}{R} + \frac{X_2}{R}$$

$$= \frac{G_1(s)\left[1 - G_2(s) G_4(s)\right] + G_2(s)\left[1 - G_1(s) G_3(s)\right]}{1 - G_1(s) G_2(s) G_3(s) G_4(s)}$$

$$= \frac{G_1(s) + G_2(s) - G_1(s) G_4(s) - G_4(s) G_2(s) G_3(s)}{1 - G_1(s) G_2(s) G_3(s) G_4(s)} \quad \textbf{Ans.}$$

Problem 1.34. For the block diagram given below, construct the equivalent signal flow graph and calculate *C/R* by use of Mason's gain formula.

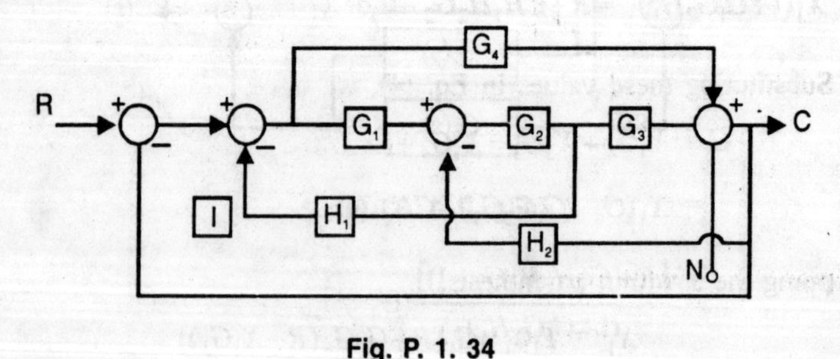

Fig. P. 1. 34

Solution:

The equivalent signal flow graph is

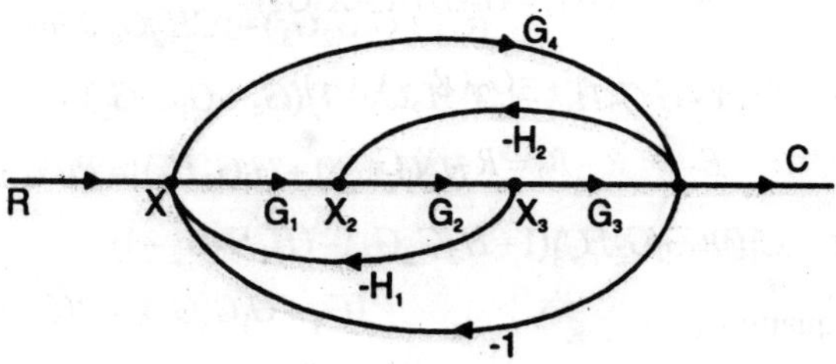

Fig. P.1.34

Let us solve for $\dfrac{C(s)}{R(s)}$ by the following 2 methods.

1. **Loop Reduction Method:** Writing the equations for the intermediate points X_1, X_2 and X_3, we have

$$X_1 = R - X_3 H_1 - C \qquad (1)$$

$$X_2 = X_1 G_1 - C H_2 \qquad (2)$$

$$X_3 = X_2 G_2 \qquad (3)$$

$$C = X_1 G_4 + X_3 G_3 \qquad (4)$$

From equations (1), (2) and (3), we have

$$X_3 = X_2 G_2 = (X_1 G_1 - C H_2) G_2$$

and $\qquad X_1 = R - X_3 H_1 - C$

and $\qquad X_1 = R - (X_1 G_1 - C H_2) H_1 G_2 - C$

Solving for X_1 in terms of R and C, we have

$$X_1(1 + G_1 G_2 H_1) = R + C H_1 H_2 G_2 - C \text{ or } X_1 = \frac{R + C(H_1 H_2 G_2 - 1)}{(1 + G_1 G_2 H_1)}$$

Substituting these values in Eq. (4), we have

$$C = X_1 G_4 + X_3 G_3 = X_1 G_4 + (X_1 G_1 G_2 - C H_2 G_2) G_3$$

$$= X_1(G_4 + G_1 G_2 G_3) - C H_2 G_2 G_3$$

$$= \frac{[R + C(H_1 H_2 G_2 - 1)]}{(1 + G_1 G_2 H_1)}(G_4 + G_1 G_2 G_3) - C H_2 G_2 G_3$$

or grouping terms of C and R separately, we have

$$C(1+G_1G_2H_1) = R(G_4+G_1G_2G_3)+C(H_1H_2G_2-1)$$
$$(G_4+G_1G_2G_3)-(CH_2G_2G_3)(1+G_1G_2H_1)$$

or $\quad C[(1+G_1G_2H_1)-(H_1H_2G_2-1)(G_4+G_1G_2G_3)+$
$$H_2G_2G_3(1+G_1G_2H_1)]=R(G_4+G_1G_2G_3)$$

or $\quad C[(1+G_1G_2H_1)(1+H_2G_2G_3)-(H_1H_2G_2-1)$
$$(G_4+G_1G_2G_3)]=R(G_4+G_1G_2G_3)$$

or $\quad C[1+G_1G_2H_1+H_2G_2G_3+G_1G_2^2G_3H_1H_2-H_1H_2G_2G_4+G_4$
$$-G_1G_2^2G_3H_1H_2+G_1G_2G_3]=R(G_4+G_1G_2G_3)$$

or $\quad \dfrac{C}{R}=\dfrac{(G_4+G_1G_2G_3)}{(1+G_4+G_1G_2H_1+H_2G_2G_3-H_1H_2G_2G_4+G_1G_2G_3}$

2. By using Mason's Rule

Forward Paths	Loop Gains
1. $G_1G_2G_3$	1. $-G_1G_2H_1$
2. G_4	2. $-G_1G_2G_3$
	3. $-G_2G_3H_2$
	4. $-G_4$
	5. $G_2G_4H_1H_2$

Then the overall gain of the system is,

$$M = \frac{C}{R}=\sum_{k=1}^{2}\frac{M_k\Delta_k}{\Delta}$$

Loop Gains

$$P_{11} = -G_1G_2H_1$$
$$P_{21} = -G_1G_2G_3$$
$$P_{31} = -G_2G_3H_2$$
$$P_{41} = -G_4$$
$$P_{51} = G_2G_4H_1H_2$$

There are no non-touching loops

$$P_{m2} = P_{m3}=P_{m4}=...=0$$

Hence $\Delta = 1-(-G_1G_2H_1-G_1G_2G_3-G_2G_3H_2-G_4+G_2G_4H_1H_2)$

$$= 1+G_1G_2H_1+G_1G_2G_3+G_2G_3H_2+G_4-G_2G_4H_1H_2$$

The transfer function C/R is given by

$$\frac{C}{R} = \frac{P_1\Delta_1+P_2\Delta_2}{\Delta}$$

(Here $\Delta_1=\Delta_2=1$ as both the forward paths are in touch with all the loops.)

$$= \frac{G_1G_2G_3+G_4}{1+G_1G_2H_1+G_1G_2G_3+G_2G_3H_2+G_4-G_2G_4H_1H_2}$$

Problem 1.35. Draw a signal flow graph for the network shown below and therefrom determine the transfer function relating the input and output voltages. (I.E.S.)

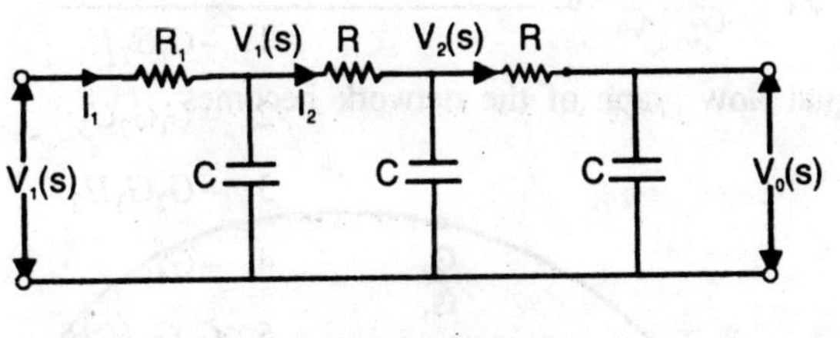

· Fig. P. 1.35

Solution:

The equations relating various voltages and currents are given by

$$V_0(s) = \frac{\dfrac{1}{Cs}}{R+\dfrac{1}{Cs}}.V_2(s) \tag{1}$$

$$V_2(s) = V_0(s)+\frac{V_2(s)}{R+\dfrac{1}{Cs}}.R$$

$$I_2(s) = \frac{V_2}{\dfrac{1}{Cs}}+\frac{V_2-V_0}{R}=V_2\left(Cs+\frac{1}{R}\right)-\frac{V_0}{R}$$

$$V_1 = V_2 + I_2 R = V_2 + V_2(RCs+1) - V_0 = V_2(2+RCs) - V_0$$

$$V_2 = \frac{V_1}{2+RCs} + \frac{V_0}{2+RCs} \tag{2}$$

$$I_1 = V_1 Cs + I_2 = Cs V_1 + \left[\left(\frac{V_1}{2+RCs} + \frac{V_o}{2+RCs} \right) \frac{RCs+1}{R} - \frac{V_0}{R} \right]$$

$$V_1 = V_i - I_1 R$$

$$= V_i - R \left[Cs + \frac{1}{2+RCs} \frac{RCs+1}{R} \right] V_1 - R \left[\frac{RCs+1}{(2+RCs)R} - \frac{1}{R} \right] V_0$$

$$V_1 = V_1 \left[1 + RCs + \frac{RCs+1}{2+RCs} \right] = V_i - \left[\frac{RCs+1}{2+RCs} - 1 \right] V_0 \tag{3}$$

$$V_1 G_1(s) = V_i - G_2(s) V_0 \text{ (say)}$$

$$V_1 = \frac{V_i}{G_1} - \frac{G_2}{G_1} V_0 \tag{4}$$

The signal flow graph of the network becomes

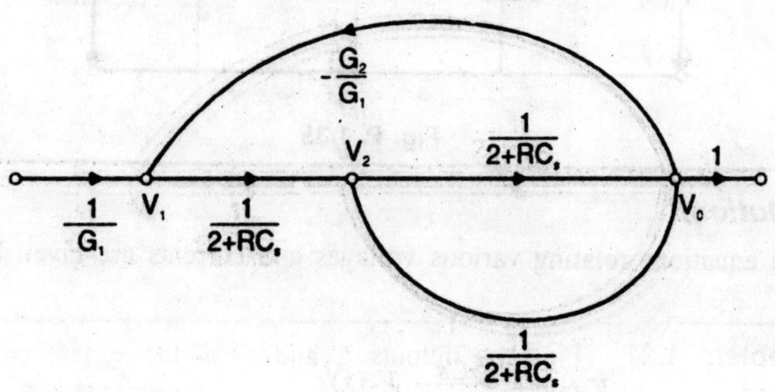

Fig. P. 1.35 (a)

In Fig. P. 1.35 (a) $G_1 = 1 + RCs + \dfrac{RCs+1}{2+RCs}$ and $G_2 = \dfrac{RCs+1}{2+RCs} - 1$

Problem 1.36.   What do you understand by the Transfer function of a system? State its properties. Find the transfer function of the lag-lead compensator network shown in Fig. P.1.36.   (I.E.S.-98)

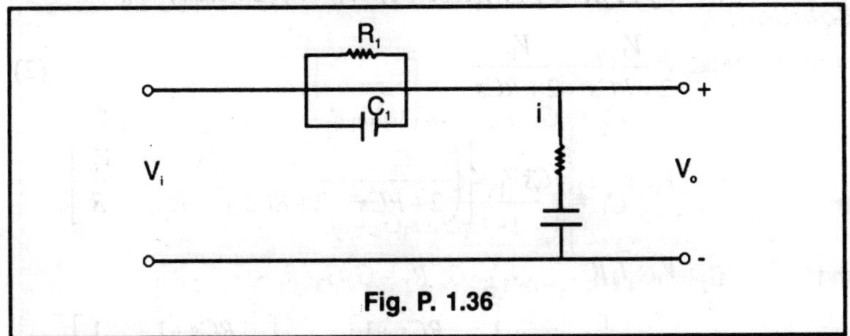

Fig. P. 1.36

Solution:

Transfer function: It is the ratio of the laplace transform of the output and input quantities assuming initial conditions zero, for linear time invariant system.

The transfer function is an expression in S-domain relating the output and the input of the LTI systems in terms of system parameters and independent of the input. It describes the input-output behaviour of the system and does not give any information concerning the internal structure of the system.

The transfer function of the given lag-lead compensator:

$$T(s) = \frac{V_0(s)}{V_i(s)} = \left(\frac{\left(s + \dfrac{1}{R_1C_1}\right)\cdot\left(s + \dfrac{1}{R_2C_2}\right)}{s^2 + \left(\dfrac{1}{R_1C_1} + \dfrac{1}{R_2C_1} + \dfrac{1}{R_2C_2}\right)s + \dfrac{1}{R_1R_2C_1C_2}} \right)$$

Problem 1.37. Find the outputs C_1 and C_2 of the system of Fig P.1.37

(I.E.S.-98)

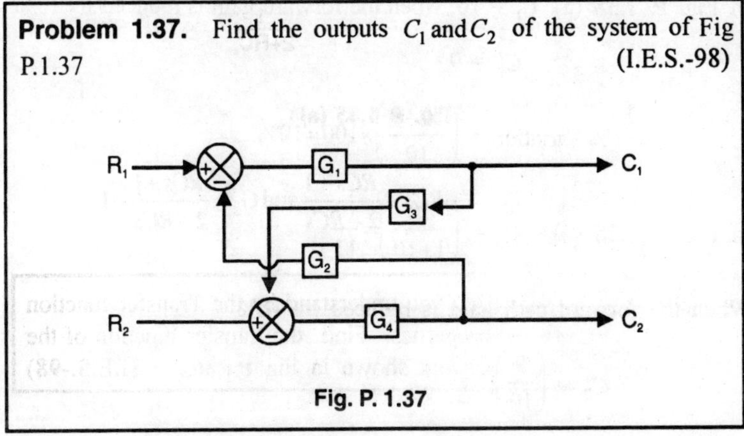

Fig. P. 1.37

Solution:

From the Fig. P. 1.37

$$C_1 = G_1(R_1 - G_2 C_1) = G_1 R_1 - G_1 G_2(R_2 - G_3 C_1) G_4$$

or

$$C_1 = \left(\frac{G_1 R_1 - G_1 G_2 G_4 R_2}{1 - G_1 G_2 G_3 G_4} \right)$$

and

$$C_2 = G_4(R_2 - C_1 G_3) = G_4 R_2 - G_4 G_3(R_1 - G_2 C_2) G_1$$

$$\therefore \quad C_2 = \left(\frac{G_4 R_2 - G_1 G_3 G_4 R_1}{1 - G_1 G_2 G_3 G_4} \right)$$

Problem 1.38. Consider the system shown in figure-I and figure-II. If the forward path gain is reduced by 10% in each system, then the variation in C_1 and C_2 will be in what percent?   (I.E.S.-99)

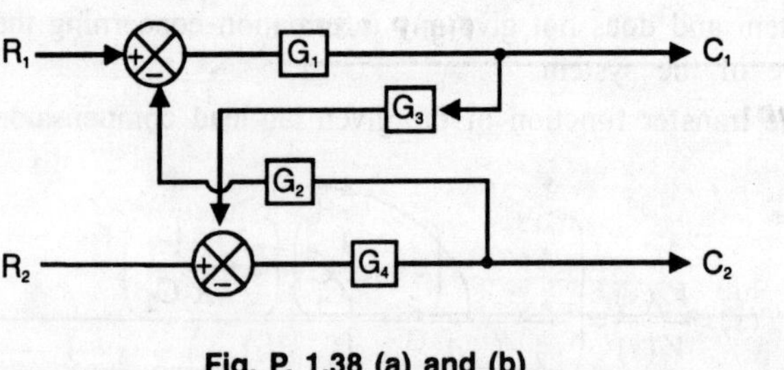

Fig. P. 1.38 (a) and (b)

Solution:

In Fig. P. 1.38 (a) $C_1 = 10$; when the forward gain is reduced by 10%

$$C_1' = 9$$

$$\therefore \quad \% \text{ Variation} = \left(\frac{10 - 9}{10} \right) \times 100 = 10\%$$

In Fig. P. 1.38 (b) $C_2 = \left(\dfrac{10}{1 + 10} \right) = \dfrac{10}{11}$

When the forward path gain is reduced by 10%

$$C_2' = \left(\frac{9}{10} \right)$$

$$\therefore \% \text{ Variation in } C_2 = \left(\frac{\dfrac{10}{11} - \dfrac{9}{10}}{\dfrac{10}{11}} \right) \times 100 = 1\%$$

Problem 1.39. Find the signal flow graph of the block diagram as below:

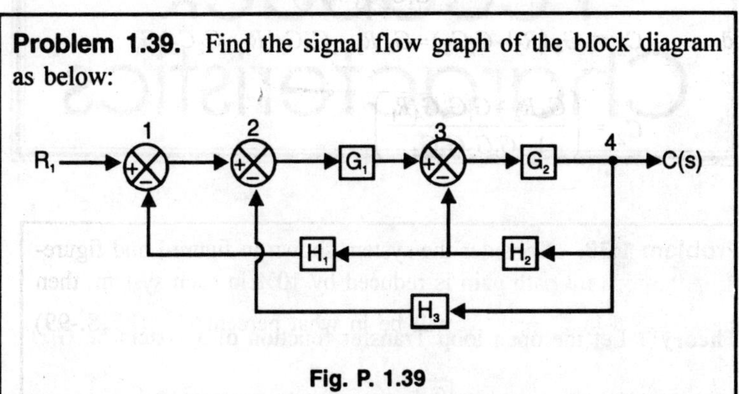

Fig. P. 1.39

Solution:

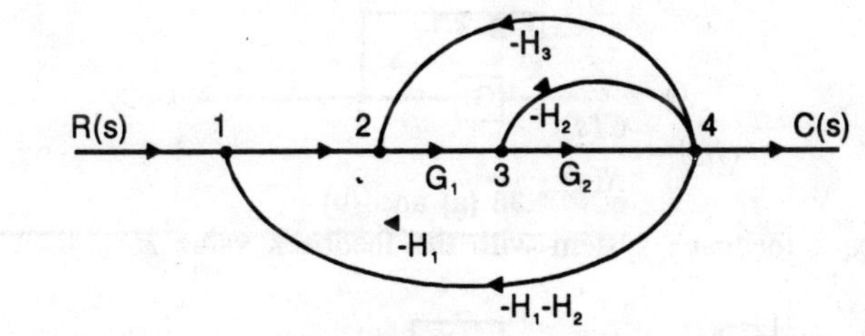

Fig. P. 1.39

Forward path: Loops:

$$G_1 G_2 \qquad\qquad -G_1 G_2 H_1 H_2$$

$$-G_2 H_2$$

$$-G_1 G_2 H_3$$

$$\Delta_1 = 1, \quad \Delta = (1 + G_2 H_2 + G_1 G_2 H_3 + G_1 G_2 H_1 H_2)$$

CHAPTER 2

Feedback Characteristics

Theory: Let the open loop Transfer function of a system be *G(s)*

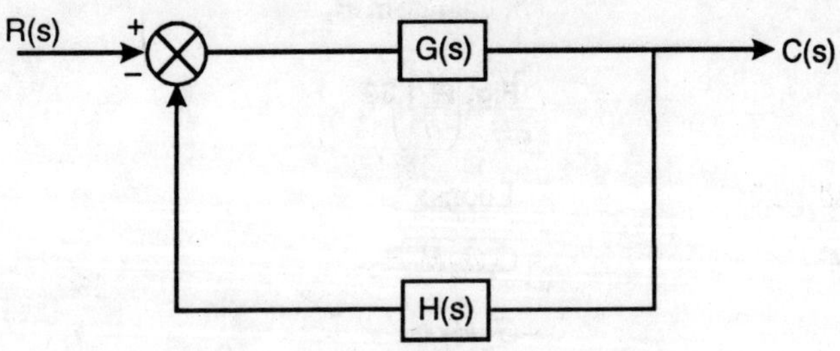

Fig. 2.1

then, $$G(s) = \frac{C(s)}{R(s)}$$

If there be a feedback system with the feedback value *H(S)* then

Fig. P. 2.2

closed loop Transfer function $G'(s) = \dfrac{G(s)}{1+G(s)H(s)}$

i.e., $$\frac{C(s)}{R(s)} = \frac{G(s)}{1+G(s)H(s)} = \frac{G}{1+GH}$$

So, with feedback, open loop transfer function (OLTF) or gain reduces by a factor of $(1 + GH)$.

CLTF $$T = \frac{G}{1+GH}$$

Sensitivity

Sensitivity with respect to forward path gain $G(S)$ is defined as

$$\text{Sensitivity} = \frac{\text{Percentage change in } T(S)}{\text{Percentage change in } G(S)}$$

$$\Rightarrow \qquad S_G^T = \frac{\partial T / T}{\partial G / G} = \frac{\partial T}{\partial G} \times \left(\frac{G}{T} \right)$$

For open loop system, $\quad T = G, \ S_G^T = 1$

For closed loop system, $\quad T = \dfrac{G}{1 + GH}$

$$S_G^T = \frac{\partial T}{\partial G} \times \frac{G}{T} = \frac{(1+GH)-GH}{(1+GH)^2} \times \frac{G}{G/1+GH}$$

$$S_G^T = \frac{1}{1+GH} \tag{1}$$

Sensitivity with respect to feedback factor H(S)

$$S_H^T = \frac{\% \text{ change in } T}{\% \text{ change in } H} = \frac{\partial T / T}{\partial H / H}$$

$$S_H^T = \frac{\partial T}{\partial H} \times \left(\frac{H}{T} \right)$$

$$\frac{\partial T}{\partial H} = \frac{-G^2}{(1+GH)^2}$$

$$S_H^T = -\frac{GH}{(1+GH)} \tag{2}$$

For $GH \gg 1 \qquad S_H^T = -1$

Thus, from (1) & (2) we conclude that a closed loop control system is more sensitive to variations in feedback path parameters than the variations in forward path parameters.

Also, from (1), we see that open loop sensitivity is reduced by a factor of $(1 + GH)$.

Problem 2.1. The block diagram of a position control system is shown in fig. Determine the sensitivity of CLTF T w.r.t G and H for ω = 1 rad/sec.

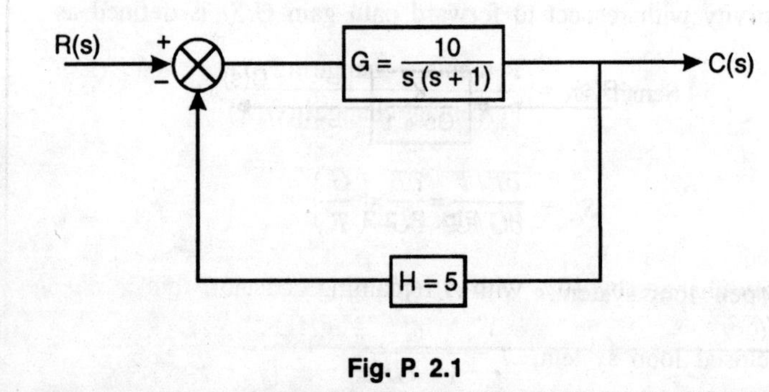

Fig. P. 2.1

Solution:

CLTF
$$T(s) = \frac{G}{1+GH}$$

$\Rightarrow$
$$T(s) = \frac{10/s(s+1)}{1+50/s(s+1)} = \frac{10}{s^2+s+50}$$

$$S_G^T = \frac{1}{1+GH} = \frac{1}{1+\dfrac{50}{s(s+1)}} = \frac{s(s+1)}{s^2+s+50}$$

$$S_G^T(j\omega) = \frac{j\omega(1+j\omega)}{-\omega^2+j\omega+50}$$

$$S_G^T(j\omega)\Big|_{\text{at }\omega=1} = \frac{\sqrt{1^2+1^2}}{\sqrt{49^2+1^2}} = 0.0288$$

$$S_H^T = -\frac{GH}{1+GH} = -\frac{50/s(s+1)}{1+\dfrac{50}{s(s+1)}} = \frac{-50}{s^2+s+50}$$

$$S_H^T(j\omega)\Big|_{\text{at }\omega=1} = \frac{-50}{49+j} = 1.0202$$

Problem 2.2. Given fig. shows open loop System with $K = 10$.

For $R(S) = \dfrac{A}{S}$ find A for $C(SS) = 1$.

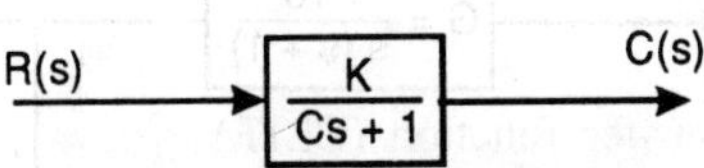

Fig. P. 2.2

K now changes by 10% with A remaining constant, find % change is $C(SS)$.

Solution:

$$G(s) = \frac{K}{Cs+1} = \frac{C(s)}{R(s)}$$

$$\Rightarrow \qquad C(s) = R(s)G(s) = \frac{KA}{s(Cs+1)}$$

Applying final value theorem to get the steady state value $C(ss)$.

$$C(ss) = \lim_{s\to 0} sC(s) = \lim_{s\to 0}\left(\frac{KA}{Cs+1}\right) = KA$$

given $\qquad C(ss) = 1$

$$\Rightarrow \qquad KA = 1$$

$$\therefore \qquad A = \left(\frac{1}{K}\right) = \frac{1}{10} = 0.1$$

Since $C(ss) \propto K$

Thus, 10% change in K will result in 10% change of $C(SS)$.

Problem 2.3. In the closed loop system shown in Fig. P.2.3, find the system sensitivty $S_{G_1}^T$ and $S_{G_2}^T$.

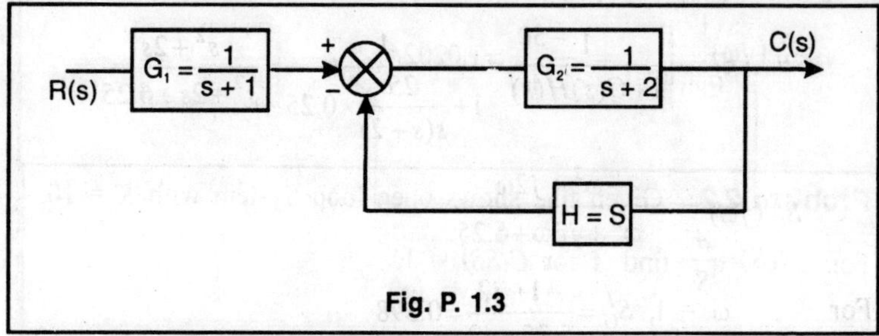

Fig. P. 1.3

Solution:

The closed loop transfer function (CLTF)

$$T(s) = \frac{G_1(s) \cdot G_2(s)}{1 + G_2(s)H(s)}$$

Now $\quad S_{G_1}^{T} = \dfrac{\partial T}{\partial G_1} \cdot \dfrac{G_1}{T} = \left(\dfrac{G_2}{1+G_2 H}\right)\left(\dfrac{(1+G_2 H)G_1}{G_1 \cdot G_2}\right) = 1$

$$S_{G_2}^{T} = \frac{\partial T}{\partial G_2} \cdot \frac{G_2}{T} = \frac{G_1(1+G_2 H) - G_1 G_2 H}{(1+G_2 H)^2} \cdot \left(\frac{1+G_2 H}{G_1}\right) = \left(\frac{1}{1+G_2 H}\right)$$

Putting the given values,

$$S_{G_2}^{T} = \frac{1}{1+\dfrac{s}{s+2}} = \frac{(s+2)}{2(s+1)}$$

Problem 2.4. Determine the sensitivity of the overall T.F. for the system shown in fig. at $\omega = 1$ rad/sec. w.r.t. G & H.

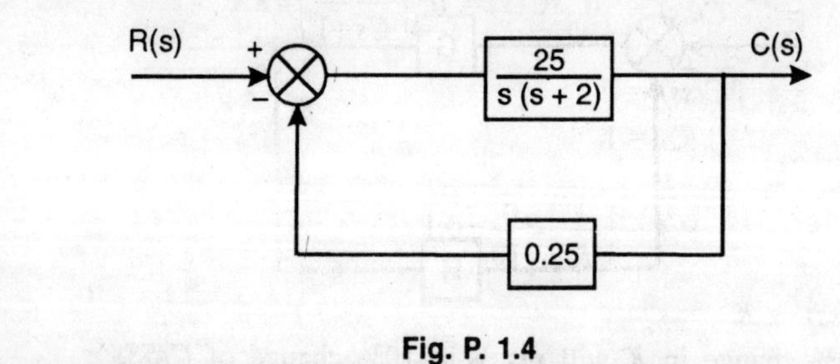

Fig. P. 1.4

Solution:

Given $\qquad G(s) = \dfrac{25}{s(s+2)}$ and

$$H(s) = 0.25$$

Now, Sensitivity of overall transfer function T w.r.t. G

$$S_G^T = \frac{1}{1+G(s)H(s)} = \frac{1}{1+\dfrac{25}{s(s+2)} \times 0.25} = \frac{s^2+2s}{s^2+2s+6.25}$$

$$S_G^T(j\omega) = \frac{-\omega^2+j2\omega}{-\omega^2+j2\omega+6.25}$$

For $\qquad \omega = 1, \; S_G^T = \dfrac{-1+j2}{5.25+j2} = 0.398$

Sensitivity of T w.r.t H,

$$S_H^T = -\frac{G(s)H(s)}{1+G(s)H(s)} = \frac{-\left(\dfrac{25}{s(s+2)}\right) \times 0.25}{1+\dfrac{25}{s(s+2)} \times 0.25} = \frac{-6.25}{s^2+2s+6.25}$$

$$S_H^T(j\omega) = \frac{-6.25}{-\omega^2+j2\omega+6.25}$$

For $\qquad \omega = 1, \; S_H^T = \dfrac{-6.25}{5.25+j2} = -1.11$

Problem 2.5. A closed loop control system for regulating the generator terminal voltage E_0 is shown in fig. The forward path gain $G = 200$ and the feedback factor $H = 0.1$. Determine

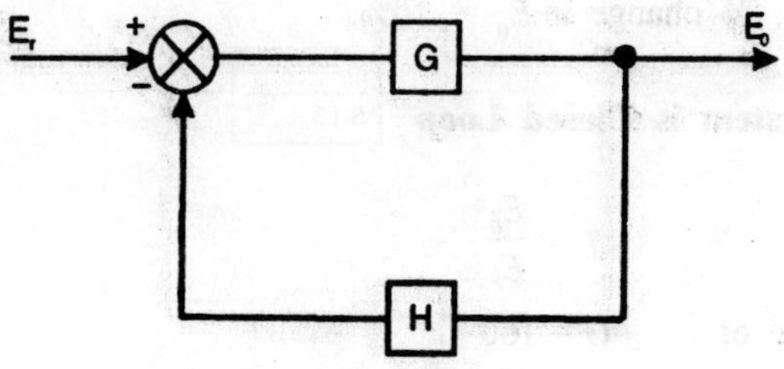

Fig. P. 1.5

(a) The reference voltage E_r for keeping $E_0 = 250$.
(b) The % change in generator terminal voltage if the forward path gain is reduced by 10%.

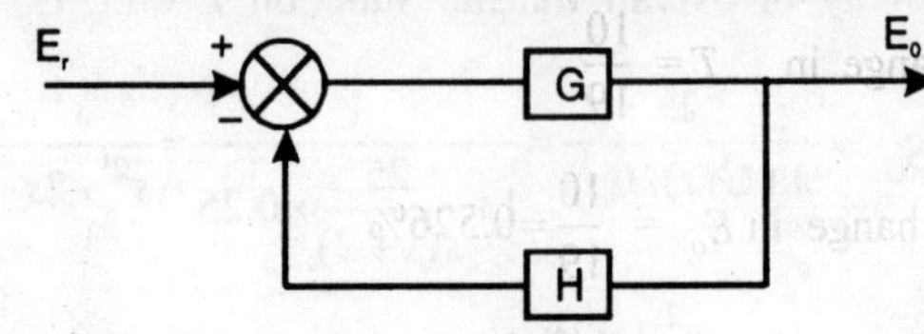

Fig. P. 2.5 (a)

Determine the results (a) and (b) if the system is made open loop.

Solution:

(a) If the system is open loop

$$\frac{E_o}{E_r} = G \Rightarrow E_r = \frac{E_o}{G} = \frac{250}{200} = 1.25V$$

If the system is closed loop

$$T = \frac{E_o}{E_r} = \frac{G}{1+GH} = \frac{200}{1+200\times0.1} = \frac{200}{21}$$

$$\Rightarrow \qquad E_r = \frac{21}{200} \cdot E_0 = \frac{21}{200} \times 250 = 26.25V$$

(b) If the system is open loop

$$G = \frac{E_o}{E_r} \Rightarrow E_o = G.E_r$$

$$\Rightarrow \text{ \% change in } E_o = \text{\% change in } G$$

i.e., % change in E_o = 10%.

If the System is Closed Loop

$$T = \frac{E_o}{E_r}$$

New value of $G = 180$

Also, $E_o \propto T \Rightarrow$ % change in E_o = % change in T

Sensitivity $S_G^T = \frac{1}{1+GH} = \frac{1}{1+180\times0.1} = \frac{1}{19}$

$$\Rightarrow \qquad \frac{1}{19} = \frac{\%\,\text{change in}\,T}{\%\,\text{change in}\,G}$$

$$\Rightarrow \quad \% \text{ change in } T = \frac{10}{19}$$

Thus, % change in $E_o = \frac{10}{19} = 0.526\%$

Problem 6. Find the magnitude of sensitivity $|S_K^T|$ at $\omega = 5$ for the system given below.

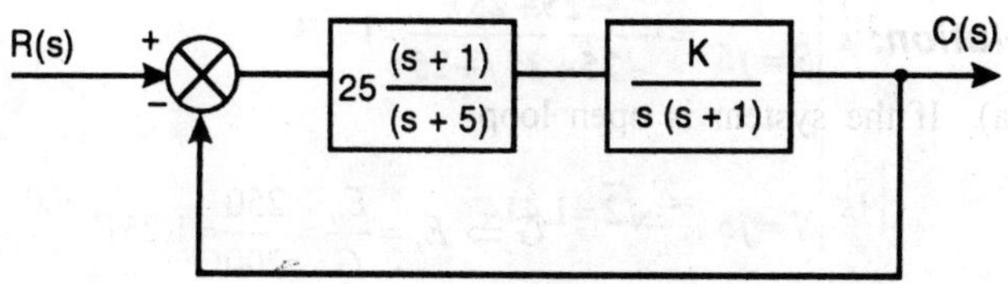

Fig. P. 2.6

Figure: Single Loop Configuration for $T(s) = \dfrac{25}{s^2 + 5s + 25}$

Solution:

The Transfer function can be written as

$$T = \frac{kG}{1 + kGH}$$

The sensitivity function S_K^T is defined as

$$S_K^T = \frac{\partial T / T}{\partial K / K} = \frac{\partial T}{\partial K} \times \frac{K}{T}$$

Now

$$\frac{\partial T}{\partial K} = \frac{(1 + KGH)G - KG(GH)}{(1 + KGH)^2} = \frac{G}{(1 + KGH)^2}$$

$$S_K^T = \frac{G}{(1 + KGH)^2} \times \frac{K}{\dfrac{KG}{(1 + KGH)}} = \frac{1}{1 + KGH}$$

Putting $K = 1$, we get $S_K^T = \dfrac{1}{1+GH}$

from Fig. P.2.6 $G = \dfrac{25}{s(s+5)}; H = 1$

Hence $S_K^T = \dfrac{1}{1+\dfrac{25}{s(s+5)}} = \dfrac{s^2+5s}{s^2+5s+25}$

$$S_K^T\Big|_{S=j5} = \frac{-25+25j}{-25+25j+25} = 1+j1$$

$$\left|S_K^T\right|_{S=j5} = \sqrt{2} = 1.414$$

Problem 2.7. Determine S_K^T for the following figure for $\omega = 5$

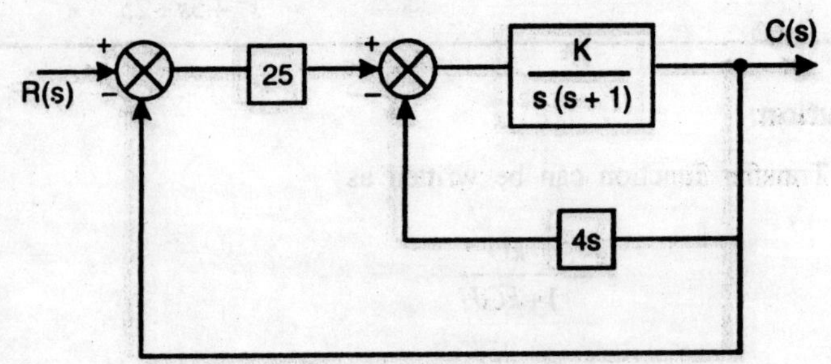

Fig. P. 2.7

Figure A two loop configuration for $T(s) = \dfrac{25}{s^2+5s+25}$

Solution:

As was derived in the previous problem

$$S_K^T = \frac{1}{1+KGH}$$

Putting $K = 1$, $S_K^T = \dfrac{1}{1+GH}$ and value of sensitivity can be calculated but we are presenting here an alternative solution for this.

$$G(s) = 25\left\{\frac{\dfrac{K}{s(s+1)}}{1+\dfrac{4Ks}{s(s+1)}}\right\} = \frac{25K}{s^2+s+4sK}$$

$$T = \frac{25K/s^2+s+4sK}{1+\dfrac{25K}{s^2+s+4sK}} = \frac{25K}{s^2+s+4sK+25K}$$

$$S_K^T = \frac{\partial T}{\partial K}\left(\frac{K}{T}\right)$$

Now

$$\frac{\partial T}{\partial K} = \frac{25(s^2+S)}{(s^2+4sK+s+25K)^2}$$

and

$$S_K^T = \frac{25(s^2+s)}{(s^2+4sK+s+25K)} \cdot \frac{K}{(25K/s^2+s+4sK+25K)^2}$$

$$= \frac{s^2+s}{s^2+4sK+s+25K}$$

Putting $K = 1$, we have

$$S_K^T = \frac{s^2+s}{s^2+5s+25}$$

$$\left. S_K^T \right|_{s=j5} = \frac{-25+5j}{25j}$$

$$\Rightarrow \qquad \left| S_K^T \right|_{s=j5} = \frac{\sqrt{650}}{25} = 1.02$$

Problem 8. A feedback control system is shown in the fig. Use the rules of block diagram algebra in order to find the closed loop transfer function $\dfrac{C(s)}{R(s)}$

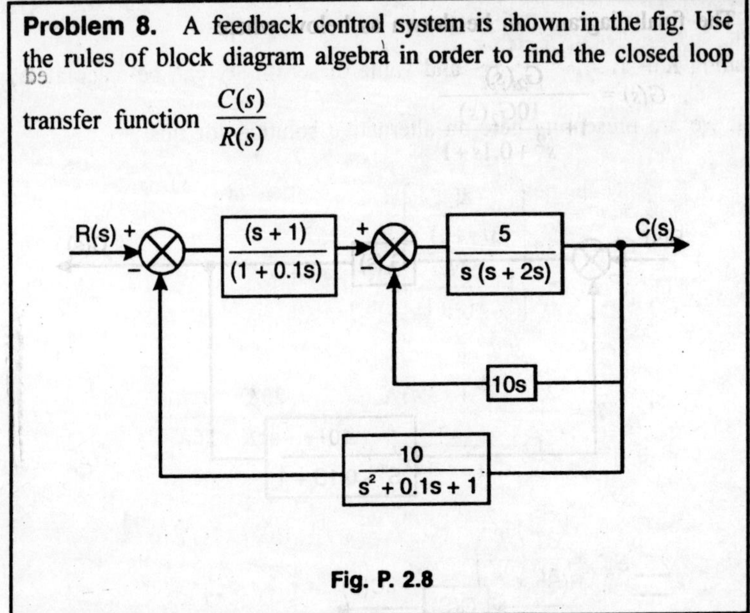

Fig. P. 2.8

Solution:

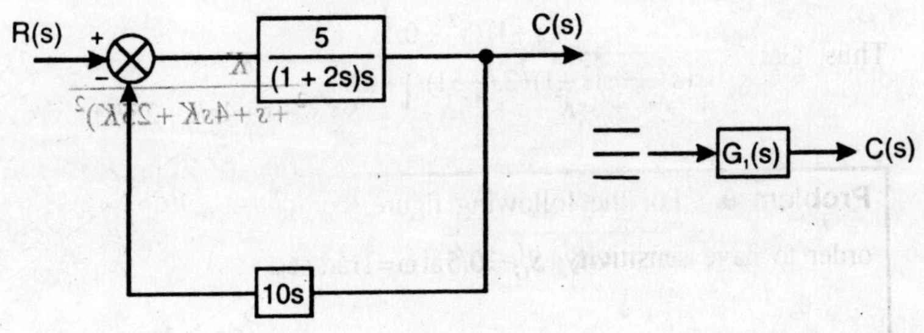

Fig. P. 2.8 (a)

$$\dot{G_1} = \frac{5/s(1+2s)}{1+\dfrac{50s}{s(1+2s)}} = \frac{5}{2s^2+51s}$$

Now $\qquad G_2(s) = \dfrac{(s+1)}{(1+0.1s)} \cdot G_1(s) = \dfrac{5(s+1)}{s(2s+51)(1+0.1s)}$

The final diagram can be drawn as below where

$$G(s) = \frac{G_2(s)}{1 + \dfrac{10G_2(s)}{s^2 + 0.1s + 1}}$$

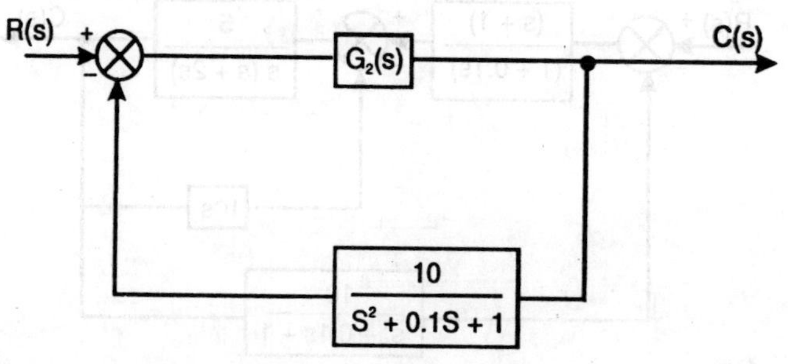

Fig. P. 2.8

Thus $G(s) = \dfrac{5(s+1)(s^2+0.1s+1)}{s(s^2+0.1s+1)(2s+51)(1+0.1s)+50(s+1)} = \dfrac{C(s)}{R(s)}$

Problem 9. For the following figure, compute the value of a in order to have sensitivity $S_G^T = 0.5$ at $\omega = 1\,\text{rad/sec}$.

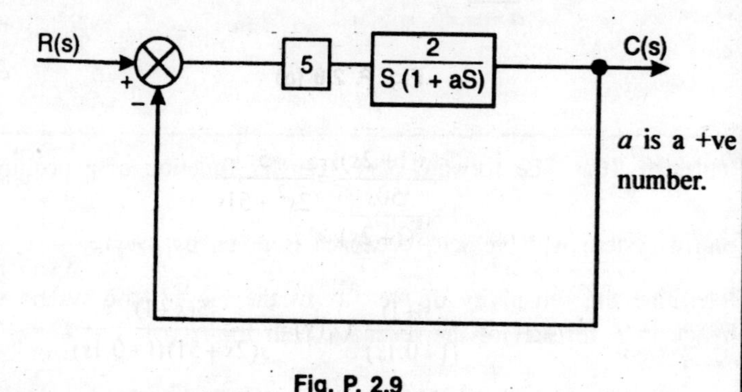

Fig. P. 2.9

Solution:

$$G(s) = \frac{10}{s(1+as)}, \quad H=1$$

$$S_G^T = \frac{1}{1+GH} = \frac{1}{1+\dfrac{10}{s(1+as)}} = \frac{s(1+as)}{as^2+s+10}$$

$$\left| S_G^T \right|_{\omega=1} = \frac{j(1+ja)}{-a+j+10}$$

$$\left| S_G^T \right|_{\omega=1} = \frac{1 \cdot \sqrt{1+a^2}}{\sqrt{(10-a)^2+1}}$$

But given $\qquad \left| S_G^T \right|_{\omega=1} = 0.5$

$$\Rightarrow \qquad 4\sqrt{1+a^2} = 0.5\sqrt{(10-a)^2+1}$$

$$\Rightarrow \qquad 4(1+a^2) = (10-a)^2+1 = 100-20a+a^2+1$$

$$\Rightarrow \qquad 3a^2+20a-97 = 0$$

$$\Rightarrow \qquad a = \frac{-20 \pm \sqrt{400+12\times97}}{6} = 3.26; \; -9.92$$

a is given to be a positive number

$$\therefore \qquad a = 3.26$$

Problem 10. The forward path Transfer function of a position control system with velocity feedback is given by $G(s) = \dfrac{K}{s(s+P)}$. Determine the sensitivity of the T.F. of the closed-loop system to change in P for T.F. of the feedback path for $H(s) = 1$ if $K = 12$, $p = 3$.

Solution:

$$G(s) = \frac{K}{s(s+P)}$$

$$T(s) = \frac{K/s(s+P)}{1+\dfrac{K}{s(s+P)}} = \frac{K}{s^2+Ps+K}$$

Sensitivity of CLTF with respect to P

$$S_P^T = \frac{\partial T}{\partial P} \times \left(\frac{P}{T}\right)$$

$$\frac{\partial T}{\partial P} = \frac{-Ks}{(s^2+Ps+K)^2} = \frac{-12s}{(s^2+Ps+12)^2}$$

$$S_P^T = \frac{-12s}{(s^2+Ps+12)^2} \times \frac{P(s^2+Ps+12)}{1^2}$$

$$= \frac{-Ps}{s^2+Ps+12} = \frac{-3s}{s^2+3s+12}$$

Problem 11. Determine the sensitivity of the C.L.T.F. of OLTF $G(s)=\dfrac{K}{s(s+P)}$ with respect to α when the feedback factor $H(s)=(1+\alpha s)$. Take $K = 12$, $P = 3$ and $\alpha = 0.14$.

Solution:

$$G(s) = \frac{K}{s(s+P)}$$

$$T(s) = \frac{K/s(s+P)}{1+\dfrac{K}{s(s+P)}\times(1+\alpha s)} = \frac{K}{s^2+Ps+K+K\alpha s}$$

$$= \frac{K}{s^2+Ps+K\alpha s+K} = \frac{K}{s^2+s(P+K\alpha)+K}$$

Putting values of $K = 12$ and $P = 3$

$$T(s) = \frac{12}{s^2 + s(3+12\alpha) + 12}$$

Now $$S_\alpha^T = \frac{\partial T}{\partial \alpha} \times \frac{\alpha}{T}$$

$$\frac{\partial T}{\partial \alpha} = \frac{12(-12)}{(s^2 + s(3+12\alpha) + 12)^2}$$

$$S_\alpha^T = \frac{-144}{(s^2 + s(3+12\alpha) + 12)^2} \times \frac{\alpha(s^2 + s(3+12\alpha) + 12)}{12}$$

$$= \frac{-12\alpha}{s^2 + s(3+12\alpha) + 12}$$

Now putting $\alpha = 0.14$,

$$S_\alpha^T = \frac{-12 \times 0.14}{s^2 + s(3 + 12 \times 0.14) + 12} = \frac{-1.68\,s}{s^2 + 4.68\,s + 12}$$

Problem 12. Find the characteristic equation for a system having differential equation $\dfrac{d^3 y}{dt^3} + 9\dfrac{d^2 y}{dt^2} + 3\dfrac{dy}{dt} + y = 29$. Also determine the sensitivity w.r.t. open loop gain.

Solution:

Writing the differential equation in Laplace's transform assuming initial conditions to be zero.

$$s^3 y(s) + 9s^2 y + 3sy + y = \frac{29}{s} u(s)$$

$\Rightarrow$ Overall Transfer function (Open Loop)

$$G(s) = \frac{Y(s)}{U(s)} = \frac{29/s}{s^3 + 9s^2 + 3s + 1} = \frac{29}{s^4 + 9s^3 + 3s^2 + s}$$

CLTF $$T(s) = \frac{G(s)}{1 + G(s)H(s)} = \frac{29}{s^4 + 9s^3 + 3s^2 + s + 29}$$

Sensitivity S_G^T of CLTF is defined as

$$S_G^T = \frac{1}{1+GH}$$

$$\Rightarrow \qquad S_G^T = \frac{1}{1+\dfrac{29}{s^4+9s^3+3s^2+s}} = \frac{s^4+9s^3+3s^2+s}{s^4+9s^3+3s^2+s+29}$$

Now, Characteristic equation of a polynominal is the denominator of the closed loop transfer function equated to zero.

Denominator of CLTF $T(s) = s^4+9s^3+3s^2+s+29$

Thus, Characteristic equation for the given system is

$$s^4+9s^3+3s^2+s+29 = 0$$

Problem 13. Given: $KK_t = 99; s = j1\,\text{rad/sec}$, find the sensitivity of the closed-loop system (shown in fig.) to variation in parameter K.

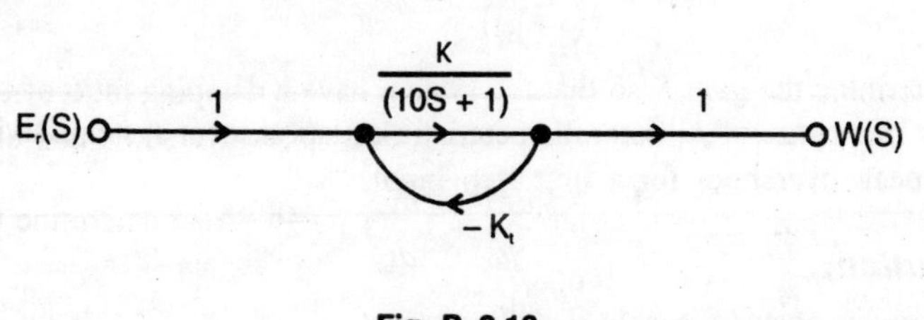

Fig. P. 2.13

Solution:

$$S_K^T = \frac{1}{1+GH} = \frac{1}{1+\dfrac{KK_t}{10s+1}} = \frac{10s+1}{10s+1+KK_t} = \frac{10s+1}{10s+100}$$

$$S_K^T \Big|_{j=1} = \frac{\sqrt{10^2+1}}{\sqrt{10^2+100^2}} = \frac{\sqrt{101}}{\sqrt{10100}} = 0.1$$

CHAPTER 3

Time Response Analysis

Problem 3.1.   A unity feedback system is characterized by an open-loop transfer function

$$G(s) = \frac{K}{s(s+10)}$$

Determine the gain K so that the system have a damping ratio of 0.5. For this value of K, determine setting time, peak overshoot and time to peak overshoot for a unit step input.

Solution:

The given open loop gain,

$$G(s) = \frac{K}{s(s+10)}$$

$\therefore$ The transfer function is, $T(s) = \left(\dfrac{K}{s^2+10s+\text{K}}\right)$

Now comparing the transfer function with the standard second order system's transfer function,

$$T(S) = \frac{w_n^2}{s^2+2\zeta w_n s+w_n^2}$$

We find that, $\zeta w_n = 5$ and $w_n^2 = K$

therfore, $\zeta = \dfrac{5}{w_n} = \left(\dfrac{5}{\sqrt{K}}\right)$ \hfill (1)

For the given value of $\zeta = 0.5$, [using eqn (1)]

$$K = \frac{25}{(0.5)^2} = 100$$

Now, the setting time, $\qquad t_s = \dfrac{4}{\zeta w_n} = \dfrac{4}{5} = 0.8\,\text{sec}$

peak overshoot, $\qquad M_p = e^{\frac{-\zeta\pi}{\sqrt{1-\zeta^2}}} = e^{\frac{-\pi\cdot(0.5)}{\sqrt{0.75}}} = 16.3\%$

Time to peak overshoot, $\qquad t_p = \dfrac{\pi}{w_n\sqrt{1-\zeta^2}} = \dfrac{3.14}{10\sqrt{1-0.5^2}} = 0.36\,\text{sec.}$

Problem 3.2. Measurements conducted on a servomechanism shown in system response to be $c\,(t) = 1 + 0.2\ e^{-60\,t} - 1.2\ e^{-10\,t}$ when subjected to a unit step input.

(a) Obtain the expression for the closed-loop transfer function.

(b) Determine the undamped natural frequency (w_n) and the damping ratio (ζ) of the system.

Solution:

(a) Taking laplace Transformation of the system response:

$$C(s) = \alpha\big[c(t)\big].$$

Therefore, $\qquad C(s) = \left(\dfrac{1}{s}\right) + \dfrac{0.2}{s+60} - \dfrac{1.2}{(s+10)}$

$$= \dfrac{1}{s} + \dfrac{1}{5} - \left(\dfrac{1}{s+60} - \dfrac{6}{s+10}\right)$$

$$= \dfrac{1}{s} - \dfrac{1}{5}\left(\dfrac{5s+350}{(s+60)(s+10)}\right)$$

$$= \dfrac{600}{s(s+10)(s+60)}$$

Then, $\qquad T(s) = \dfrac{C(s)}{R(s)}$

Here $\qquad R(s) = 1/s$

$$\therefore \qquad T(s) = \frac{600}{(s+60)(s+10)} = \left(\frac{600}{s^2+70s+600}\right)$$

(b) Now comparing it with the standard second order system's transfer function, we have,

$$T(s) = \frac{w_n^2}{s^2+2\zeta w_n s + w_n^2}$$

Here $\qquad w_n^2 = 600, \therefore w_n = \sqrt{600} = 24.50\,\text{rad/sec.}$ **Ans.**

and $\qquad 2\zeta w_n = 70$ the damping ratio, $\zeta = \left(\dfrac{35}{24.50}\right) = 1.43$ **Ans.**

Problem 3.3.   The open loop transfer function of a servo-system with unity feedback is,

$$G(s) = \frac{10}{s(0.1s+1)}$$

Evaluate the static error constants (k_p, k_v, k_a) for the system. Obtain the steady error of the system when subjected to an input given by the polynomial,

$$r(t) = a_0 + a_1 t + \frac{a_2}{2} t^2$$

Solution:

The given open loop transfer function,

$$G(s) = \left(\frac{10}{s(0.1s+1)}\right)$$

For the unit step input, $R(s) = 1/s$

$$e_{ss} = \lim_{s \to 0} SE(s) = \lim_{s \to 0} \frac{sR(s)}{1+G(s)} = \lim_{s \to 0} \frac{1}{1+\dfrac{10}{s(0.1s+1)}} = 0$$

Then, $\qquad K_p = \dfrac{1}{e_{ss}} = \infty.$ (Position error constant)

Now the unit-ramp input;

$$R(s) = 1/s^2$$

$$e_{ss} = \lim_{s\to 0} sE(s) = \lim_{s\to 0} \frac{sR(s)}{1+G(s)} = \lim_{s\to 0}\left(\frac{\frac{1}{s^2}\cdot s}{1+\dfrac{1}{s(0.1s+1)}}\right) = 1$$

then, $K_V = 1/e_{ss} = 1$ (velocity error constant)

For the unit-parabolic input;

$$R(s) = \left(1/s^3\right)$$

$$e_{ss} = \lim_{s\to 0} sE(s) = \lim_{s\to 0} \frac{sR(s)}{1+G(s)} = \lim_{s\to 0}\left(\frac{s\cdot 1/s^3}{1+\dfrac{10}{s(0.1s+1)}}\right) = \dot{\alpha}$$

Then, Acceleration error constant, $K_a = 1/e_{ss} = 0$.

When $r(t) = a_0 + a_1 t + \dfrac{a_2}{2}t^2$

then, $R(s) = \left(\dfrac{a_0}{s} + \dfrac{a_1}{s^2} + \dfrac{a_2}{s^3}\right)$, $e_{ss} = \lim_{s\to 0}\left(\dfrac{sR(s)}{1+G(s)}\right)$

then, $e_{ss} = \lim_{s\to 0}\left(\dfrac{S\cdot\left(\dfrac{a_0}{s} + \dfrac{a_1}{s^2} + \dfrac{a_2}{s^3}\right)}{1+\dfrac{10}{s(0.1s+1)}}\right) = \lim_{s\to 0}\left(\dfrac{\left(a_0 s^2 + a_1 s + a_2\right)}{s\big(s(0.1s+1)+10\big)}\right) = \infty.$

Problem 3.4. Consider the system shown below:

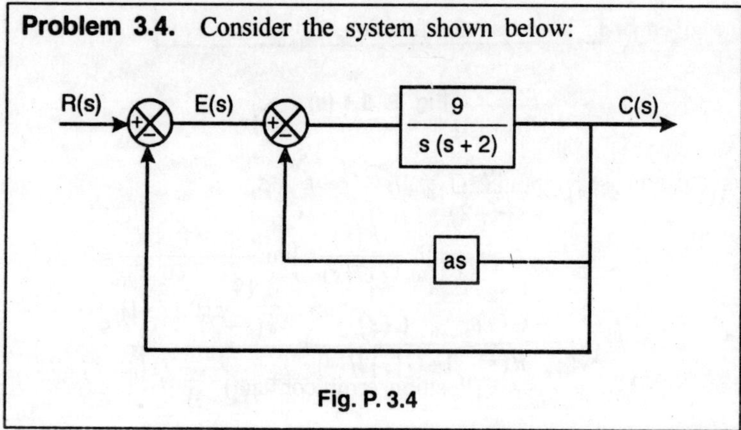

Fig. P. 3.4

(i) In the absence of derivative feedback ($a = 0$), determine the damping factor and natural frequency, also determine the steady state error resulting from a unit ramp output.

(ii) Determine the derivative feedback constant. Which will increase the damping factor of the system to 0.7. What is the steady state error to unit ramp input with this setting of the derivative feeback constant?

(iii) Illustrate how the steady state of the system with the derivative feedback to unit-ramp can be reduced to the same value as in part:

    (a) While the damping factor is maintained at 0.7.

    (b) The open loop transfer function of unity feedback system is given by

$$G(s) = \frac{k}{s(T_1 s + 1)(T_2 s + 1)}$$

Derive an expression for gain k in terms of T_1, T_2 and the specified gain margin G_m.

[IES 92]

Solution:

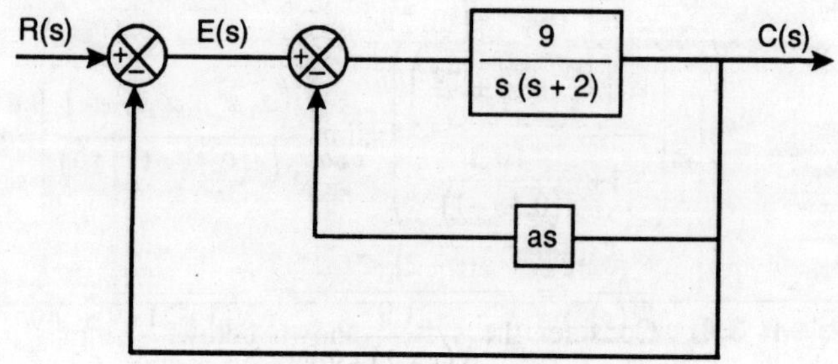

Fig. P. 3.4 (a)

(i) $$G(s) = \frac{9}{s(s+2)}; a=0$$

$$M(s) = \frac{C(s)}{R(s)} = \frac{C(s)}{1+G(s)H(s)} = \frac{\dfrac{9}{s(s+2)}}{1+\dfrac{9}{s(s+2)}} = \frac{9}{s(s+2)+9}$$

$$\therefore \quad s(s+2)+9 = 0; \text{ or } s^2+2s+9=0$$

Comparing with $s^2+2\zeta w_n s+w_n^2=0$,

$\therefore$ Natural frequency $w_n = \sqrt{9}=3$ and damping factor $= \dfrac{2}{2w_n}$

$$= \dfrac{2}{2\times3}=\dfrac{1}{3} \text{ i.e. } \zeta=1/3$$

Steady-state error $e_{SS} = \underset{s\to0}{\text{Lim}}\, sE(s)$

$$E(s) = \dfrac{R(s)}{1+G(s)} \text{ for unity feedback loop}$$

For ramp unit; $r(t) = tu(t)$, $R(s)=\dfrac{1}{s^2}$

$$\therefore \quad e_{ss} = \underset{s\to0}{\text{Lim}}\dfrac{s\dfrac{1}{s^2}}{1+\dfrac{9}{s(s+2)}}=\underset{s\to0}{\text{Lim}}\dfrac{1}{s+\dfrac{9}{s+2}}=\dfrac{2}{9}$$

(ii) with $a\neq0, \dfrac{C(s)}{E(s)} = \dfrac{\dfrac{9}{s(s+2)}}{1+\dfrac{9as}{s(s+2)}}=\dfrac{9}{s(s+2)+9as}$

$$\dfrac{C(s)}{R(s)} = \dfrac{\dfrac{9}{s(s+2)+9as}}{1+\dfrac{9}{s(s+2)+9as}}=\dfrac{9}{s(s+2)+9as+9}\oplus$$

$$\therefore \quad s(s+2)+9as+9 = 0$$

or, $\qquad s^2+2s+9as+9 = 0$

$$s^2+(2+9a)s+9 = 0$$

On comparing with $\quad s^2+2\zeta w_n s+w_n^2 = 0$

$$w_n = 3, \quad \zeta=\dfrac{2+9a}{2\times3}=0.7, a=\dfrac{2.2}{9}=0.2444$$

$$e_{ss} = \underset{s \to 0}{\text{Lim}}\; s \cdot \frac{1/s^2}{1 + \dfrac{9}{s(s+2)+9as}}$$

$$= \underset{s \to 0}{\text{Lim}}\; \frac{1}{s + \dfrac{9}{(s+2)+9a}} = \frac{2+9a}{9}$$

$$\frac{4.2}{9} = 0.4666$$

(iii)   System should have $\zeta = 0.7$ and $e_{ss} = \dfrac{2}{9}$

Now replacing 9 in $\dfrac{9}{s(s+2)}$ by k, $\dfrac{9}{s(s+2)} \to \dfrac{k}{s(s+2)}$

With this, $G(s) = \dfrac{k}{s(s+2)+kas}$

$$\frac{C(s)}{R(s)} = \frac{k}{s(s+2)+kas+9}$$

$$\therefore \; s(s+2)+ka+k = 0, \text{and } s^2 + s(2+ka)+k = 0$$

On comparing, $\overset{\cdot}{w_n} = \sqrt{k}\;; \zeta = \dfrac{2+k_a}{2\sqrt{k}}$

$$e_{ss} = \underset{s \to 0}{\text{Lim}}\; s \frac{1/s^2}{1 + \dfrac{k}{s(s+2)+kas}}$$

$$= \underset{s \to 0}{\text{Lim}}\; \frac{1}{s + \dfrac{k}{(s+2)+ka}} = \frac{2+ka}{k}$$

Thus $\dfrac{2+ka}{2\sqrt{k}} = 0.7$ and $\dfrac{2+ka}{k} = \dfrac{2}{9}$

$$\therefore \quad \frac{\sqrt{k}}{2} = 0.7 \times \frac{9}{2} \text{ or, } \sqrt{k} = 6.3 \text{ or, } k = 39.69$$

and $\dfrac{2+39.69a}{2 \times 6.3} = 0.7 \therefore a = 0.1718$

(b)
$$G(s) = \frac{k}{s(T_1 s+1)(T_2 s+1)}$$

$$= \frac{k}{s\left(1+T_1 T_2 s^2 + s(T_1+T_2)\right)}$$

$$G(s)H(s) = \frac{k}{s\left(1+T_1 T_2 s^2 + s(T_1+T_2)\right)}$$

For unity feedback loop

$$G(j\omega)H(j\omega) = \frac{k}{j\omega\left[1-T_1 T_2 \omega^2 + j\omega(T_1+T_2)\right]}$$

$$= \frac{k/\omega}{-\omega(T_1+T_2)+j(1-T_1 T_2 \omega)} \;.$$

$$= \frac{k/\omega\left[\omega(T_1+T_2)+j\left(1-T_1 T_2 \omega^2\right)\right]}{\omega^2(T_1+T_2)^2+\left(1-T_1 T_2 \omega^2\right)^2}$$

For finding phase crossover frequency ω_c, we put imaginary part of $G(j\omega)H(j\omega)=0$,

$$\left(1-T_1 T_2 \omega_c^2\right) = 0 \text{ or } \omega_c^2=\frac{1}{T_1 T_2}$$

Thus $G(j\omega_c)H(j\omega_c) = \dfrac{k/\omega_c \times \omega_c (T_1+T_2)}{\omega_c^2 (T_1+T_2)^2}$

$$= \frac{k}{\omega_c^2 (T_1+T_2)}=\frac{-kT_1 T_2}{(T_1+T_2)}$$

$$\left|G(j\omega_c)H(j\omega_c)\right| = \frac{kT_1 T_2}{T_1+T_2}$$

Gain margin $G_m = 20\log_{10}\dfrac{1}{\left|G(j\omega_c)H(j\omega_c)\right|}$

$$= 20\log_{10}\frac{T_1+T_2}{kT_1 T_2}$$

or, $\quad 10^{G_m/20} = \dfrac{T_1+T_2}{kT_1 T_2}$

or, $\quad k = \dfrac{T_1+T_2}{T_1 T_2 10^{G_m/20}}$ where G_m in dB

                                                          Control System

---

Problem 3.5. The loop transfer function (OLTE) of a unity feedback control system is given by

$$G(s) = \left(\frac{k}{s(s+1)}\right)$$

If the gain k is increased to infinity then what will be the damping ratio?

Solution:

Given
$$G(s) = \left(\frac{k}{s(s+1)}\right)$$

The characteristic equation corresponding to unity feedback control loop is

$$s(s + 1) + k = 0$$

i.e.
$$s^2 + s + k = 0$$

This can be written in the form of $s^2 + 2\zeta\omega_n s + \omega_n^2 = 0$, where ζ is the damping ratio. Comparing we get,

$$\omega_n = \sqrt{k}\;;\; \zeta = \frac{1}{2\sqrt{k}}$$

Thus if $k \to \infty$, $\zeta \to 0$

Problem 3.6. A first order system and its response to a unit step input are shown in the Figure given below. Determine the parameters a and k.

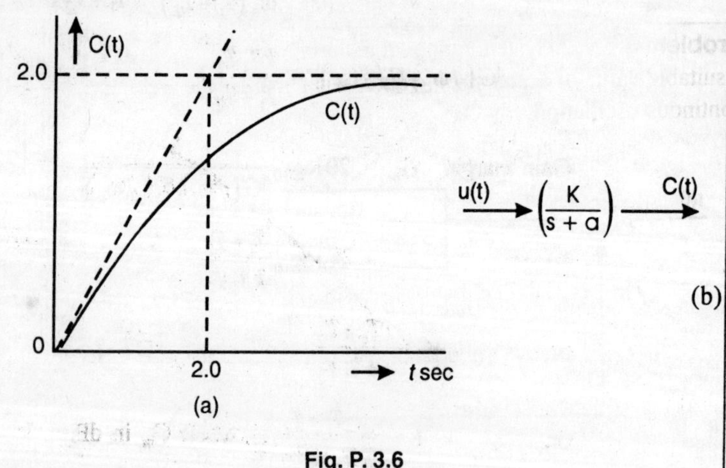

Fig. P. 3.6

Solution:

The system reponse,

$$C(s) = \left(\frac{k}{s+a}\right)U(s) = \frac{1}{s}\cdot\left(\frac{k}{s+a}\right)$$

Now, By the final value theorem,

$$C(\infty) = \lim_{s\to 0} sC(s) = \lim_{s\to 0}\left(\frac{k}{s+a}\right) = \left(\frac{k}{a}\right) = 2 \tag{1}$$

Now, From the figure, the slope,

$$\lim_{t\to 0}\left(\frac{d}{dt}c(t)\right) = \frac{2}{0.2} = 10 \tag{2}$$

taking laplace theorem,

$$\mathcal{L}\left(\frac{d}{dt}c(t)\right) = sc(s) = \left(\frac{k}{s+a}\right)$$

$$\frac{d}{dt}c(t) = \mathcal{L}^{-1}\left(\frac{k}{s+a}\right) = ke^{-at} \tag{3}$$

From (2) & (3), we get, $k = 10$ **Ans.** $\tag{4}$

$\therefore$ From (1) & (4) $a = \left(\frac{10}{2}\right) = 5.$ **Ans.**

Problem 3.7. A system is shown in the figure. If allowed to choose a suitable value of scalar parameter k, what will be the frequency for continous oscillation.

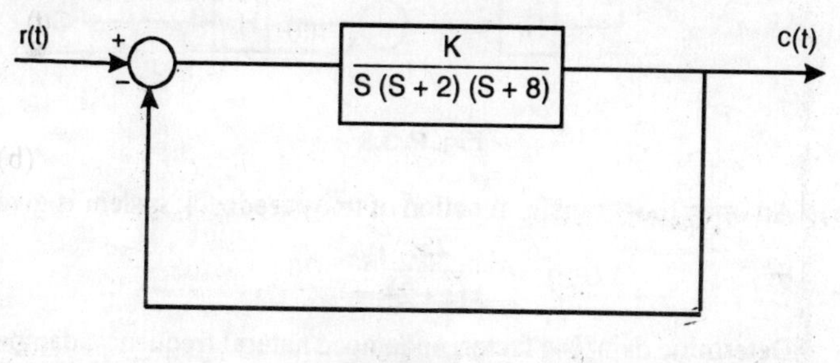

Fig. P. 3.7

Solution:

The condition for a system with feedback loop to oscillate is $\alpha\beta = -1$. where $\alpha = G(s)$ and β = Feedback path gain.

The characteristic equation for above system is

$$1 + \frac{k}{s(s+2)(s+8)} = 0$$

$$s(s+2)(s+8) + k = 0$$

$$s\left[s^2 + 10s + 16\right] + k = 0$$

$$j\omega\left(-\omega^2 + 10\,j\omega + 16\right) + k = 0$$

For the system to oscillate j terms should be set to zero.

$$\therefore \qquad -j\omega^3 + 16\,j\omega = 0 \ \text{ or } \ \omega = 4 \text{ rad/s}$$

Also, $\qquad -10\omega^2 + k = 0$

or $\qquad\qquad k = 10\omega^2 = 10\times16 = 160$

Problem 3.8. (a) Use block diagram reduction methods to obtain the equivalent transfer function from R to C.

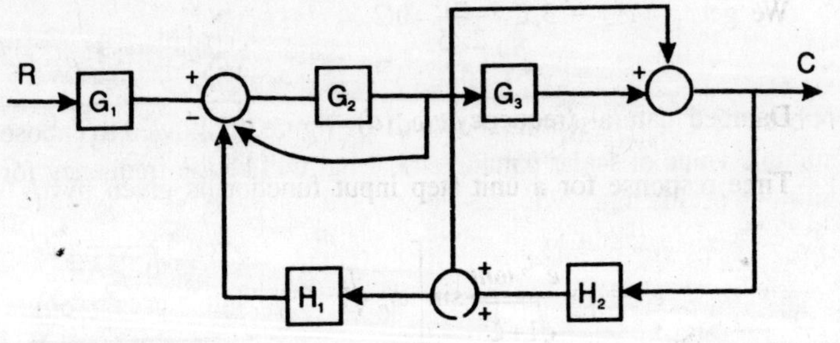

Fig. P. 3.8

(b) An open loop transfer function of unity feedback system is given

by $\qquad G(s) = \dfrac{25}{s(s+2)}$

Determine damping factor, undamped natural frequency, damped natural frequency and time response for a unit step input.

Solution:

(a)
$$\frac{C}{R} = \frac{G}{1+GH} = \frac{\dfrac{G_1 G_2 (1+G_3)}{(1+G_2)}}{1+\dfrac{G_1 G_2 (1+G_3)}{1+G_2}\dfrac{H_1}{G_1}\dfrac{1+H_2(1+G_2)}{(1+G_3)}}$$

$$= \frac{G_1 G_2 (1+G_3)}{(1+G_2)-(1+G_2 H_1 (1+H_2))}$$

(b)
$$G(s) = \frac{25}{s(s+2)}, H(s)=1$$

$$1 + G(s)\,H(s) = 1+\frac{25}{s(s+2)} = \frac{s(s+2)+25}{s(s+2)}$$

Characteristic equation is $1 + G(s)\,H(s) = 0$

$\therefore$
$$s(s + 2) + 25 = 0$$

or,
$$s^2 +2s+25 = 0$$

Comparing with
$$s^2 +2\zeta\omega_n s+\omega_n^2 = 0$$

We get
$$\omega_n = 5,\ \zeta=\frac{2}{2\times5}=0.2$$

Damped natural frequency, $\omega_n (d) = \omega_n \sqrt{1+\zeta^2}=4.899$

Time response for a unit step input function is given by

$$e^t = 1+\frac{e^{-\zeta\omega n t}}{\sqrt{1+\zeta^2}}\sin\left[\omega_n \sqrt{1-\zeta^2}\,t-\tan^{-1}\frac{\sqrt{1-\zeta^2}}{-\zeta}\right]$$

$$= 1+1.0209\,e^{-t} \sin(4\times99t-101.5°)$$

Problem 3.9. A servomechanism is used to control the angular position θ_0 of a mass through a command signal θ_1. The moment of inertia of moving parts referred to the load shaft is 200 kg-m^2 and the motor torque at the load is $6.88 \times 10^{+4}$ N-m per rad, of error.

The damping torque coefficient referred to the load shaft is 5×10^3 N-m/rad/sec.

(i) Find the time response of the servomechanism to a step input of 1 rad and determine the frequency of transient oscillation, The time to rise to the peak overshoot and the value of the peak overshoot.

(ii) Determine the steady state error when the command signal is a constant angular velocity of 1 revolution/minute.

(iii) Determine the steady state error which exists when a steady torque of 1200 N-m applied at load shaft.   [IES - 94]

Solution:

From
$$T = J\frac{d^2\theta}{dt^2} + D\frac{d\theta}{dt}$$

$$6.88\times10^4\left(\theta_i - \theta_0\right) = 200\frac{d^2\theta_0}{dt^2} + 5\times10^3\frac{d\theta_0}{dt}$$

Where θ_i is the command signal and θ_0 is the output variable load.

$$\therefore \qquad 68.8\left(\theta_i - \theta_0\right) = 0.2\frac{d^2\theta_0}{dt^2} + 5\frac{d\theta_0}{dt}$$

$$68.8\left(\theta_i - \theta_0\right) = 0.2\frac{d^2\theta_0}{dt^2} + 5\frac{d\theta_0}{dt}$$

Laplace transformation, $68.8\left[\theta_i(s) - \theta_0(s)\right] = 0.2s^2\theta_0(s) + 5s\theta_0(s)$

or,
$$\frac{\theta_0(s)}{\theta_i(s)} = \left[1 + \frac{5}{68.8} + \frac{0.2}{68.8}s^2\right] = \frac{68.8}{0.2s^2 + 5s + 68.8}$$

$$= \frac{344}{s^2 + 25s + 344}$$

$\therefore$ Characteristic equation of the system becomes, $s^2 + 25s + 344 = 0$

Comparing with $s^2 + 2\zeta\omega_n s + \omega_n^2 = 0$

$$\omega_n = \sqrt{344} = 18.547; \quad \zeta = \frac{25}{2\times18.547} = 0.674$$

$$\zeta\omega_n = 0.674 \times 18.547 = 12.5$$

$$\sqrt{1-\zeta^2} = \sqrt{1-0.674^2} = 0.7387$$

$$\sqrt{1-\zeta^2}\,\omega_n = 13.7\frac{\sqrt{1-\zeta^2}}{\zeta} = 1.096;$$

$$\frac{\tan^{-1}\sqrt{1-\zeta^2}}{\zeta} = 47.6°$$

Unit step response of a second order system as above, is given by

$$C(t) = 1 + \frac{e^{-\zeta\omega_n t}}{\sqrt{1-\zeta^2}}\sin\left(\omega_n\sqrt{1-\zeta^2}\,t - \tan^{-1}\frac{\sqrt{1-\zeta^2}}{\zeta}\right)$$

$$= 1 + \frac{e^{-12.5t}}{0.7387}\sin\left(13.7t + 47.6°\right)$$

$$= 1 + 1.3537\,e^{-12.5t}\sin\left(13.7t + 47.6°\right)$$

$$t_{max} = \frac{\pi}{\omega_n\sqrt{1-\zeta^2}} = \frac{\pi}{13.7} = 0.2293$$

Max. overshoot $= e^{-\pi\zeta}/\sqrt{1-\zeta^2} = e^{-\pi\times 0.9124} = 0.0596$

(i) Thus the frequency of transient oscillation is 13.7 rad/s, the time to rise to the peak overshoot is 0.2293 and the value of the peak overshoot is 0.0569.

(ii) For a unit ramp input $\zeta(t) = tu(t)$, the output is

$$\zeta_0(t) = L^{-1}\left[\frac{\omega_n^2}{s^2\left(s^2 + 2\zeta\omega_n s + \omega_n^2\right)}\right]$$

$$= t - \frac{2\zeta}{\omega_n} + \frac{1}{\omega_n\sqrt{1-\zeta^2}}e^{-\zeta\omega_n t}\sin\left[\omega_n\sqrt{1-\zeta^2}\,t - \phi\right]$$

Where $\quad \phi = 2\tan^{-1}\dfrac{\sqrt{1-\zeta^2}}{-\zeta}$

When t approached infinity, $q_0 = t - \dfrac{2\zeta}{\omega_n}$

Therefore the steady state error for such an input is $\dfrac{2\zeta}{\omega_n}$.

Now the input is a constant angular velocity of I rpm which corresponds to an input of

$$\theta = \frac{2\pi t}{60} u(t) \text{ radians,}$$

$$\therefore \quad ss \text{ error} = \frac{2\pi}{60} \times \frac{2\pi}{\omega_n} = \frac{2\pi}{60} \times \frac{2 \times 0.674}{18.547}$$

$$= 7.6 \times 10^{-3} \text{ rad} = 0.436°$$

(iii) At a steady torque of 1200 *N-m*. Since applied torque $t = 6.88 \times 10^4 \,(\theta_i - \theta_0)$

$$\text{We have} \quad \theta_i - \theta_0 = \frac{1200}{6.88 \times 10^4} = \frac{0.12}{6.88} = 0.1744 \text{ rad}$$

$$= 0.999°$$

Problem 3.10.   The system shown in the given fig. has a unit step input. In order that the steady state error is 0.1. Find the value of K required.

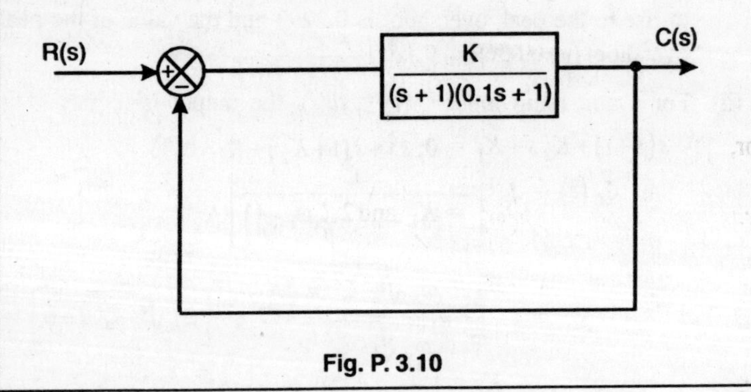

Fig. P. 3.10

Solution:

For a step input, steady state error $= \dfrac{1}{1 + K_p}$; where K_p is positional error constant.

$$k_p = \lim_{s \to 0} G(s) \cdot H(s) = \lim_{s \to 0} \frac{1}{(s+1)(0.1s+1)+K}$$

$$\therefore \quad \frac{1}{1+K_p} = \frac{1}{1+K} = 0.1 \quad \therefore \quad K = 9.0$$

Problem 3.11. The system shown in figure has second order system response with a damping ratio of 0.6 and frequency of damped oscillations of 10 rad/sec. Find the values of K_1 and K_2.

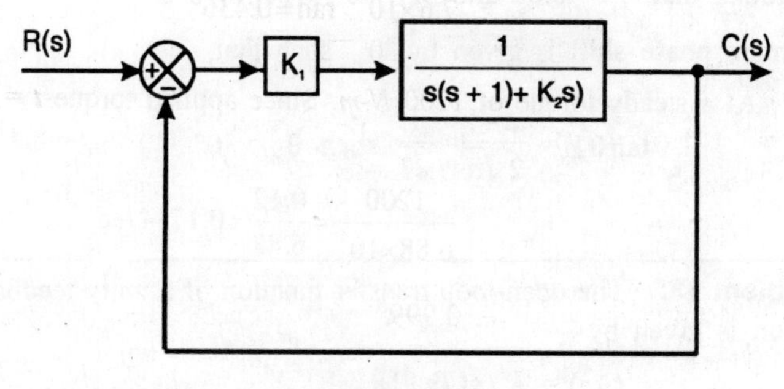

Fig. P. 3.11

Solution:

The characteristic equation of the system shown,

$$1+G(s)\cdot H(s) = 0, i.e. \ 1+\frac{K_1}{s(s+1)+K_2 s}=0$$

or,

$$s(s+1)+K_2 s+K_1 = 0; \ s^2+s(1+K_2)+K_1=0$$

$$\therefore \quad \omega_n^2 = K_1 \ \text{and} \ 2\,\zeta\omega_n=(1+K_2)$$

$$\omega_d = \omega_n\sqrt{1-\zeta^2} = \sqrt{K_1}\cdot\sqrt{1-0.6^2} = 0.8\sqrt{K_1}$$

$$\therefore \quad K_1 = \left(\frac{10}{.8}\right)^2 = 156.25$$

Then, $\qquad\qquad \omega_n = 12.5$

Therefore, $\qquad 1+K_2 = 2 \times 12.5 \times 0.6 \ \therefore \ K_2 = 14$ **Ans.**

Problem 3.12.　What will be the maximum phase shift that could be obtained by a lead compensator with transfer function ?

$$G_c(s) = \frac{4(1+0.15s)}{(1+0.05s)}$$

Solution:

For the phase shift, only $\dfrac{(1+0.15s)}{(1+0.05s)}$ is important

It is found that, $T = 0.05$ and $aT = 0.15$ i.e., $a = 3$.

Maximum phase shift is given by, θ_m such that,

$$\tan\theta_m = \frac{a-1}{2\sqrt{a}} = \frac{1}{\sqrt{3}} \quad \text{then} \quad \theta_m = 30°$$

Problem 13.　The open-loop transfer function of a unity feedback system is given by

$$G(s) = K/s(Ts+1)$$

where K and T are positive constants.

　　By what factor should the amplifier gain be reduced so that the peak overshoot of unit-step response of the system is reduced from 75% to 25%?

Solution:

The open loop transfer function is given by

$$G(s) = \frac{K}{s(Ts+1)}$$

Where K and T are (+) ve constants. Then the closed loop T. F. (CLTF),

$$T(s) = \frac{K}{Ts^2+s+K} = \left(\frac{K/T}{s^2+\dfrac{1}{T}s+\dfrac{K}{T}}\right)$$

Now, comparing with the standard second order form,

$$T(s) = \left(\frac{\omega_n^2}{s^2+2\zeta\omega_n s+\omega_n^2}\right)$$

$$2\zeta\omega_n = \left(\frac{1}{T}\right) \text{ and } \omega_n = \left(\frac{\sqrt{K}}{T}\right)$$

Then

$$\zeta = \left(\frac{1}{2\sqrt{KT}}\right)$$

Now peak overshoot, $M_p = e^{-\frac{\zeta\pi}{\sqrt{1-\zeta^2}}}$

Given, $M_{p_1} = 0.75$ and $M_{p_2} = 0.25$

Therefore, $e^{-\frac{\zeta_1\pi}{\sqrt{1-\zeta_1^2}}} = 0.75$

$\therefore \qquad \frac{\zeta_1}{\sqrt{1-\zeta_1^2}} = 0.09; \; \frac{1-\zeta_1^2}{\zeta_1^2} = 8.1\times10^{-3}$

$\therefore \qquad \zeta_1 = 0.091$

Again, where $M_{p_2} = 0.25 = e^{\frac{\zeta_2\pi}{\sqrt{1-\zeta_2^2}}}$

or, $\frac{\zeta_2}{\sqrt{1-\zeta_2^2}} = 0.44; \; \frac{1-\zeta_2^2}{\zeta_2^2} = 5.16$

$\therefore \qquad \zeta_2 = 0.4$

Now, $\left(\frac{\zeta_1}{\zeta_2}\right) = \sqrt{\frac{K_2}{K_1}}$

$\therefore \qquad \frac{K_1}{K_2} = \left(\frac{\zeta_2}{\zeta_1}\right)^2 = \left(\frac{0.40}{0.09}\right)^2 \cong 20$

Therefore the amplifier gain should be reduced by 20.

Problem 3.14.   Figure shows a system employing proportional plus error-rate control. Determine the value of the error-rate factor K_e so that the damping ratio is 0.5. Determine the values of settling time, maximum overshoot and steady-state error (for unit-ramp input) with and without error-rate control. Comment upon the effect of error-rate control on system dynamics.

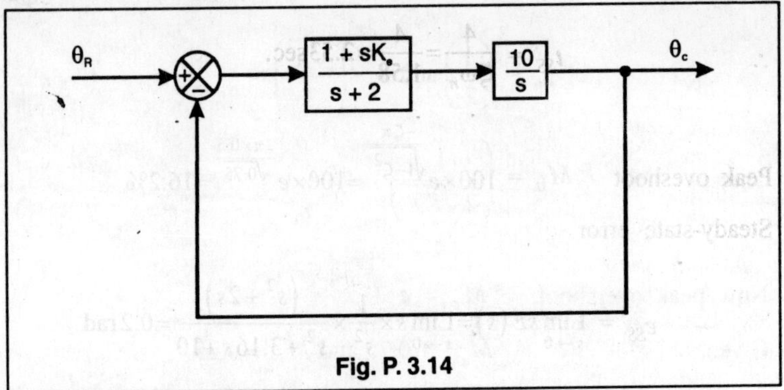

Fig. P. 3.14

Solution:

From the block diagram given

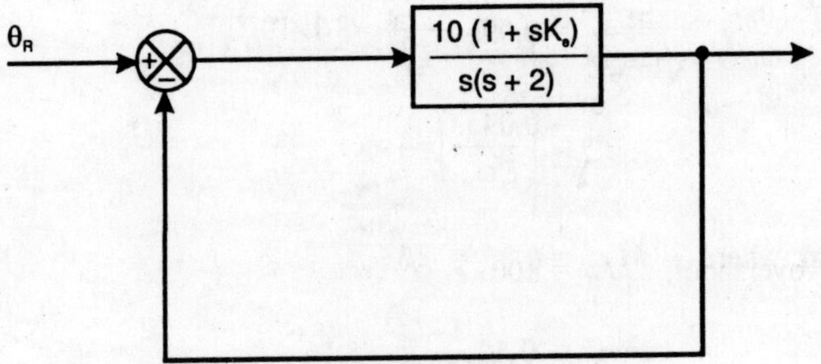

Fig. P. 3.14 (a)

$\therefore$ *T. F.* is, $T(s) = \dfrac{\theta_c(s)}{\theta_R(s)} = \dfrac{10(1+s\,K_e)}{s^2+(2+10K_e)s+10}$

The characteristic equation is, $s^2+(2+10K_e)s+10 = 0$

Which yields, $\omega_n = \sqrt{10}$ and $2\zeta\omega_n = (2+10K_e)$

Given the value of damping ratio, $\zeta = 0.5$

$$2\times0.5\times\sqrt{10} = 2+10K_e \quad (\therefore K_e = 0.116) \text{ **Ans.**}$$

Situation I: with error-rate control;

Characteristic equation is, $s^2+(2+10K_e)s+10 = 0$

or, $s^2+3.16s+10 = 0$ $(\zeta\omega_n = 3.16/2 = 1.58)$

$$\therefore \quad t_{SS} = \frac{4}{\zeta\omega_n} = \frac{4}{1.58} = 2.53 \text{ sec.}$$

Peak oveshoot $\quad M_p = 100 \times e^{\dfrac{-\zeta\pi}{\sqrt{1-\zeta^2}}} = 100 \times e^{\dfrac{-\pi \times 0.5}{\sqrt{0.75}}} = 16.2\%$

Steady-state error

$$e_{SS} = \lim_{s \to 0} sE(s) = \lim_{s \to 0} s \times \frac{1}{s^2} \times \frac{\left(s^2 + 2s\right)}{s^2 + 3.16s + 10} = 0.2 \text{ rad}$$

Situation II: Without error-rate control; i.e. $K_e = 0$

The characteristic equation:

$$s^2 + 2s + 10 = 0$$

$$\therefore \quad \zeta\omega_n = 1 \quad \therefore \quad \zeta = \frac{1}{\omega_n} = \frac{1}{\sqrt{10}} = 0.316$$

$$\therefore \quad t_{ss} = \left(\frac{4}{\zeta\omega_n}\right) = 4 \text{ sec.}$$

peak overshoot, $\quad M_p = 100 \times e^{\dfrac{-\pi \times 0.316}{\sqrt{0.9}}} = 35.2\%$

$$e_{SS} = s \times \frac{1}{s^2} \times \frac{s^2 + 2s}{s^2 + 2s + 10} = 0.2 \text{ rad}$$

Comment: Hence the presence of error-rate control reduces the peak-overshoot and setting time but steady state error remains unchanged.

Problem 3.15. A feedback system employing output-rate damping is shown in Fig.

Fig. P. 3.15

(a) In the absence of derivation feedback ($K_0 = 0$), determine the damping factor and natural frequency of the system. What is the steady-state error resulting from unit-ramp.

(b) Determine the derivative feedback constant K_0, which will increase the damping factor of the system to 0.6. What is the steady-state error resulting from unit-ramp input with this setting of the derivative feedback constant?

(c) Illustrate how the steady-state error of the system with derivative feedback to unit-ramp input can be reduced to same value as in part (a), while the damping factor is maintained at 0.6.

Solution:

(a) From the block diagram above in the question:

$$\text{T.F.} \quad T(s) = \frac{\theta_c}{\theta_R} = \frac{K_A}{s^2 + s(K_0 + 2) + K_A}$$

$$E(s) = \theta_R(s) = \frac{s^2 + s(K_0 + 2)}{s^2 + s(K_0 + 2) + K_A}$$

The steady error constant with $(K_0 = 0)$,

$$e_{ss} = \underset{s \to 0}{\text{Lim}} \; sE(s)$$

$$= \underset{s \to 0}{\text{Lim}} \; s \cdot \frac{1}{s^2} \cdot \left(\frac{s^2 + 2s}{s^2 + 2s + K_A} \right) = \frac{2}{K_A} = 0.2 \, \text{rad.}$$

Also, the T.F. becomes,

$$T(s) = \frac{K_A}{s^2 + 2s + K_A} = \frac{10}{s^2 + 2s + 10}$$

Here the undamped natural fequency, $\omega_n = \sqrt{10}$ and $\zeta\omega_n = 1$

$\therefore$ the damping ratio (factor) $= \dfrac{1}{\sqrt{10}} = 0.316$

When $K_0 \neq 0$

$$T(s) = \frac{K_A}{s^2 + (K_0 + 2)s + K_A} = \left(\frac{10}{s^2 + (K_0 + 2)s + 10} \right)$$

Here,
$$2\zeta\omega_n = (K_0+2); \quad 2\sqrt{10}\times 0.6=(K_0+2)$$

or,
$$K_0 \cong 3.8 - 2 \cong 1.8$$

The steady-state error,

$$e_{ss} = \lim_{s\to 0} s\cdot\frac{1}{s^2}\times\frac{s+(K_0+2)}{s^2+s(K_0+2)+K_A}$$

$$= \frac{(K_0+2)}{(10)}=0.38\,\text{rad}$$

(c) Here e_{ss} is as in part (a) = 0.2

Then,
$$\frac{(K_0+2)}{(K_A)} = 0.2$$

Hence,
$$K_0+2 = 0.2\,K_A \tag{1}$$

Also,
$$2\zeta\omega_n = (K_0+2) \tag{2}$$

$$\left.\begin{array}{c}\zeta=0.6\\[4pt]\omega_n=\sqrt{K_A}\end{array}\right] \tag{3}$$

Then, from equation 1, 2 & 3

$$2\times 0.6\sqrt{K_A} = 0.2\,K_A$$

$$\therefore \qquad \sqrt{K_A} = \left(\frac{1.2}{0.2}\right)$$

Hence,
$$K_A = 36$$

and now,
$$K_0 = 0.2\times K_A-2=7.2-2=5.2$$

Problem 3.16. A unity feedback system is characterized by the open-loop transfer function

$$G(s) = 1/s(0.5s+1)(0.2s+1)$$

Determine the steady-state errors for unit-step, unit-ramp and unit-acceleration inputs. Also determine the damping ratio and natural frequency of the dominant roots.

Solution:

Given OLTF, $\quad G(s) = \dfrac{1}{s(0.5s+1)(0.2s+1)}$

For unity feedback system, $H(s) = 1$

For unit step input, $R(s) = 1/s$

$$e_{ss} = \lim_{s \to 0} s \frac{R(s)}{1+G(s)}$$

$$= \lim_{s \to 0} s \frac{1}{s} \cdot \frac{1}{1 + \dfrac{1}{s(0.5s+1)(0.2s+1)}} = 0.$$

for the unit ramp input, $R(s) = 1/s^2$

Then, steady error constant, $e_{ss} = \lim_{s \to 0} \dfrac{sR(s)}{1+G(s)}$

Hence, $\qquad e_{ss} = \lim_{s \to 0} s \cdot \dfrac{1}{s} \cdot \dfrac{s(0.5s+1)(0.2s+1)}{1+s^2(0.5s+1)(0.2s+1)}$

Also For the unit acceleration input, $R(s) = \dfrac{1}{s^3}$

Then the steady state error, $e_{ss} = \lim_{s \to 0} s \cdot \dfrac{R(s)}{1+C(s)}$

Therfore, $\qquad e_{ss} = \lim_{s \to 0} s \cdot \dfrac{1}{s^3} \left(\dfrac{1}{1 + \dfrac{1}{s(0.5s+1)(0.2s+1)}} \right) = \infty$

Problem 3.17. The open loop transfer function of a unity feedback system is given by,

$$G(s) = \frac{K}{s(1+Ts)}$$

where K and T are positive constants. By what factor should the amplifier gain be reduced so that the peak overshoot of the unit step response of the system is reduced from 75% to 25%. (I.E.S-96)

Solution:

$$G(s) = \frac{K}{s(1+Ts)}$$

Let the equation be, $1+ G(s) H(s) = 0$

or, $\qquad 1 + G(s) = 0$ or, $1+\dfrac{K}{s(1+Ts)}=0$

or, $\qquad s(1+Ts)+K = 0$

or, $\qquad s^2+\dfrac{1}{T}s+\dfrac{K}{T} = 0 \qquad\qquad$ (1)

$$s^2+2\zeta\omega_n s+\omega_n^2 = 0 \qquad\qquad (2)$$

Comparing (1) and (2) we get $\omega_n=\sqrt{\dfrac{K}{T}}\quad \zeta=\dfrac{1}{2\omega_n T}$

$$\zeta = 0.5\frac{1}{\sqrt{\dfrac{K}{T}}\cdot T}=\frac{0.5}{\sqrt{KT}}=\frac{1}{\sqrt{4KT}}$$

Percent peak overshoot $= 100e^{-\pi\zeta/\sqrt{1-\zeta^2}}$

Therefore per unit overshoot $= e^{-\pi\zeta/\sqrt{1-\zeta^2}}$

Case I: Peak overshoot per unit $= 0.75$

Now, $\qquad \dfrac{-\pi\zeta}{\sqrt{1-\zeta^2}} = \dfrac{-\pi\dfrac{1}{\sqrt{4KT}}}{\sqrt{1-\dfrac{1}{4KT}}}=\dfrac{-\pi}{\sqrt{4KT-1}}$

or, $\qquad e^{-\pi\sqrt{4KT-1}} = 0.75$

$$\frac{-\pi}{\sqrt{4KT-1}} = \ln 0.75 = -0.2877$$

or $\qquad 4\,KT - 1 = 119.254, \quad K = \dfrac{120.254}{4T}$

Case II: $\quad e^{-\pi/\sqrt{4K'T-1}} = 0.25$

or, $\qquad 4\,K'T - 1 = 5.13557$

Hence $\qquad K' = \dfrac{6.13557}{4T}$

$$\frac{K}{K'} = \frac{120.254}{6.13557} = 19.6$$

Thus, K should be reduced by a factor of 19.6 in order to reduce peak overshoot from 75% to 25%.

Problem 3.18. A unity feedback position control system has a forward path transfer function, $G(s) = \dfrac{K}{s}$. For unit step input, compute the value of K that minimizes ISE (integral square error).

(I.E.S-96)

Solution:

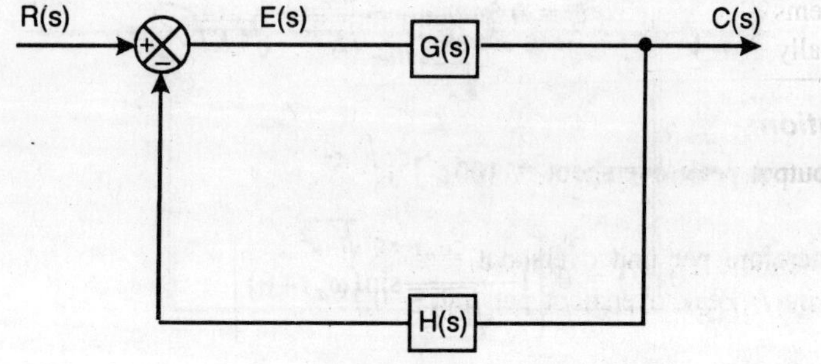

Fig. P. 3.18

$$E(s) = R(s) - H(s)\,C(s)$$

$$= R(s) - \frac{H(s)G(s)R(s)}{1+G(s)H(s)}$$

$$= \frac{R(s)}{1+G(s)H(s)} = \frac{R(s)}{1+G(s)} \quad \text{for unity feedback.}$$

Now $\qquad R(s) = \dfrac{1}{s}, \; G(s) = \dfrac{K}{s}$

$\therefore \qquad E(s) = \dfrac{1/s}{1+(K/s)} = \dfrac{1}{(s+K)}$

Hence error $\qquad e(t) = e^{-Kt}, \;$ or $\; e^2(t) = e^{-2Kt}$

$$\int_0^\infty e^{-2Kt}\, dt = \left. \dfrac{e^{-2Kt}}{-2K} \right|_0^\infty$$

$$= \dfrac{1}{-2K}\left[e^{-\infty} - e^0\right] = -\dfrac{1}{2K} \cdot -1 = \dfrac{1}{2K}$$

The integral of square of the error function does not show any minimum. K should be as large as possible for this integral to be of small magnitude.

Problem 3.19. A servo system for the position control of rotatable mass is stablised by viscous damping which is three-quarters of that is needed for critical damping. The undamped natural frequency of the system is 12 Hz. Derive an expression for the output of the systems if the input control is suddenly moved to a new position being initially at rest. Hence find the maximum overshoot.

Solution:

The output response can be obtained as

$$\theta_0(t) = \theta_i\left[1 - \dfrac{e^{-\zeta\omega_n t.}}{\sqrt{1-\zeta^2}}\sin(\omega_d t + \theta)\right]$$

Now $\qquad\qquad \zeta = 0.75 \text{ (given)}$

$\therefore \qquad\qquad \theta = \tan^{-1}\dfrac{\sqrt{1-\zeta^2}}{\zeta} = \tan^{-1}\dfrac{\sqrt{1-0.75^2}}{0.75} = 41.41°$

But $\qquad\qquad 41.41° = \dfrac{41.4 \times \pi}{180} = 0.72 \text{ radians.}$

$$\omega_d = \omega_n\sqrt{1-\zeta^2}$$

Now $\omega_n = 12\,\text{Hz} = 12\,\text{cycles/sec} = 2\pi \times 12 = 75.4\,\text{rad/s}$

$$\therefore \quad \omega_d = 75.4\sqrt{1-0.75^2} = 50$$

$$\frac{1}{\sqrt{1-\zeta^2}} = \frac{1}{\sqrt{1-0.75^2}} = 1.5$$

$$\zeta\omega_n = 0.75 \times 75.4 = 56.6$$

$$\therefore \quad \theta_0(t) = \theta_i\left[1 - 1.5e^{-56.6t}\sin(50t + 0.72)\right]$$

Maximum overshoot.

$$M_p = e^{\dfrac{-\pi\zeta}{\sqrt{1-\zeta^2}}} = e^{\dfrac{-\pi \times 0.75}{\sqrt{1-0.75^2}}} = 2.8\%$$

Problem 3.20. An underdamped second order system having a Transfer function of

$$M(s) = \frac{K\omega_n^2}{s^2 + 2\zeta\omega_n + \omega_n^2}$$

has a frequency response plot shown in the fig. below. Then calculate the system gain K and damping ratio ζ.

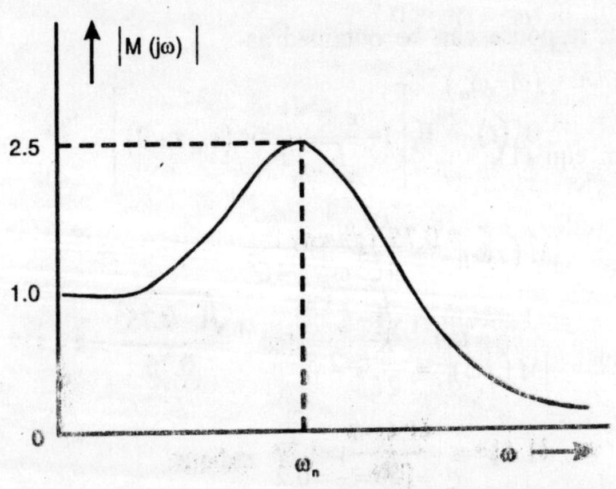

Fig. P. 3.20

Solution:

T.F. is given as
$$M(s) = \frac{K\omega_n^2}{s^2 + 2\zeta\omega_n s + \omega_n^2}$$

Then,
$$M(j\omega) = \frac{K\omega_n^2}{-\omega^2 + 2j\zeta\omega_n\omega + \omega_n^2}$$

Hence,
$$|M(j\omega)|^2 = \left(\frac{K^2\omega_n^4}{\left(\omega_n^2 - \omega^2\right) + 4\zeta^2\omega_n^2\omega^2}\right) \qquad (1)$$

From the frequency response,
$$|M(j0)| = 1$$

Hence,
$$|M(j0)|^2 = 1$$

or,
$$\frac{K^2\omega_n^4}{\omega_n^4} = K^2 = 1 \quad \therefore K = 1$$

The peak value of $M(j\omega)$ takes place when the denominator of the equation (1) becomes minimum.

It is only possible when,
$$\omega_n^2 - \omega^2 = 0$$

$$\therefore \qquad \left(\omega = \omega_n\right)$$

$\therefore$ From eqn (1),

$$|M(j\omega)|^2 = \frac{K^2\omega_n^4}{4\zeta^2\omega_n^4} = \frac{K^2}{4\zeta^2}$$

$$\therefore \qquad |M(j\omega)| = \frac{K}{2\zeta} = 2.5$$

$$\therefore \qquad \zeta = \frac{K}{5} = \frac{1}{5} = 0.2$$

Hence $K = 1$ and $\zeta = 0.2$ **Ans.**

Problem 3.21. Determine the values of M, B and K from the response curve of the system shown in fig.

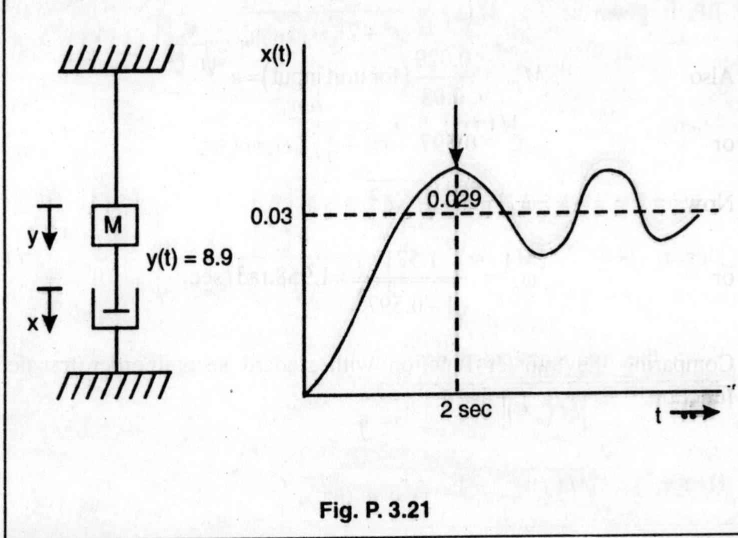

Fig. P. 3.21

Solution:

The transfer function of the system is

$$\frac{X(s)}{Y(s)} = \frac{1}{Ms^2 + Bs + K}$$

$$Y(t) = 8.9, \quad \therefore Y(s) = \frac{8.9}{s}$$

$$\therefore \qquad X(s) = \frac{0.9}{s\left(Ms^2 + Bs + K\right)}$$

Also $\qquad X(\infty) = 0.03$ from the response curve

$$\therefore \qquad X(\infty) = \lim_{s \to 0} sX(s) = \lim_{s \to 0} \frac{8.9}{Ms^2 + Bs + K} = \frac{8.9}{K}$$

or. $\qquad K = 296.67$ N/m

Also $\qquad t_p = 2\,\text{sec} = \dfrac{\pi}{\omega_d}$

or $$\omega_d = \frac{\pi}{2} = 1.571 \, \text{rad/sec}$$

Also $$M_p = \frac{0.029}{0.03} \, (\text{for unit input}) = e^{-\dfrac{\pi\zeta}{\sqrt{1-\zeta^2}}}$$

or $$\zeta = 0.597$$

Now $$\omega_d = \omega_n \sqrt{1-\zeta^2}$$

or $$\omega_n = \frac{1.571}{\sqrt{1-0.597^2}} = 1.958 \, \text{rad/sec}$$

Comparing the transfer function with stadard second order transfer function.

$$\frac{1}{Ms^2 + Bs + K} = \frac{\omega_n^2}{s^2 + 2\zeta\omega_n + \omega_n^2}$$

we get $$\omega_n^2 = \frac{K}{M}$$

or $$M = \frac{K}{\omega_n^2} = \frac{296.67}{(1.958)^2} = 77.38 \, \text{kg}$$

and $$2\zeta\omega_n = \frac{B}{M}$$

or $$B = 2\zeta\omega_n M = 2 \times 0.597 \times 1.958 \times 77.38$$
$$= 180.91 \, \text{kg/m/sec}$$

Hence $$K = 296.67 \, \text{N/m}$$
$$M = 77.38 \, \text{kg}$$
$$B = 180.91 \, \text{kg/m/sec}.$$

Problem 3.22. Measerments conducted on a servo Mechanism show the error response to be

$$\frac{e}{r} = 1.66 \, e^{-8t} \sin\left(6t + 37°\right)$$

When the input is a sudden displacement r. Find the natural frequency of oscillations, damping ratio and damped frequency of oscillations.

Solution:

The error response for a step input has the expression

$$e = re^{-\zeta\omega_n t}\sin(\omega_d t + \theta)$$

Comparing we get $r = 1.66$, $\zeta\omega_n = 8$

$$\omega_d = \omega_n\sqrt{1-\zeta^2} = 6$$

Solving we get $\zeta = 0.8$ and $\omega_n = 10\,\text{rad/sec}$.

Problem 3.23.   The closed-loop poles of a system is shown in fig. Find the unit step response of the system and the settling time for 2% tolerance.

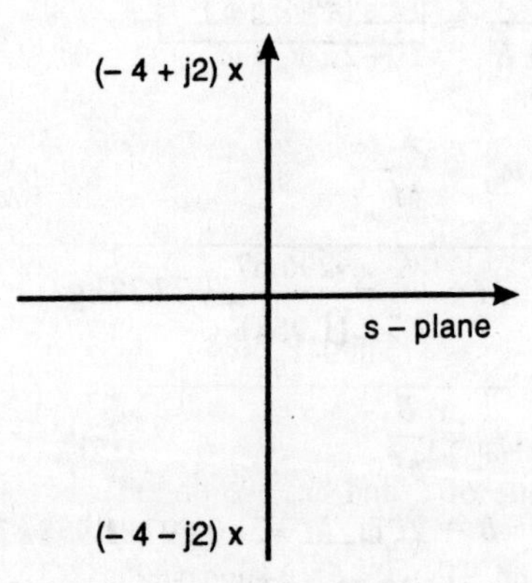

Fig. P. 3.23

Solution:

$$-\zeta\omega_n = -4 \quad \text{and} \quad \omega_n\sqrt{1-\zeta^2} = 2$$

Solving we get, $\omega_n = 4.46$ rad/sec and $\zeta = 0.9$

Time response is $C(t) = 1 - \dfrac{e^{\zeta\omega_n t}}{\sqrt{1-\zeta^2}}\sin\left[\omega_n\sqrt{1-\zeta^2}\,t + \tan^{-1}\dfrac{\sqrt{1-\zeta^2}}{\zeta}\right]$

$$= 1 - \frac{e^{-4t}}{0.44}\sin\left[2t + \tan^{-1}\frac{0.44}{0.9}\right]$$

The characteristic equation is $\left(S+4+j\sqrt{2}\right)\left(S+4-j\sqrt{2}\right) = S^2 + 8S + 20$

$$\text{Settling time} = \frac{3.91}{\zeta\omega_n} = \frac{3.91}{4} = 0.98 \text{ sec}$$

Problem 3.24. Consider the system shown in figure, where $\zeta = 0.6$ and $\omega_n = 5$ rad/sec. Obtain the rise time t_r, peak time t_p, maximum overshoot M_p, and settling time t_s when the system is subjected to a unit-step input.

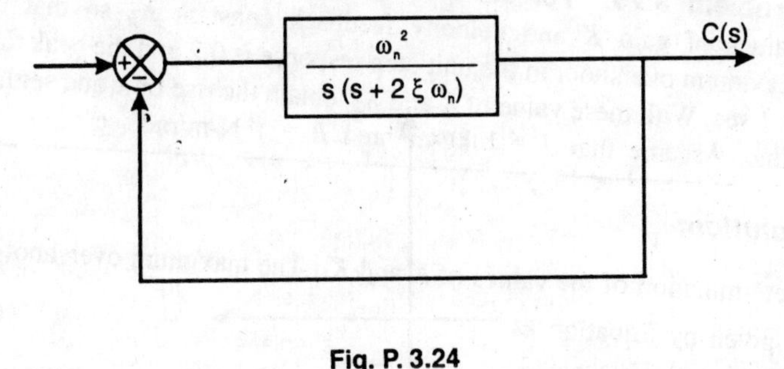

Fig. P. 3.24

Solution:

From the given value of ζ and ω_n, we obtain

$$\omega_d = \omega_n\sqrt{1-\zeta^2} = 4 \text{ and } \sigma = \zeta\omega_n = 3.$$

Rise time t_r: The rise time is $t_r = \dfrac{\pi-\beta}{\omega_d} = \dfrac{3.14-\beta}{4}$

Where β is given by $\quad \beta = \tan^{-1}\dfrac{\omega_d}{\sigma} = \tan^{-1}\dfrac{4}{3} = 0.93\,\text{rad}$

The rise time t_r is thus $t_r = \dfrac{3.14-0.93}{4} = 0.55\,\text{sec}$

Peak time t_p: The peak time is $\quad t_p = \dfrac{\pi}{\omega_d} = \dfrac{3.14}{4} = 0.785\,\text{sec}$

Maximum overshoot M_p : The maximum overshoot is

$$M_p = e^{-(\sigma/\omega_d)\pi} = e^{-(3/4)\times 3.14} = 0.095$$

The maximum percent overshoot is thus 9.5%

Settling time t_s: For the 2% criterion, the settling time is

$$t_s = \frac{4}{\sigma} = \frac{4}{3} = 1.33\,\text{sec}$$

For the 5% criterion, $\qquad t_s = \dfrac{3}{\sigma} = \dfrac{3}{3} = 1\,\text{sec}$

Problem 3.25. For the system shown in Figure determine the values of gain K and velocity feedback constant K_h so that the maximum overshoot in the unit-step response is 0.2 and the peak time is 1 sec. With these value of K and K_h, obtain the rise time and settling time. Assume that $J = 1$ kg-m^2 and $B = 1$ N-m/rad/sec.

Solution:

Determination of the values of K and K_h: The maximum overshoot M_p is given by Equation as

$$M_p = e^{-\left(\zeta/\sqrt{1-\zeta^2}\right)\pi}$$

This value must be 0.2 thus,

$$e^{-\left(\zeta/\sqrt{1-\zeta^2}\right)\pi} = 0.2$$

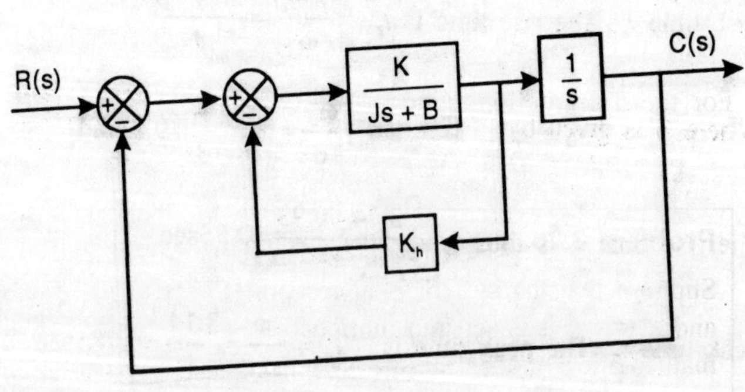

Fig. P. 3.25

or
$$\frac{\zeta \pi}{\sqrt{1-\zeta^2}} = 1.61$$

Which yields $\zeta = 0.456$

The peak time t_p is specified as 1 sec; therefore, from Equation

$$t_p = \frac{\pi}{\omega_d} = 1 \quad \text{or} \quad \omega_d = 3.14$$

Since ζ is 0.456, ω_n is
$$\omega_n = \frac{\omega_d}{\sqrt{1-\zeta^2}} = 3.53$$

Since the natural frequency ω_n is equal to $\sqrt{K/J}$,

$$K = J\omega_n^2 = \omega_n^2 = 12.5 \, \text{N-m}$$

Then, K_h is, from Equation

$$K_h = \frac{2\sqrt{KJ}\zeta - B}{K} = \frac{2\sqrt{K}\zeta - 1}{K} = 0.178 \, \text{sec}$$

Rise time t_r: is given by

$$t_r = \frac{\pi - \beta}{\omega_d}$$

Where
$$\beta = \tan^{-1}\frac{\omega_d}{\sigma} = \tan^{-1} 1.95 = 1.10$$

Thus, t_r is $\quad t_r = 0.65 \, \text{sec}$

Settling time t_s: For the 2% criterion

$$t_s = \frac{4}{\sigma} = 2.48 \, \text{sec}$$

For the 5% criterion, $t_s = \dfrac{3}{\sigma} = 1.86 \, \text{sec}$

Problem 3.26. Consider the mechanical system shown in Figure. Suppose that the system is at rest initially [$x(0) = 0$, $\dot{x}(0) = 0$], and at $t = 0$ it is set into motion by a unit-impulse force. Obtain a mathematical model for the system. Then find the motion of the system.

Solution:

The system is excited by a unit-impulse input. Hence

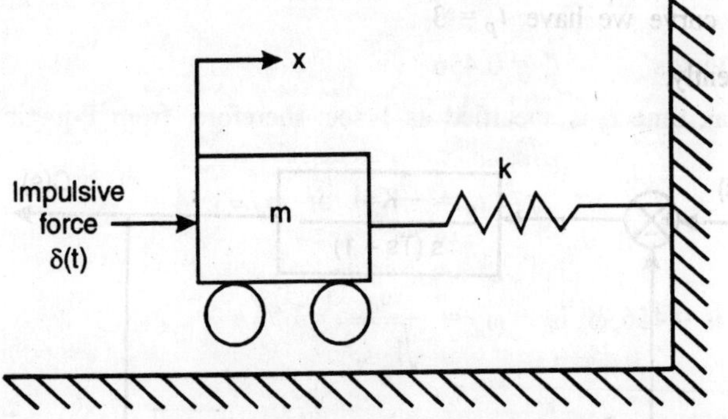

Fig. P. 3.26 Mechanical system.

$$m\ddot{x}+kx \;=\; \delta(t)$$

This is a mathematical model for the system.

Taking the Laplace transform of both sides of this equation gives

$$m\left[s^2 X(s)-sx(0)-\dot{x}(0)\right]+kX(s)= 1$$

By substituting the conditions $x(0)=0$ and $\dot{x}(0)=0$ into this last equation and solving for $X(s)$, we obtain

$$X(s) = \frac{1}{ms^2 +k}$$

The inverse Laplace transform of $X(s)$ becomes

$$X(t) = \frac{1}{\sqrt{mk}}\sin\frac{\sqrt{k}}{m}t$$

The oscillation is a simple harmonic motion. The amplitude of the oscillation is $1/\sqrt{mk}$.

Problem 27. When the system shown in Figure P 1.27 (*a*) is subjected to a unit-step input, the system output responds as shown in Fig P 1.27 (*b*). Determine the values of K and T from the reponse curve.

Solution:

The maximum overshoot of 25.4% corresponds to $\zeta = 0.4$. From the response curve we have $t_p = 3$

Consequently,

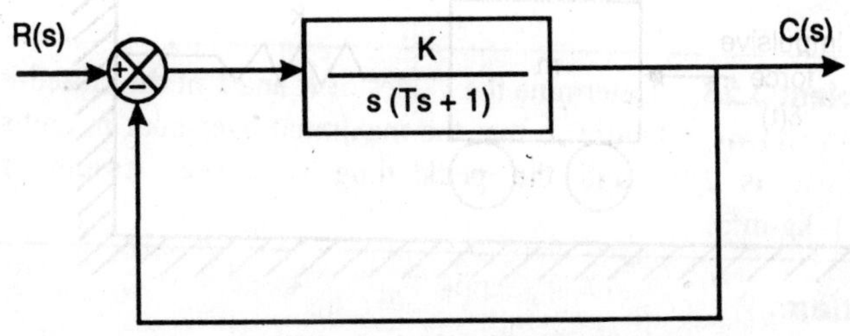

Fig. P. 3.27 (a) Closed-loop system

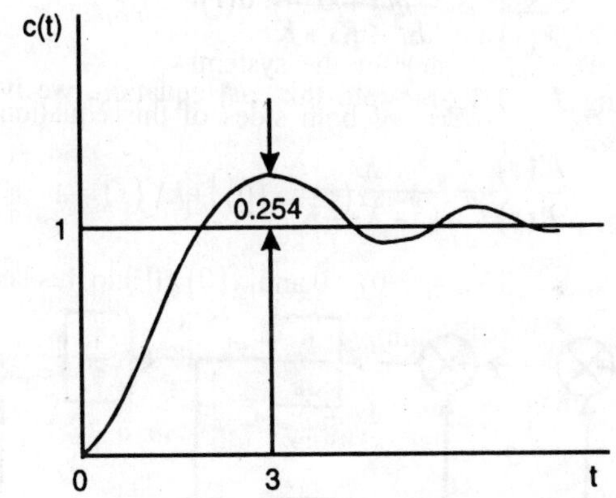

Fig. P. 3.27 (b) Unit-step response curve

$$t_p = \frac{\pi}{\omega_d} = \frac{\pi}{\omega_n \sqrt{1-\zeta^2}} = \frac{\pi}{\omega_n \sqrt{1-0.4^2}} = 3$$

It follows that $\quad \omega_n = 1.14$

From the block diagram we have $\quad \dfrac{C(s)}{R(s)} = \dfrac{K}{Ts^2 + s + K}$

From which $\quad \omega_n = \sqrt{\dfrac{K}{T}}, \; 2\zeta\omega_n = \dfrac{1}{T}$

Therefore, the values of T and K are determined as

$$T = \frac{1}{2\zeta\omega_n} = \frac{1}{2\times0.4\times1.14} = 1.09$$

$$K = \omega_n^2 T = 1.14^2 \times 1.09 = 1.42$$

Problem 3.28. Determine the values of K and k of the closed-loop system shown in Figure so that the maximum overshoot in unit-step response is 25% and the peak time is 2 sec. Assume that $J = 1$ kg-m^2.

Solution:

The closed-loop transfer function is

$$\frac{C(s)}{R(s)} = \frac{K}{Js^2 + Ks + K}$$

By substituting $J = 1$ kg-m^2 into this last equation, we have

$$\frac{C(s)}{R(s)} = \frac{K}{s^2 + Ks + K}$$

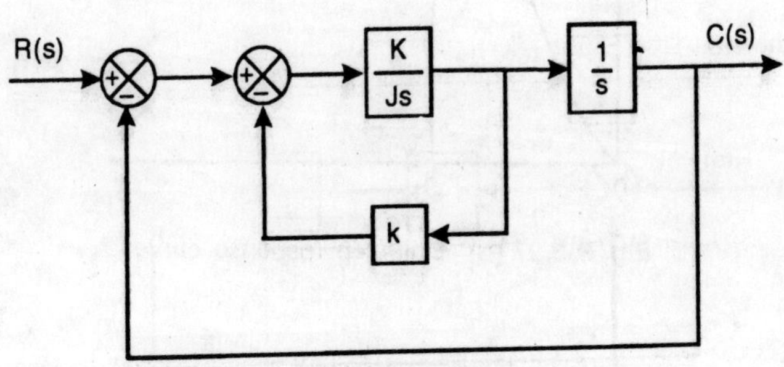

Fig. P. 3.28

Note that $\qquad \omega_n = \sqrt{K},\ 2\zeta\omega_n = Kk$

The maximum overshoot = $e^{-\zeta\pi/\sqrt{1-\zeta^2}}$

which is specified as 25%. Hence $e^{-\zeta\pi/\sqrt{1-\zeta^2}} = 0.25$

From which $\dfrac{\zeta\pi}{\sqrt{1-\zeta^2}} = 1.386$ or $\zeta = 0.404$

The peak time t_p is specified as 2 sec. And so

$$t_p = \frac{\pi}{\omega_d} = 2 \text{ or } \omega_d = 1.57$$

Then the undamped natural frequency ω_n is

$$\omega_n = \frac{\omega_d}{\sqrt{1-\zeta^2}} = \frac{1.57}{\sqrt{1-0.404^2}} = 1.72$$

Therefore, we obtain $K = \omega_n^2 = 1.72^2 = 2.95\,\text{N-m}$

$$k = \frac{2\zeta\omega_n}{K} = \frac{2\times0.404\times1.72}{2.95} = 0.471\,\text{sec}$$

Problem 3.29.   What is the unit-step response of the system shown in Fig. P.3.29.

Solution:

The closed-loop transfer function is $\dfrac{C(s)}{R(s)} = \dfrac{10s+10}{s^2+10s+10}$

For the unit-step input $[R(s) = 1/s]$, we have

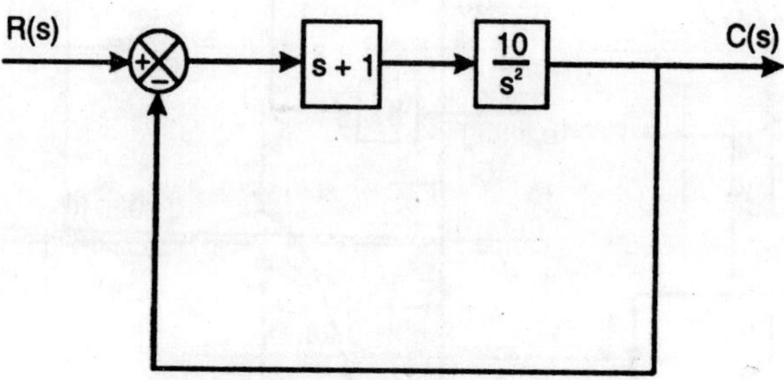

Fig. P. 3.29 Closed-loop system.

$$C(s) = \frac{10s+10}{s^2+10s+10}\frac{1}{s} = \frac{10s+10}{\left(s+5+\sqrt{15}\right)\left(s+5-\sqrt{15}\right)s}$$

$$= \frac{-4-\sqrt{15}}{3+\sqrt{15}} \frac{1}{s+5+15} + \frac{-4+\sqrt{15}}{3-\sqrt{15}} \frac{1}{s+5-\sqrt{15}} + \frac{1}{s}$$

The inverse Laplace transform of $C(s)$ gives

$$c(t) = -\frac{4+\sqrt{15}}{3+\sqrt{15}} e^{-\left(5+\sqrt{15}\right)t} + \frac{4-\sqrt{15}}{-3+\sqrt{15}} e^{-\left(5+\sqrt{15}\right)t} + 1$$

$$= -1.1455\, e^{-8.87t} + 0.1455\, e^{-1.13t} + 1$$

Clearly, the output will not exhibit any oscillation. The response curve exponentially approaches the final value $c(\infty) = 1$.

Problem 3.30. Figure (a) shows a mechanical vibratory system. When 2 Ib of force (step input) is applied to the system, the mass oscillates, as shows in Figure (b). Determine m, b and k of the system from this response curve. The displacement x is measured from the equilibrium position.

Solution:

The transfer function of this system is $\dfrac{X(s)}{P(s)} = \dfrac{1}{ms^2 + bs + K}$

Since $\qquad P(s) = \dfrac{2}{s}$

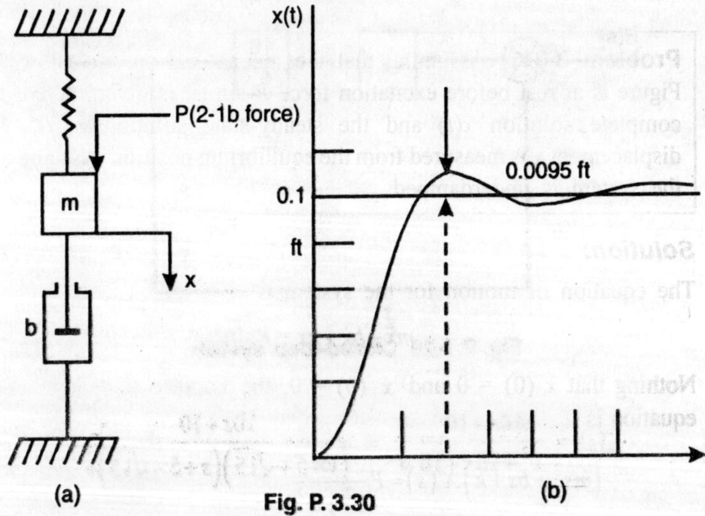

Fig. P. 3.30

We obtain

$$X(s) = \frac{2}{s\left(ms^2 + bs + k\right)}$$

It follows that the steady-state value of x is

$$x(\infty) = \operatorname*{Lim}_{s \to 0} s\,X(s) = \frac{2}{k} = 0.1\,\text{ft}$$

Hence, $k = 20$ lb$_f$/ft Note that $M_p = 9.5\%$ corresponds to $\zeta = 0.6$. The peak time t_p is given by

$$t_p = \frac{\pi}{\omega_d} = \frac{\pi}{\omega_d\sqrt{1-\zeta^2}} = \frac{\pi}{0.8\,\omega_n}$$

The experimental curve shows that $t_p = 2$ sec. Therefore,

$$\omega_n = \frac{3.14}{2 \times 0.8} = 1.96\,\text{rad/sec}$$

Since $\omega_n^2 = k/m = 20/m$, we obtain

$$m = \frac{20}{\omega_n^2} = \frac{20}{1.96^2} = 5.2\,\text{slugs} = 166\,\text{lb}$$

(Note that 1 slug = 1 lbf-sec^2/ft.) Then b is determined from $2\zeta\omega_n = \dfrac{b}{m}$

or $\qquad b = 2\zeta\omega_n m = 2 \times 0.6 \times 1.96 \times 5.2 = 12.2$ lbf/ft/sec

---

Problem 3.31. Assuming that the mechanical system shown in Figure is at rest before excitation force $P \sin \omega t$ is given, derive the complete solution $x(t)$ and the steady-state solution $x_{ss}(t)$. The displacement x is measured from the equilibrium position. Assume that the system is underdamped.

Solution:

The equation of motion for the system is

$$m\ddot{x} + b\dot{x} + kx = P\sin\omega t$$

Nothing that $x\,(0) = 0$ and $\dot{x}\,(0) = 0$, the Laplace transform of this equation is

$$\left(ms^2 + bs + k\right)X(s) = P\frac{\omega}{s^2 + \omega^2}$$

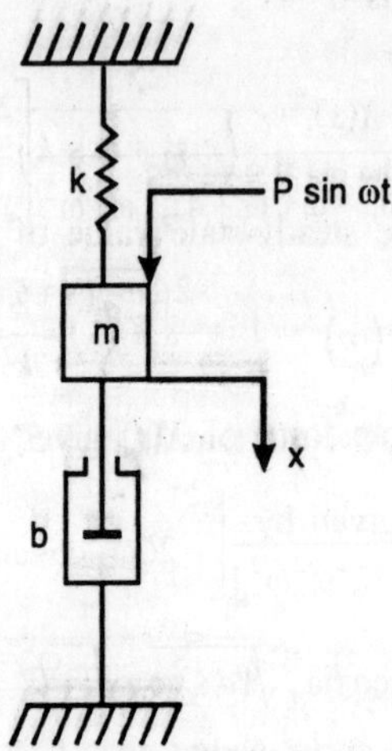

Fig. P. 3.31

For
$$X(s) = \frac{P\omega}{\left(s^2 + \omega^2\right)\left(ms^2 + bs + k\right)}$$

Since the system is underdamped, $X(s)$ can be written as follows:

$$X(s) = \frac{P\omega}{m}\frac{1}{s^2 + \omega^2}\frac{1}{s^2 + 2\zeta\omega_n s + \omega_n^2} \quad \text{Where } 0 < \zeta < 1$$

Where
$$\omega_n = \sqrt{k/m} \text{ and } \zeta = b/\left(2\sqrt{mk}\right).X(s)$$

Can be expanded as

$$X(s) = \frac{P\omega}{m}\left(\frac{as + c}{s^2 + \omega^2} + \frac{-as + d}{s^2 + 2\zeta\omega_n s + \omega_n^2}\right)$$

By simple calculations it can be found that

$$a = \frac{-2\zeta\omega_n}{\left(\omega_n^2 - \omega^2\right)^2 + 4\zeta^2\omega_n^2\omega^2}, \quad c = \frac{\left(\omega_n^2 - \omega^2\right)^2}{\left(\omega_n^2 - \omega^2\right)^2 + 4\zeta^2\omega_n^2\omega^2}$$

$$d = \frac{4\zeta^2\omega_n^2 - \left(\omega_n^2 - \omega^2\right)}{\left(\omega_n^2 - \omega^2\right)^2 + 4\zeta^2\omega_n^2\omega^2}$$

Hence
$$X(s) = \frac{P\omega}{m} \frac{1}{\left(\omega_n^2 - \omega^2\right)^2 + 4\zeta^2 \omega_n^2 \omega^2} \left[\frac{-2\zeta\omega_n + \left(\omega_n^2 - \omega^2\right)}{s^2 + \omega^2} \right.$$

$$\left. + \frac{2\zeta\omega_n\left(s + \zeta\omega_n\right) + 2\zeta^2 \omega_n^2 - \left(\omega_n^2 - \omega^2\right)}{s^2 + 2\zeta\omega_n s + \omega_n^2} \right]$$

The inverse Laplace transform of $X(s)$ gives

$$x(t) = \frac{P\omega}{m[\omega_n^2 - \omega^2)^2 + 4\zeta^2 \omega_n^2 \omega^2]} \left[-2\zeta\omega_n \cos\omega t + \frac{\left(\omega_n^2 - \omega^2\right)}{\omega} \sin\omega t \right.$$

$$+ 2\zeta\omega_n e^{-\zeta\omega_n t} \cos\omega_n \sqrt{1-\zeta^2} t + \frac{2\zeta\omega_n^2 - \left(\omega_n^2 - \omega_n\right)}{\omega_n \sqrt{1-\zeta^2}} e^{-\zeta\omega_n t}$$

$$\left. \sin\omega_n \sqrt{1-\zeta^2} t \right]$$

At steady state $(t \to \infty)$ the terms involving $e^{-\zeta\omega_n t}$ approach zero. Thus at steady state

$$x_{SS}(t) = \frac{P\omega}{m\left[\left(\omega_n^2 - \omega^2\right)^2 + 4\zeta^2 \omega_n^2 \omega^2\right]} \left(-2\zeta\omega_n \cos\omega t + \frac{\omega_n^2 - \omega_2}{\omega} \sin\omega t\right)$$

$$= \frac{P\omega}{\left(k - m\omega^2\right)^2 + b^2 \omega^2} \left(-b\cos\omega t + \frac{k - m\omega^2}{\omega} \sin\omega t\right)$$

$$= \frac{P\omega}{\sqrt{\left(k - m\omega^2\right)^2 + b^2 \omega^2}} \sin\left(\omega t - \tan^{-1}\frac{b\omega}{k - m\omega^2}\right)$$

Problem 3.32.   In the system shown in Figure the numerical values of m, b, and k are given as $m = 1$ kg. $b = 2$ N-sec/m, and $k = 100$ N/m. The mass is displaced 0.05 m and released without initial velocity. Find the frequency observed in the vibration. In addition, find the amplitude four cycles later. The displacement x is measured from the equilibrium position.

Solution:

The equation of motion for the system is

$$m\ddot{x} + b\dot{x} + kx = 0$$

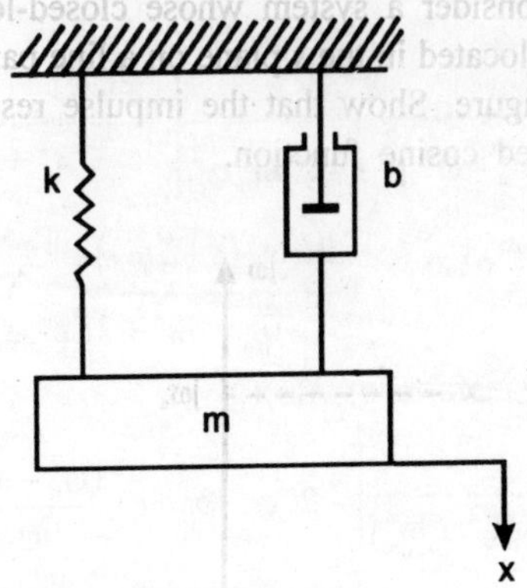

Fig. P. 3.32

Substituting the numerical values for m, b, and k into this equation gives

$$\ddot{x} + 2\dot{x} + 100x = 0$$

Where the initial conditions are $x(0) = 0.05$ and $\dot{x}(0) = 0$. From this last equation the undamped natural frequency ω_n and the damping ratio ζ are found to be $\quad \omega_n = 10, \; \zeta = 0.1$

The frequency actually observed in the vibration is the damped natural frequency ω_d.

$$\omega_d = \omega_n \sqrt{1-\zeta^2} = 10\sqrt{1-0.01} = 9.95 \, \text{rad/sec}$$

In the present analysis, $\dot{x}(0)$ is given as zero. Thus, solution $x(t)$ can

be written as $\quad x(t) = x(0)e^{-\zeta\omega_n t}\left(\cos\omega_d t + \dfrac{\zeta}{\sqrt{1-\zeta^2}}\sin\omega_d t \right)$

It follows that at $\quad t = nT$, where $T = 2\pi/\omega_d$,

$$x(nT) = x(0)e^{-\zeta\omega_n \, nT}$$

Consequently, the amplitude four cycles later becomes

$$x(4T) = x(0)e^{-\zeta\omega_n \, 4T} = x(0)e^{-(0.1)(10)(4)(0.6315)}$$

$$= 0.05\,e^{-2.526} = 0.05 \times 0.07998 = 0.004 \, \text{m}$$

Problem 3.33. Consider a system whose closed-loop poles and closed-loop zero are located in the s plane on a line parallel to the jw axis, as shown in Figure. Show that the impulse response of such a system in a damped cosine function.

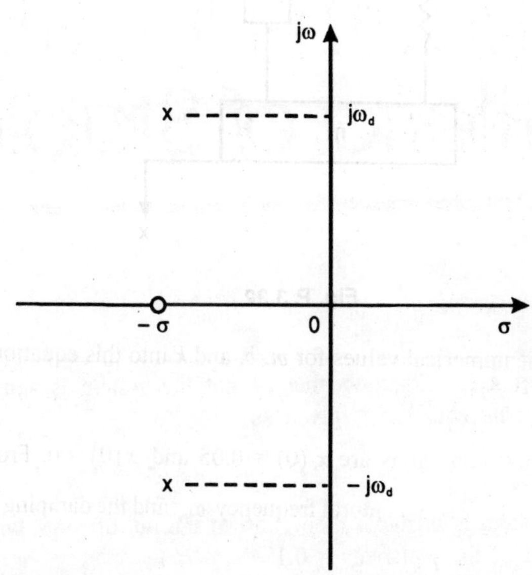

Fig. P. 3.33 Closed-loop pole-zero configuration of system whose impulse response is a damped cosine function

Solution:

The closed-loop transfer function is

$$\frac{C(s)}{R(s)} = \frac{K(s+\sigma)}{(s+\sigma\, j\omega_d)(s+\sigma-j\omega_d)}$$

For a unit-impulse input, R(s) = 1 and

$$C(s) = \frac{K(s+\sigma)}{(s+\sigma)^2 + \omega_d^2}$$

The inverse Laplace transform of $C(s)$ is

$$c(t) = Ke^{-\sigma t}\cos\omega_d t, \quad for\, t \geq 0$$

Which is a damped cosine function.

CHAPTER 4

Routh-Hurwitz Criteria of Stability

Problem 4.1. State whether or not the system is stable whose characteristic equation is given as

$$s^4 + 2s^3 + 3s^2 + 4s + 5 = 0$$

If the system is unstable then confirm the no. of roots lying in the right-half of the s-plane.

Solution:

Let us construct the array of coefficients

$$
\begin{array}{c|ccc}
s^4 & 1 & 3 & 5 \\
s^3 & 2 & 4 & 0 \\
s^2 & \dfrac{6-4}{2} & \dfrac{10-0}{2} & \\
s^1 & -6 & & \\
s^0 & 5 & &
\end{array}
$$

We check the no. of sign changes in the first column and there will be that many roots lying in the RHS of s-plane.

In the above array,

Since the sign of coeff. of s^2 in first column is $+ve$, that of s^1 is $-ve$ and again that of s^0 is $+ve$, thus there are two sign changes in the coefficients. Therefore, two roots are lying in the *RHS* of s-plane and hence the system is unstable.

Problem 4.2. The characteristic equation of a closed-loop system is given by $s^4 + 6s^3 + 11s^2 + 6s + K = 0$. Find the range of K for closed-loop stable behaviour.

Solution:

Characteristic eqn is $s^4 + 6s^3 + 11s^2 + 6s + K = 0$

Applying Routh-Hurwitz criterion

$$
\begin{array}{cccc}
s^4 & 1 & 11 & K \\
s^3 & 6 & 6 & 0 \\
s^2 & 10 & K & \\
s^1 & 6-0.6K & & \\
s^0 & K & &
\end{array}
$$

As we know that for stability of the system all the coefficients of the first column of the Routh array must be positive.

Thus, $\qquad K \geq 0$ and $6 - 0.6K \geq 0$

$\Rightarrow \qquad 0.1K \leq 1 \Rightarrow K \leq 10$

Thus, range of K is $0 \leq K \leq 10$

Problem 4.3. What is the frequency of oscillation of the system shown in the fig. if the system is made to oscillate continuously by choosing a suitable value of the scalar parameter K.

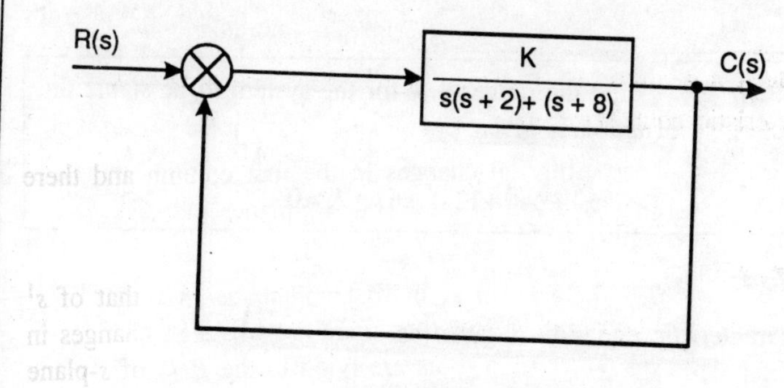

Fig. P. 4.3

Solution:

Overall T. F. of closed loop system

$$T(s) = \frac{G(s)}{1+G(s)H(s)}$$

The characteristic equation is

$$1 + G\,H = 0$$

$$\Rightarrow \qquad 1+\frac{K}{s(s+2)(s+8)} = 0$$

$$\Rightarrow \qquad s\left(s^2+10s+16\right)+K = 0$$

$$\Rightarrow \qquad s^3+10s^2+16s+K = 0$$

Putting $s = j\omega$,

$$-j\omega^3-10\omega^2+16\,j\omega+K = 0$$

$$\Rightarrow \quad \left(K-10\omega^2\right)+j\left(16\omega-\omega^3\right) = 0 \qquad\qquad (1)$$

For the system to oscillate continuously the j term should be zero.

i.e. $\qquad\qquad 16\omega-\omega^3 = 0 \Rightarrow \omega^2 = 16$

i.e. frequency of oscillation $\omega = 4$ rad/sec.

Also, from (1), $\quad K-10\omega^2 = 0$

$$\Rightarrow \qquad\qquad K = 10\omega^2 = 10 \times 16 = 160$$

Problem 4.4. Find the range of K for the system to be stable the characteristic equation is given by

$$s^4+6s^3+12s^2+6s+K = 0$$

Solution:

The characteristic equation is given as

$$s^4+6s^3+12s^2+6s+K = 0$$

Applying Routh-Hurwitz criterion

$$
\begin{array}{c|ccc}
s^4 & 1 & 12 & K \\
s^3 & 6 & 6 & \\
s^2 & 11 & K & \\
s^1 & \dfrac{66-6K}{11} & & \\
s^0 & K & &
\end{array}
$$

For the system to be stable,

$$K \geq 0 \text{ and } \frac{66-6K}{11} \geq 0$$

$$\Rightarrow \qquad 6K \leq 66 \Rightarrow K \leq 11$$

Thus, range of K is $0 \leq K \leq 11$

N.B.: at $K = 0$ and $K = 11$, the system is just stable, i.e. for limitedly stable.

Problem 4.5. The OLTF of a system os given by $G(s) = \dfrac{K}{s(s+2)(s+4)}$. Determine the value of K which will cause systained oscillations in the closed loop unity feedback system.   [IES'1996]

Solution:

Given OLTF $\qquad G(s) = \dfrac{K}{s(s+2)(s+4)}$

Characteristic equation of the CLTF will be

$$s\left(s^2+6s+8\right)+K = 0$$

$$\Rightarrow \qquad s^3+6s^2+8s+K = 0$$

From Routh-Hurwitz criterion

$$
\begin{array}{c|cc}
s^3 & 1 & 8 \\
s^2 & 6 & K \\
s^1 & \dfrac{48-K}{6} & \\
s^0 & K &
\end{array}
$$

So for the system to have sustained oscillations

$$\frac{48-K}{6} = 0 \implies K = 48.$$

Problem 4.6. The characteristic equation of a system is given as

$$s^2 - (K+2)s + (2K+5) = 0$$

(a) Find the value of K for which the system is (i) Stable (ii) Unstable (iii) Limitedly stable.

(b) For the stable case for what values of K, is the system (i) underdamped (ii) overdamped (iii) critically damped.

Solution:

Given characteristic equation is

$$s^2 - (K+2)s + (2K+5) = 0$$

Applying the Routh-Hurwitz criteria

$$
\begin{array}{ccc}
s^2 & 1 & (2K+5) \\
s^1 & -(K+2) & \\
s^0 & (2K+5) &
\end{array}
$$

Now (a) (i) For the system to be stable

$$-(K+2) > 0 \implies K+2 < 0 \implies K < -2$$

also, $(2K+5) > 0 \implies 2K > -5 \implies K > -2.5$

i.e. $-2.5 < K < -2$

(ii) For the system to be unstable

$$K > -2 \text{ and } K < -2.5$$

(iii) For the system to be limitedly stable

$$K = -2 \text{ and } K = -2.5$$

(b) To find the condition of overdamping and underdamping, we compare the characteristic equation with

$$s^2 + 2\xi\omega_n + \omega_n^2 = 0$$

Thus, $\omega_n = \sqrt{2K+5}$ and $2\xi\omega_n = -(K+2)$

$$\Rightarrow \qquad \xi = \frac{-(K+2)}{2\sqrt{2K+5}}$$

(i) For the response to be underdamped

$$\xi\,(\text{damping ratio}) < 1$$

$$\Rightarrow \quad -(K+2) < 2\sqrt{2K+5}$$

$$\Rightarrow \quad (K+2)^2 > 4(2K+5)$$

$$\Rightarrow \quad K^2 + 4K + 4 > 8K + 20 \quad \Rightarrow K^2 - 4K - 16 > 0 \qquad (1)$$

We find the value of K from the quadratic equation as $K =$ 6.47 and $-$ 2.47, and from (1)

For underdamping, $K > 6.47$ and $K > -2.47$

But $K > 6.47$ leads to the unstable region

Thus, $-2.47 < K < -2$

(ii) For the system's response to be overdamped,

$\xi > 1$ and so range of K is $-2.5 < K < -2.47$

(iii) For the response to be critically damped

$\xi = 1$ i.e. $K = -2.47$

N.B.: In these types of problems first get the expression of damping ratio ξ from given characteristic equation and therefrom apply the relevant conditions.

Problem 4.7. The OLTF of a unity feedback control system is given by $G(s) = \dfrac{K}{(s+2)(s+4)(s^2+6s+25)}$. By applying Routh-Hurwitz criteria determine:

(i) the range of K for which the closed loop system will be stable;

(ii) the value of K which causes the sustained oscillation and then find the corresponding oscillation frequency.

Solution:

Characteristic eqn. is given by $1 + GH = 0$ i.e.

$$1+\frac{K}{(s+2)(s+4)\left(s^2+6s+25\right)}=0$$

$$\Rightarrow \quad \left(s^2+6s+8\right)\left(s^2+6s+25\right)+K=0$$

$$\Rightarrow \quad s^4+12s^3+69s^2+198s+(K+200)=0$$

Routh-table

s^4	1	69	$(K+200)$
s^3	12	198	
s^2	52.5	$(K+200)$	
s^1	$\dfrac{10395-12(K+200)}{52.5}$		
s^0	$(K+200)$		

(i) For the system to be stable $(K+200)>0 \Rightarrow K>-200$

Also, $\dfrac{10395-12(K+200)}{52.5}>0 \Rightarrow \dfrac{10395}{12}>K+200$

$\Rightarrow K+200 < 866.25 \Rightarrow K<666.25$

The range of K for the system to be stable is $-200 < K < 666.25$

(ii) The system will have sustained oscillations at the limiting value of K. Also, the frequency of oscillation is calculated by the S^2-coefficients from the table i.e. frequency of oscillation ω is calculated from

$$52.5S^2+(K+200)=0 \qquad (1)$$

Now, the limiting value of $K = -200$ will give ω to be zero, so this is discarded for the case and $K = 666.25$

From (1), $\quad -52.5\omega^2+866.25=0$

$\Rightarrow \quad \omega^2 = 16.5 \text{ and } \omega=\pm4.062$

Thus, the system will have sustained oscillations at $K = 666.25$ with frequency $\omega=4.062$ rad/sec.

Problem 4.8. Predict the nature of roots of the characteristic equation given below

$$s^3 + 2s^2 + s + 2 = 0$$

Solution:

This is one of the special cases where the first column has a zero in a row i.e. one coefficient in first column is zero. In this type of problems the zero term is replaced by a very small positive number ϵ and the rest of the array is evaluated.

If the sign of the coefficient above the zero (ϵ) is the same as that below it, it indicates that there are a pair of imaginary roots otherwise it is indication of one sign changes.

$$
\begin{array}{c|cc}
s^3 & 1 & 1 \\
s^2 & 2 & 1 \\
s^1 & 0 \approx \epsilon & \\
s^0 & 2 & \\
\end{array}
$$

Thus, the system has pair of imaginarry roots $s = \pm j$

Problem 4.9. Check the stability of the following characteristic equation

$$s^3 - 3s + 2 = 0$$

Solution:

$$
\begin{array}{c|cc}
s^3 & 1 & -3 \\
s^2 & 0 \approx \epsilon & 2 \\
s^1 & 3 - \dfrac{2}{\epsilon} & \\
s^0 & 2 & \\
\end{array}
$$

There are two sign changes in the first column which indicates that two roots are in the RHS plane. Hence the system is unstable.

> **Problem 4.10.** State whether the following characteristic *eqn* has a root in RHS plane or not.
>
> $$s^5 + 2s^4 + 24s^3 + 48s^2 - 25s - 50 = 0$$

Solution:

This is also a special case where all the coefficients in one row are zero. It indicates that these are roots of equal magnitude lying radially opposite in the s-plane, that is two real roots with equal magnitudes and opposite signs and/or two conjugate imaginary roots.

In such a case, the evaluation of the rest of the array can be continued by forming an auxiliary polynomial with the coefficients of the last row and by using the coefficients of the derivative of this polynomial in the next row. Such roots with equal magnitude and opposite in sign can be found by solving the auxiliary polynomial.

$$
\begin{array}{llll}
s^5 & 1 & 24 & -25 \\
s^4 & 2 & 48 & -50 \;\leftarrow \text{auxiliary polynomial } p(s) \\
s^3 & 0 & 0 &
\end{array}
$$

$$p(s) = 2s^4 + 48s^2 - 50$$

$$\frac{d\,p(s)}{ds} = 8s^3 + 96s$$

$$
\begin{array}{lrrr}
s^5 & 1 & 24 & -25 \\
s^4 & 2 & 48 & -50 \\
s^3 & 8 & 96 & \\
s^2 & 24 & -50 & \\
s^1 & 112.7 & 0 & \\
s^0 & -50 & &
\end{array}
$$

Since there is one sign change in the first column of the new array, the is one root with a +ve real part and hence system is unstable. Pair of roots of equal magnitude but opposite sign is obtained as,

$$2s^4 + 48s^2 - 50 = 0$$

$$\Rightarrow \qquad s^2 = 1, \; s^2 = -25 \qquad \text{or, } s = \pm1, \; s = \pm j5$$

Problem 4.11. Determine the range of K for stability for the

CLTF $G(s) = \dfrac{K}{s\left(s^2 + s + 1\right)(s+2) + K}$

Solution:

Characteristic *eqn* is $s^4 + 3s^3 + 3s^2 + 2s + K = 0$

The array of coefficients is

$$
\begin{array}{c|ccc}
s^4 & 1 & 3 & K \\
s^3 & 3 & 2 & \\
s^2 & \dfrac{7}{3} & K & \\
s^1 & 2 - \dfrac{9}{7}K & & \\
s^0 & K & &
\end{array}
$$

For stability K must be +ve and all coefficients in the first column must be +ve.

$$\therefore \qquad 2 - \frac{9}{7}K > 0 \implies K < \frac{14}{9}$$

$\therefore$ Range of K is $\dfrac{14}{9} > K > 0.$

Problem 4.12. Find the no. of roots in the left-half of S-plane for the equation $s^3 - 4s^2 + s + 6 = 0$ \qquad (IAS' 2001, Gate' 2004)

Solution:

Characteristic *eqn* is $s^3 - 4s^2 + s + 6 = 0$

Applying Routh-Hurwitz criterion,

$$
\begin{array}{c|cc}
s^3 & 1 & 1 \\
s^2 & -4 & 6 \\
s^1 & 2.5 & \\
s^0 & 6 &
\end{array}
$$

Since there are two sign changes in the first column therefore, these are two roots in the right-half of S-plane.

Order of polynomial $= 3$

i.e. total no. of roots $= 3$

Thus, the no. of roots lying in the left-half of the S-plane $= 3 - 2 = 1$

Problem 4.13.   Find the no. of roots lying in the left   half of s-plane for the equation $s^5 + 2s^4 - 3s^3 + s^2 + 6s + 1 = 0$

Solution:

The Routh array will be as follows:

$$
\begin{array}{c|ccc}
s^5 & 1 & -3 & 6 \\
s^4 & 2 & 1 & 1 \\
s^3 & -3.5 & 5.5 & \\
s^2 & 4.143 & 1 & \\
s^1 & 6.345 & & \\
s^0 & 1 & &
\end{array}
$$

Since there are two sign changes in the first column so there are two roots in the RHS of s-plane.

The no. of roots in the left half of s-plane $= 5 - 2 = 3$

Problem 4.14.   For thé characteristic equation given below, determine the stability of a closed-loop control system.

$$s^5 + s^4 + 2s^3 + 2s^2 + 11s + 10 = 0$$

Solution:

Routh array is shown below:

$$
\begin{array}{c|ccc}
s^5 & 1 & 2 & 11 \\
s^4 & 1 & 2 & 10 \\
s^3 & 0 & 1 &
\end{array}
$$

Since one coefficient of first column is zero so Routh criterian fails here and to solve this replace zero by ϵ.

$$
\begin{array}{cccc}
s^5 & 1 & 2 & 11 \\
s^4 & 1 & 2 & 10 \\
s^3 & \epsilon & 1 & \\
s^2 & \dfrac{2\epsilon-1}{\epsilon} & 10 & \\
s^1 & \left(1-\dfrac{10\epsilon^2}{2\epsilon-1}\right) & & \\
s^0 & 10 & &
\end{array}
$$

To check the sign, we apply the limit as follow:

$$
\lim_{\epsilon\to 0}\frac{2\epsilon-1}{\epsilon} = -\infty = -\text{ve}
$$

Also,

$$
\lim_{\epsilon\to 0}\left(1-\frac{10\epsilon^2}{2\epsilon-1}\right) = 1 = +\text{ve}
$$

Thus, there are two sign changes in the first column of the Routh array indicating that two roots are in the Right-half of s-plane and hence the system is unstable.

Problem 4.15. (a) Using Routh-Hurwitz criterion determine the relation between K and T so that unity feedback control system whose OLTF given below is stable

$$
G(s) = \frac{K}{s\left[s(s+10)+T\right]}
$$

(b) Determine the modified relation between K and T it all the roots of characteristic equation as determined in (a) are to lie to the left of the line $s = -1$ in s-plane.

Solution:

(a) The characteristic equation for the system is

$$
s\left[s(s+10)+T\right]+K = 0
$$

$$\Rightarrow \quad s^3 + 10\,s^2 + Ts + K = 0$$

Constructing the Routh's array

$$
\begin{array}{c|cc}
s^3 & 1 & T \\
s^2 & 10 & K \\
s^1 & \dfrac{10T - K}{10} & \\
s^0 & K &
\end{array}
$$

For the system to be stable

$$K > 0 \text{ and } \frac{10T - K}{10} > 0 \Rightarrow K < 10\,T$$

Thus, the required relation is $0 < K < 10\,T$

(b) For all the roots to lie in the left half of the line $s = -1$, the modified characteristic equation is obtained by putting $s = (Z + 1)$ thus, the modified characteristic equation is

$$(Z+1)^3 + 10(Z+1)^2 + T(Z+1) + K = 0$$

$$\Rightarrow \quad Z^3 + 13\,Z^2 + (T+23)Z + (T+K+11) = 0 \qquad (1)$$

The same Routh-Hurwitz criterion is applied to the equation (1)

$$
\begin{array}{c|cc}
Z^3 & 1 & (T+23) \\
Z^2 & 13 & (T+K+11) \\
Z^1 & \dfrac{(12T - K - 288)}{13} & \\
Z^0 & T+K+11 &
\end{array}
$$

For the roots to lie to the left of line $s = -1$, $\dfrac{12T - K - 288}{13} > 0$

$$\Rightarrow \quad K < (12\,T - 288) \qquad (a)$$

Also $T + K + 11 > 0$

$$\Rightarrow \quad K < -(T + 11) \qquad (b)$$

From (a) & (b), we find

$$(12\,T - 288) > K < -(T + 11)$$

Problem 4.16. Determine the range of values of K ($K > 0$) so that the characteristic equation

$$s^3 + 3(K+1)s^2 + (7K+5)s + (4K+7) = 0$$

has roots more negative than $s = -1$

Solution:

The characteristic equation is

$$s^3 + 3(K+1)s^2 + (7K+5)s + (4K+7) = 0$$

If s_1 is a root of this equation, the condition is that

$$s_1 < -1 \text{ or } (s_1 + 1) < 0.$$

Let, $s' = (s+1)$ Then the condition becomes $s' < 0$ and the modified characteristic equation becomes

$$(s'-1)^3 + 3(K+1)(s'-1)^2 + (7K+5)(s'-1) + 4K + 7 = 0$$

$$\Rightarrow \quad s'^3 - 3s'^2 + 3s' - 1 + 3(K+1)\left(s'^2 - 2s' + 1\right) + (7K+5)$$

$$(s'-1) + 4K + 7 = 0$$

or $\quad s'^3 + 3Ks'^2 + s'(K+2) + 4 = 0$

Constructing the Routh's array

$$
\begin{array}{ccc}
s'^3 & 1 & (K+2) \\[2mm]
s'^2 & 3K & 4 \\[2mm]
s'^1 & \dfrac{4 - 3K^2 - 6K}{-3K} & \\[2mm]
s'^0 & 4 &
\end{array}
$$

The condition that all roots $s' < 0$ is that there should be no sign changes in the column.

Hence, $\dfrac{4 - 3K^2 - 6K}{-3K} > 0$

$$\Rightarrow \quad \dfrac{3K^2 + 6K - 4}{3K} > 0 \Rightarrow 3K^2 + 6K - 4 > 0$$

Solving the quadratic equation,

$$K = \frac{-6 \pm \sqrt{36 + 4 \times 3 \times 4}}{6} = 0.527, -2.53$$

Since $K > 0$ (given),

$\therefore$ Required value of $K > 0.527$.

Problem 4.17. The characteristic equation for a feedback control system is given by

$$s^3 + 2Ks^2 + 5s^2 + 10s + 15 = 10$$

Determine the range of K for which the system is stable.

Solution:

For the given characteristic equation, Routh's array is given below

$$
\begin{array}{c c c}
s^3 & 1 & 10 \\[4pt]
s^2 & (2K+5) & 15 \\[4pt]
s^1 & \dfrac{(20K+35)}{2K+5} & \\[8pt]
s^0 & 15 &
\end{array}
$$

For the given system to be stable, $\dfrac{20K+35}{2K+5} > 0$

$\Rightarrow \quad 20\,K + 35 > 0 \Rightarrow K > \dfrac{-35}{20}$

$\Rightarrow \qquad K > -1.75$ (a)

Also $\quad 2\,K + 5 > 0 \Rightarrow K > -2.5$ (b)

Conditions (a) and (b) merges into $K > -1.75$

Problem 4.18. A system is designed to give satisfactroy performance when a particular amplifier gain K has the value 2.5. Determine by how much this gain can vary before the system becomes unstable if the characteristic eqn is $s^3 + (4 + K)s^2 + 6s + 16 + 8K = 0$.

Solution:

For the given characteristic equation Rouwth's array will be as below:

$$
\begin{array}{c|cc}
s^3 & 1 & 6 \\
s^2 & (4+K) & (16+8K) \\
s^1 & \dfrac{8-2K}{4+K} & - \\
s^0 & 16+8K & -
\end{array}
$$

For the given system to be stable,

$$4 + K > 0 \Rightarrow K < -4 \tag{1}$$

and

$$\frac{8-2K}{4+K} > 0 \Rightarrow 2K < 8 \Rightarrow K < 4 \tag{2}$$

also, $16 + 8K > 0$

$$\Rightarrow \quad K < -2 \tag{3}$$

from inequalities (1), (2) & (3), we conclude that $K < -4$

N.B.: In the inequality $K < -4$, all the conditions are satisfied.

Problem 4.19. For the characteristic equation given below find the value of a if the system· has sustained oscillations of frequency ω_n rad/sec.

$$s^3 + (2a+7)s^2 + 5s + 1 = 0$$

Also find the value of ω_n.

Solution:

The Routh's array for the given characteric equation is shown below:

$$
\begin{array}{c|cc}
s^3 & 1 & 5 \\
s^2 & (2a+7) & 1 \\
s^1 & \dfrac{10a-34}{2a+7} & - \\
s^0 & 1 & -
\end{array}
$$

For the given system to have sustained oscillations,

$$\frac{10a-34}{2a+7}=0 \Rightarrow a = 3.4$$

Oscillation frequency can be determind from $(2a+7)s^2 +1=0$

$$\Rightarrow \qquad -(2\dot{a}+7)\omega_n^2 +1 = 0$$

$$\Rightarrow \qquad \omega_n^2 = \frac{1}{13.8}, \Rightarrow \omega_n =0.27\,\text{rad/sec. } \textbf{Ans.}$$

Problem 4.20. For the system in fig. find the condition

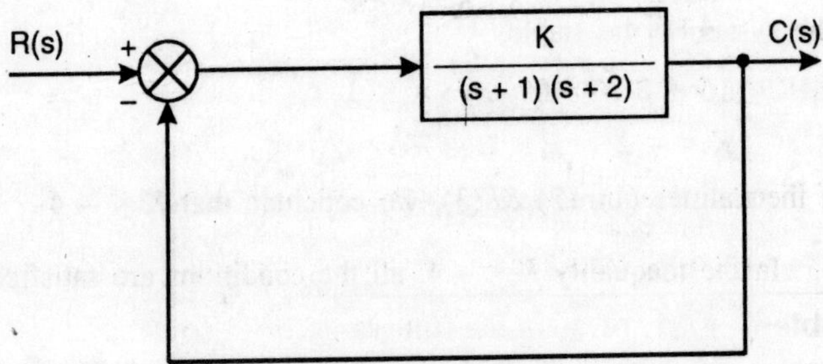

Fig. P. 4.20

For K so that the system is stable if there is a transportation lag of dead time equal to 1 sec. (transportation lag $= e^{-Ts}$). Assume the concerned frequencies to be low.

Solution:

From the given fig. the characteristic equation is given by

$$1 + G(s).H(s) = 0$$

Since, there is a transportation lag, so modified

$$G(s) = \frac{Ke^{-s}}{(s+1)(s+2)}$$

For small or low frequency, $e^{-s} \approx 1-s$

$$\therefore \qquad G(s)\,H(s) = \frac{K(1-s)}{(s+1)(s+2)}$$

and the characteristic equation becomes,

$$1+\frac{K(1-s)}{(s+1)(s+2)}=0$$

$$\Rightarrow \quad s^2+3s+2+K-Ks=0$$

$$\Rightarrow \quad s^2+s(3-K)+(2+K)=0$$

Applying Routh's criterion,

$$
\begin{array}{ccc}
s^2 & 1 & (2+K)\\
s^1 & (3-K) & 0\\
s^0 & (2+K) & 0
\end{array}
$$

For the system to be stable,

$$3-K>0 \Rightarrow K<3$$

also, $\qquad 2+K>0 \Rightarrow K<-2$

Thus, the range required is $K<-2$.

Problem 4.21. Determine the largest time constant of the characteristic equation given below:

$$s^3+4s^2+6s+4=0$$

Solution:

Given equation is $s^3+4s^2+6s+4=0$ we check that $s=-2$ satisfies the given equation and therefore $(s+2)$ is a factor of the equation

Now,

$$
\begin{array}{r}
(s+2))s^3+4s^2+6s+4(s^2+2s+2\\
\underline{-s^3\pm2\in s^2}\\
2s^2+6s+4\\
\underline{-2s^2\pm4s}\\
2s+4\\
\underline{-2s\pm4}\\
0
\end{array}
$$

Thus, the characteristic equation can be factorised as

$$(s+2)\left(s^2+2s+2\right)=0$$

$(s + 2)$ in time-constant form will be $(0.5\,s + 1)$ So, one time constant = 0.5 sec.

For the quadratic equation $s^2+2s+2=0$, comparing with the equation

$s^2+2\xi\omega_n+\omega_n^2=0$ has time-constant = $\xi\omega_n$

So, for other factor, time-constant = 1 sec.

So, the largest time-constant = 1 sec.

Problem 4.22. A unity feedback control system has an OLTF

$$G(s)=\frac{K(s+13)}{s(s+1)(s+7)}.$$

Find the value of K at which the system just attains the stability.

Solution:

The characteristic equation will be as

$$1+\frac{K(s+13)}{s(s+1)(s+7)}=0$$

$$\Rightarrow \quad s^3+8s^2+7s+Ks+13K=0$$

$$\Rightarrow s^3+8s^2+s+(7+K)+13K=0$$

Constructing the Routh's array.

$$
\begin{array}{c|cc}
s^3 & 1 & (7+K) \\[4pt]
s^2 & 8 & 13K \\[4pt]
s^1 & \dfrac{56-5K}{8} & - \\[4pt]
s^0 & 3K & -
\end{array}
$$

For the system to just attain the stability,

$$56 - 5\,K = 0$$

$$\Rightarrow \qquad 5\,K = 56 \Rightarrow K = \frac{56}{5}=11.2$$

Problem 4.23. The characteristic equation of a closed-loop-system is given by $s^4+6s^3+11s^2+6s+K=0$. Find the range of K for closed-loop stable behaviour.

Solution:

Given $\quad s^4+6s^3+11s^2+6s+K=0$

Hurwitz critarion is applied,

$$
\begin{array}{c c c c}
s^4 & 1 & 11 & K \\
s^3 & 6 & 6 & - \\
s^2 & 10 & K & - \\
s^1 & 6-0.6K & & \\
s^0 & K & &
\end{array}
$$

For stable closed loop, $K \geq 0$ and also $6(1-0.1K) \geq 0$

Thus $0 \leq K \leq 10$

Problem 4.24. What is the frequency of oscillation, to make the system shown in figure 4.24 to oscillate countinously, by a suitable choice of the scalar parameter K.

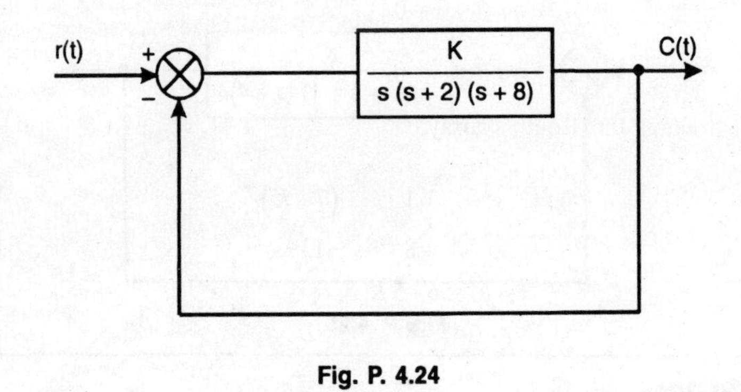

Fig. P. 4.24

Solution:

The characteristic equation will be,

$$s(s+2)(s+8)+K = 0$$

or $\qquad s\left(s^2+10s+16\right)+K = 0 \qquad\qquad (1)$

Also, for a system with feed back loop to oscillate, $\alpha\beta = -1$

Where $\alpha = G(s)$ and β = feebdack path gain

$\therefore \qquad \dfrac{K\times 1}{s\left(s+2\right)\left(s+8\right)} = -1$

or, $\qquad s\,(s + 2)\,(s + 8) + K = 0 \qquad\qquad (2)$

Which is same as the characteristic *eqn* in equation (1)

Replacing the term s by $(j\omega)$, we have,

$$j\omega\left(-\omega^2+10\,j\omega+16\right)+K = 0$$

For the system to oscillate, j terms should be zero

$\qquad -j\omega^3+16\,j\omega=0 \quad \therefore \quad \omega = 4 \text{ rad/sec.}$

Then $(K = 160)$

Problem 4.25. Consider the closed loop system shown in fig below. Determine the ranges of K for which the system is stable.

(I.E.S-95)

$$\dfrac{K}{s\,(s^2 + s + 1)\,(s + 4)}$$

Fig. P. 4.25

Solution:

Here $\qquad G(s) = \dfrac{K}{s\left(s+s+1\right)\left(s+4\right)}$

$\qquad\qquad H(s) = 1$

Hence the characteristic equation is,

$$s\left(s^2 + s + 1\right)(s+4) + K = 0$$

Now applying Routh-Hurwitz criterion to determine the range of K.

$$
\begin{array}{c|cccc}
s^4 & 1 & 5 & K \\
s^3 & 5 & 4 & - \\
s^2 & \dfrac{21}{5} & K & - \\
s^1 & 4 - \dfrac{25}{21}K & - & - \\
s^0 & K & - & - \\
\end{array}
$$

For the system to be stable, $4 - \dfrac{25}{21}K > 0,\ K < \dfrac{84}{25}$

Hence $0 < K < \dfrac{84}{25}$ **Ans.**

Problem 4.26. Find the range of K, for the system to be stable. The characteristic equation is given by

$$s^4 + 6s^3 + 11s^2 + 6s + K$$

Solution:

Applying Routh-Hurwitz criterion,

$$
\begin{array}{c|cccc}
s^4 & 1 & 11 & K \\
s^3 & 6 & 6 & - \\
s^2 & 10 & K & - \\
s^1 & \dfrac{60 - 6K}{10} & - & - \\
s^0 & K & - & - \\
\end{array}
$$

For the system to be stable, $K > 0$ and also all of the terms of the first column should be of same sign.

$$\frac{60 - 6K}{10} > 0;\ \text{or}\ 6K < 60,\ K < 10$$

$\therefore\ 0 < K < 10$

Problem 4.27. The open loop transfer function of a system is given by $G(s) = \dfrac{K}{s(s+2)(s+4)}$. Determine the value of K which will cause sustained oscillations in the closed loop unity feedback system.

(I.E.S-96)

Solution:

Given, O.L.T.F is, $G(s) = \dfrac{K}{s(s+2)(s+4)}$

For C.L.T.F, the characteristic *eqn* is,

$$s(s+2)(s+4)+K = 0$$

or,

$$s^3 + 6s^2 + 8s + K = 0$$

Constructing Routh-Hurwitz table

$$
\begin{array}{c|cc}
s^3 & 1 & 8 \\
s^2 & 6 & K \\
s^1 & 8 - K/6 & - \\
s^0 & K &
\end{array}
$$

First of all $K > 0$ and for sustained oscillations, $8 - \dfrac{K}{6} = 0$

$\therefore \qquad K = 48$ **Ans.**

Problem 4.28. The open loop transfer function of a unity feedback control system is $G(s) \cdot H(s) = \dfrac{30}{s(s+1)(s+T)}$;

Where T is a variable parameter. The closed loop system will be stable, for what values of T?

[I.E.S 1997]

Solution:

$$G(s) \cdot H(s) = \frac{30}{s(s+1)(s+T)}$$

Hence the characteristic equation,

$$s(s+1)(s+T)+30 = 0$$

or,　$s^3 + s^2(1+T)+sT+30 = 0$

Now constructing Routh table,

$$
\begin{array}{c|cc}
s^3 & 1 & T \\
s^2 & (1+T) & 30 \\
s^1 & T-\dfrac{30}{1+T} & - \\
s^0 & 30 &
\end{array}
$$

For the system to be stable

$$1 + T > 0 \text{ or } T > -1 \text{ and } T-\frac{30}{1+T}>0$$

or,　$\dfrac{T(1+T)-30}{(1+T)}>0;$ or, $\dfrac{(T+6)(T-5)}{(1+T)}>0$

$$T > -6 \text{ and } T > 5$$

Therefore T should greater than 5.

Problem 4.29. Define the term (a) Limitedly stable, (b) absolutely stable and conditionally stable.

Solution:

(a) Limitedly stable: If all the roots of the characteristic equation have negative real parts, with one or more non repeated roots are one $j\omega$-axis, the system is limitedly or marginally stable.

(b) Absolutely stable: If a system is stable for all values of the given parameter with respect to that parameter.

(c) Conditionally stable: Conditionally stable with respect to a parameter, if the system is stable for only certain bonded ranges of the values of this parameter.

Problem 4.30. Consider the following characteristic equation

$$3s^4 + 10s^3 + 5s^2 + 5s + 2 = 0$$

How many roots are in the R.H. plane?

Solution:

The Routh array is,

$$
\begin{array}{c|ccc}
s^4 & 3 & 5 & 2 \\
s^3 & 10 & 5 & - \\
    & 2 & 1 & - \\
s^2 & 3.5 & 2 & - \\
s^1 & -1/7 & & \\
s^0 & 2 & &
\end{array}
\qquad \left(\text{dividing the } s^3\text{-row by 5 throughout}\right)
$$

Examining the first column of the Routh array, it is found that there are two change of sign. Therefore the system is unstable having two poles in the R.H. plane.

Problem 4.31. The characteristic of a system is given as below,

$$s^2 - (K+2)s + (2K+5) = 0$$

(a) Find the value of K for which the system is,
   (1) stable, (2) limitedly stable (3) unstable.

(b) For the stable case for what values of K, is the system
   (1) underdamped (2) overdamped

Solution:

The characteristic *eqn* is,

$$s^2 - (K+2)s + (2K+5) = 0$$

Applying Routh criterion,

$$
\begin{array}{ccc}
s^2 & 1 & 2K+5 \\
s^1 & -(K+2) & - \\
s^0 & 2K+5 &
\end{array}
$$

(a) (i) For the system to be stable, there are two conditions,

$$- (K + 2) > 0 \text{ and } (2K + 5) > 0$$

or $\quad K < -2$ and $K > -2.5$ $\therefore$ $- 2 < K > -25$

(ii) For the system to be limitedly stable,

$$K = -2 \text{ and } K = -2.5$$

(iii) For the system to be unstable

$$K > -2 \text{ and } K < -2.5$$

(b) The roots of the characteristic *eqn*,

$$s_1, s_2 = \frac{1}{2}\left\{(K+2) \pm \sqrt{(K+2)^2 - 4(2K+5)}\right\}$$

For critically damped,

$$(K+2)^2 - 4(2K+5) = 0$$

$$\therefore \qquad K = 6.47, - 2.47$$

As per stability condition, $K = 6.47$ is unstable so, for critical damping $K = -2.47$

(i) underdamped case $-2 > K > -2.47$

(ii) overdamped case, $-2.47 > K > -2.5$

Problem 4.32. Consider the following equation (AMIE)

$$s^5 + s^4 + 2s^3 + 2s^2 + 3s + s = 0.$$

Test condition for stability.

Solution:

$$
\begin{array}{c|ccc}
s^5 & 1 & 2 & 3 \\
s^4 & 1 & 2 & 5 \\
s^3 & 0 & -2 & \\
s^2 & & & \\
s^1 & & & \\
s^0 & 5 & &
\end{array}
$$

From the Routh-array, it is seen that the first element in the 3rd row is 0.

This difficulty caused by zero term in the first column of the Routh-array can be avoided by simply placing $\dfrac{1}{Z}$ in place of s, and then the char. *eqn* becomes,

$$5Z^5 + 3Z^4 + 2Z^3 + 2Z^2 + Z + 1 = 0$$

Then the Routh array is,

$$
\begin{array}{c|ccc}
Z^5 & 5 & 2 & 1 \\
Z^4 & 3 & 2 & 1 \\
Z^3 & -4/3 & -2/3 & - \\
Z^2 & 1/2 & 1 & - \\
Z^1 & 2 & - & - \\
Z^0 & 1 & - & -
\end{array}
$$

There are two sign changes in the first column and hence the system is unstable with two roots in the right S-plane.

Problem 4.33.   Test the stability of the system with the following characteristic equation by Routh's test:

$$s^6 + 2s^5 + 8s^4 + 12s^3 + 20s^2 + 16s + 16 = 0$$

Solution:

The Routh array is,

$$
\begin{array}{c|cccc}
s^6 & 1 & 8 & 20 & 16 \\
s^5 & 1 & 12 & 16 & - \\
s^5 & 1 & 6 & 8 & - \\
s^4 & 2 & 12 & 16 & - \\
s^4 & 1 & 6 & 8 & - \\
s^3 & 0 & 0 & - & -
\end{array}
$$

Since the terms in the s^3-row are all zero. The Routh test breaks down. Next the auxiliary polynomial is formed from coefficients of the s^4-row

which are $\qquad A(s) = s^4 + 6s^2 + 8$

It's derivative, $\qquad \dfrac{dA}{ds} = 4s^3 + 12s \in$

Now the zeros of s^3-row are replaced by the coefficients 4 & 12. The Routh array is,

$$
\begin{array}{c|cccc}
s^6 & 1 & 8 & 20 & 16 \\
s^5 & 1 & 6 & 8 \\
s^4 & 1 & 6 & 8 \\
s^3 & 4 & 12 \\
s^2 & 3 & 8 \\
s^1 & 1/3 \\
s^0 & 8
\end{array}
$$

Here, we see that there is no sign change in the first column of the new array.

The roots of the auxiliary polynomial $s^4 + 6s^2 + 8 = 0$ are given by

$$s = \pm j\sqrt{2} \quad \text{and} \quad s = \pm j2$$

No root lies in the R. H. plane. Hence the system under consideration is limitedly stable.

Problem 4.34. Consider the characteristic equation of a certain system $s^3 + 7s^2 + 25s + 39 = 0$ check if all the roots of this equation have real parts more negative than -1.

Solution:

Putting $\quad Z = Z - 1$

The characteristic equation is Z-variable becomes,

$$Z^3 + 4Z^2 + 14Z + 20 = 0$$

Now forming the Routh array,

$$
\begin{array}{c|cc}
Z^3 & 1 & 14 \\
Z^2 & 4 & 20 \\
Z^1 & 9 \\
Z^0 & 20
\end{array}
$$

Since the sign of all the terms in first column are (+) ve, all the roots of 2-characteristic equation lie in the left half which implies that all the roots of original equation lie to left half of $s= -1$ in the s-plane.

It is not enough that the system is stable; it should also possess the required relative stability which is specified in terms of peak overshoot, settling time etc. Thus, the relative stability is directly related to location of C.L. poles.

Problem 4.35.   The characteristic equation of a feedback control system is given by $2s^5 + 2s^4 + 5s^3 + 5s^2 + 3s + 5 = 0$

Use Routh's criterion to determine the stability of the system.

(I.E.S.-EE 92)

Solution:

Routh's criterion is,

$$
\begin{array}{c|ccc}
s^5 & 2 & 5 & 3 \\[2ex]
s^4 & 2 & 5 & 5 \\[2ex]
s^3 & \in \to 0 & -2 & 0 \\[2ex]
s^2 & \dfrac{5\in+4}{\in} & 5 & 0 \\[3ex]
s^1 & \dfrac{\dfrac{-2(5\in+4)}{\in}-5\in}{5\in+4/\in} & 0 & - \\[3ex]
s^0 & 5 & &
\end{array}
$$

The fifth term of the first column as $\delta \to 0; \dfrac{-2(4)}{4} = (-)\text{ve}$. There are two sign changes. Therefore two roots lie on the right half of S-plane.

$\therefore$ The system is unstable.

Problem 4.36.   The characteristic *eqn* for a feedback control system is given by $s^3 + 2Ks^2 + 5s^2 + 10s + 15 = 0$

Determine the range of K for which the system is stable.

(I.E.S-EE 87)

Solution:

The Routh-array is given by,

$$
\begin{array}{ccc}
s^3 & 1 & 10 \\
s^2 & (20K+5) & 15 \\
s^1 & \dfrac{200K+35}{200K+50} & 0 \\
s^0 & 15 & -
\end{array}
$$

For the system to be stable, $20K + 5 > 0$, $K > -5/20 = -1/4$

and $\qquad \dfrac{200K+35}{200K+50} > 0;\ K > \dfrac{-35}{200} = -0.175\ \ K > -1/4$

Problem 4.37. A system is designed to give satisfactory performance when a particular amplifier gain K has the value 2.5. Determine by how much this gain can vary before the system becomes unstable if the characteristic equation is, $\qquad$ (I.E.S-EE)

$$s^3 + (4+K)s^2 + 6s + 16 + 8K = 0$$

Solution:

The given characteristic equation is,

$$s^3 + (4+K)s^2 + 6s + 16 + 8K = 0$$

The Routh-array is

$$
\begin{array}{c|cc}
s^3 & 1 & 6 \\
s^2 & (4+K) & (16+8) \\
s^1 & \dfrac{8-2K}{4+K} & 0 \\
s^0 & 16+8K & 0
\end{array}
$$

For stability $\qquad 4 + K > 0\ ;\ K > -4$ $\qquad\qquad$ (1)

$\qquad 8 - 2K > 0\ ;\ 2K < 8,\ K < 4$ $\qquad\qquad$ (2)

and $\qquad 16 + 8K > 0\ ;\ K > -2$ $\qquad\qquad$ (3)

Hence, the range of K for stability from eqn (1), (2) & (3)

$$(\therefore -2 < K < 4)$$

Since at present $K = 2.5$, K can vary by $-4.5 < \Delta K < 1.5$

Problem 4.38. The open loop T.F. of a unity feedback control system is given by $G(s) = \dfrac{K}{(s+2)(s+4)(s^2+6s+2s)}$

By applying Routh-Hurwitz criterion, determine:

  (i) the range of K for which the closed loop system will be stable;

  (ii) the values of K which causes sustained oscillations.

       Find corresponding oscillation frequencies?

(I.E.S-EE 90)

Solution:

The characteristic *eqn* becomes, $1 + G(s).H(s) = 0$ i.e.

$$1 + \frac{K}{(s+2)(s+4)(s^2+6s+25)}$$

or, $$(s+2)(s+4)(s^2+6s+2s) + K = 0$$

or, $$s^4 + 12s^3 + 69s^2 + 198s + (K+200) = 0$$

The Routh table is,

s^4	1	69	$K+200$
s^3	12	198	0
s^2	$\dfrac{630}{12}$	$K+200$	0
s^1	$\left[198 - \dfrac{(K+200)144}{630}\right]$	0	0
s^0	$(K+200)$	0	0

(1) For the system to be stable, $K+200 > 0$ and $198 - \dfrac{144(K+200)}{630} > 0$

$$\therefore \quad K < \frac{630 \times 198}{144} - 200 = 666.25$$

or, $-200 < K < 666.25$

(2) The system will have sustained oscillation when,

$$K + 200 = 0 \text{ or, } K = -200$$

The corresponding oscillation frequency is,

$$\left(198 - \frac{(K+200)144}{630}\right)s = 0 \text{ which gives, } s = 0$$

$$\therefore \ \omega = 0$$

For the values of oscillation freq.

$$198 - \frac{144}{630}(K+200) = 0$$

$$\therefore \qquad\qquad K = 666.25$$

The corresponding auxiliary equation is,

$$\frac{630}{12}s^2 + K + 200 = 0$$

$$52.5\,s^2 + 866.25 = 0, \ \omega^2 = 16.5$$

$$\therefore \qquad\qquad \omega = \pm 4.062 \,\text{rad/sec.}$$

CHAPTER 5

Mathematical Modelling of Dynamic Systems

Problem 5.1. Figure P. 5.1 shows a schematic diagram of an automobile suspension system. As the car moves along the road, the vertical displacements at the tires act as the motion excitation to the automobile suspension system.

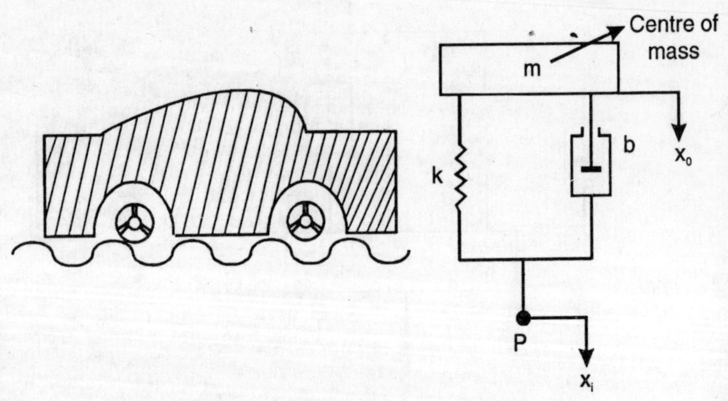

Fig. P. 5.1

A simplified version of the suspension system is also shown in fig. Considering the motion of the body only in the vertical direction, obtain the transfer function $\dfrac{X_o(s)}{X_i(s)}$

Solution:

The Equation of motion for the system shown is

$$m\ddot{x}_0 + b(\dot{x}_o - \dot{x}_i) + k(\dot{x}_o - \dot{x}_i) = 0$$

$$\Rightarrow \qquad m\ddot{x}_0 + b\dot{x}_o - b\dot{x}_i + kx_o - kx_i = 0$$

$$\Rightarrow \qquad m\ddot{x}_0 + b\dot{x}_o + kx_o = b\dot{x}_i + kx_i$$

Taking the Laplace transform of the above equation, assuming zero initial coniditions,

We get, $\qquad ms^2 X_o(s) + bsX_o(s) + kX_o(s) = bsX_i(s) + kX_i(s)$

$$\Rightarrow \qquad (ms^2 + bs + k)X_o(s) = (bs + k)X_i(s)$$

Transfer function required is

$$\therefore \qquad \frac{X_o(s)}{X_i(s)} = \frac{bs + k}{ms^2 + bs + k}$$

Problem 5.2. Obtain the transfer function $\dfrac{Y(s)}{U(s)}$ of the system shown in Fig. P. 5.2.

Fig. P. 5.2

Solution:

This is also a simplified version of an automobile or motorcycle suspension system.

Writing Equation of motion for the two masses.

For mass m_1,

$$m_1\ddot{x}+k_1(x-u)+k_2(x-y)+b(\dot{x}-\dot{y})=0 \tag{1}$$

Alternatively from Newton's second law of motion,

$$k_2(y-x)+b(\dot{y}-\dot{x})+k_1(u-x) = m_1\ddot{x}$$

for mass m_2,

$$m_2\ddot{y}+k_2(y-x)+b(\dot{y}-\dot{x}) = 0 \tag{2}$$

from (i), we have

$$m_1\ddot{x}+b\dot{x}+(k_1+k_2)x = b\dot{y}+k_2y+k_1u$$

from (ii), we have

$$m_2\ddot{y}+b\dot{y}+k_2y = b\dot{x}+k_2x$$

Taking Laplace's transforms of these two equations, assuming zero initial conditions,

$$\left[m_1s^2+bs+(k_1+k_2)\right]X(s) = (bs+k_2)Y(s)+k_1U(s)$$

and

$$\left[m_2s^2+bs+k_2\right]Y(s) = (bs+k_2)X(s)$$

Eliminating $X(s)$ from the last two equations, we get

$$\left(m_1s^2+bs+k_1+k_2\right)\frac{m_2s^2+bs+k_2}{bs+k_2}Y(s)=(bs+k_2)Y(s)+k_1U(s)$$

Which yields

$$\frac{Y(s)}{U(s)}=\frac{k_1(bs+k_2)}{m_1m_2s^4+(m_1+m_2)bs^3+\left[k_1m_2+(m_1+m_2)k_2\right]s^2+k_1bs+k_1k_2}$$

Problem 5.3. Consider the electrical circuit shown in Fig. P. 5.3

Obtain the transfer function $\dfrac{E_o(s)}{E_i(s)}$ by use of the block diagram approach.

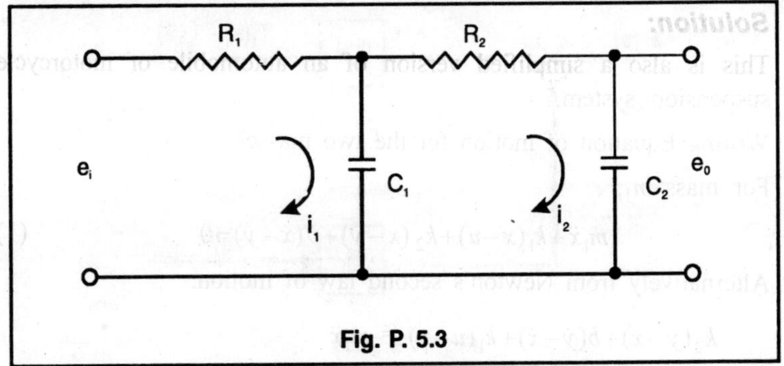

Fig. P. 5.3

Solution:

Equations for the circuits are

$$\frac{1}{C_1}\int(i_1-i_2)\,dt + R_1 i_1 = e_i \tag{1}$$

$$\frac{1}{C_1}\int(i_2-i_1)\,dt + R_2 i_2 + \frac{1}{C_2}\int i_2\,dt = 0 \tag{2}$$

and

$$\frac{1}{C_2}\int i_2\,dt = e_o \tag{3}$$

The Laplace transform of equations (1), (2) and (3) with zero initial conditions, we get

$$\frac{1}{C_1 s}\big[I_1(s)-I_2(s)\big] + R_1 I_1(s) = E_i(s) \tag{4}$$

$$\frac{1}{C_1 s}\big[I_2(s)-I_1(s)\big] + R_2 I_2(s) + \frac{1}{C_2 s}I_2(s) = 0 \tag{5}$$

and

$$\frac{1}{C_2 s}I_2(s) = E_o(s) \tag{6}$$

Equation (4) can be re-written as

$$C_1 s\big[E_i(s)-R_1 I_1(s)\big] = I_1(s)-I_2(s) \tag{7}$$

Equation (5) can be modified as,

$$I_2(s) = \frac{C_2 s}{R_2 C_2 s + 1}\cdot\frac{1}{C_1 s}\big[I_1(s)-I_2(s)\big] \tag{8}$$

Equation (7) gives the block diagram of fig. 5.3 (a). Also equation (8) yields the block diagram of fig. 5.3 (b) and equation (6) gives the block diagram shown in fig. 5.3 (c).

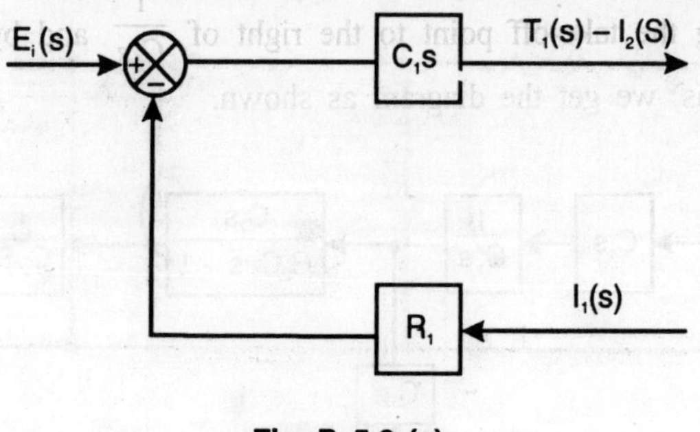

Fig. P. 5.3 (a)

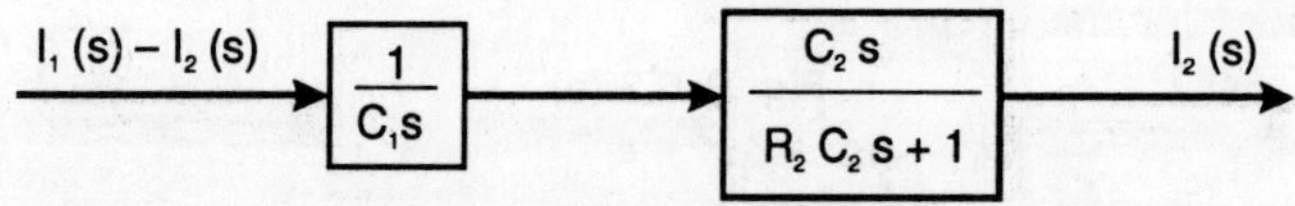

Fig. P. 5.3 (b)

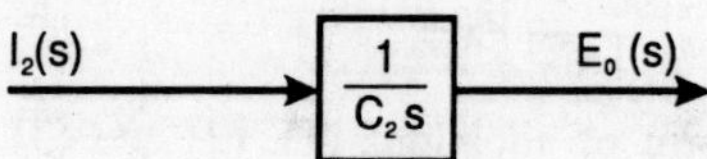

Fig. P. 5.3 (c)

The combination of all the above three block diagrams is shown in the figure below and this block diagram can be successively modified to get final result as under:

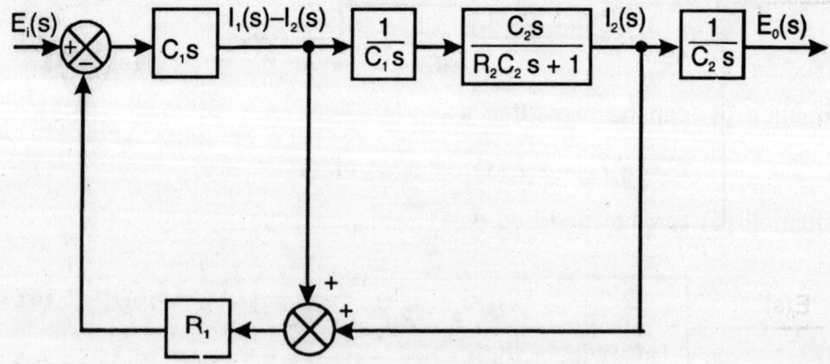

Fig. P. 5.3 (d)

Now, taking the take-off point to the right of $\dfrac{1}{C_1 s}$ and by relevant modifications, we get the diagram as shown.

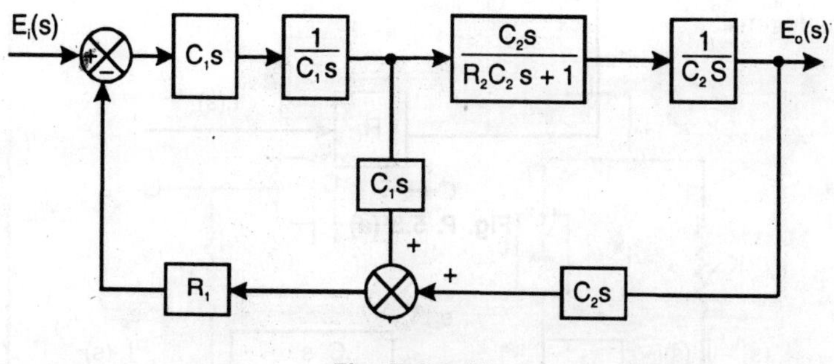

Fig. P. 5.3 (e)

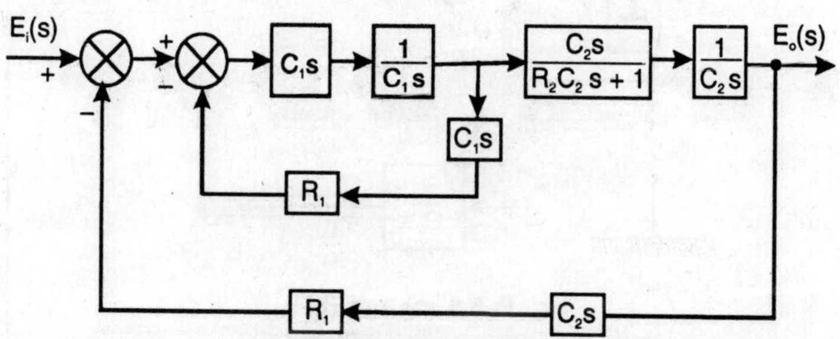

Fig. P. 5.3 (f)

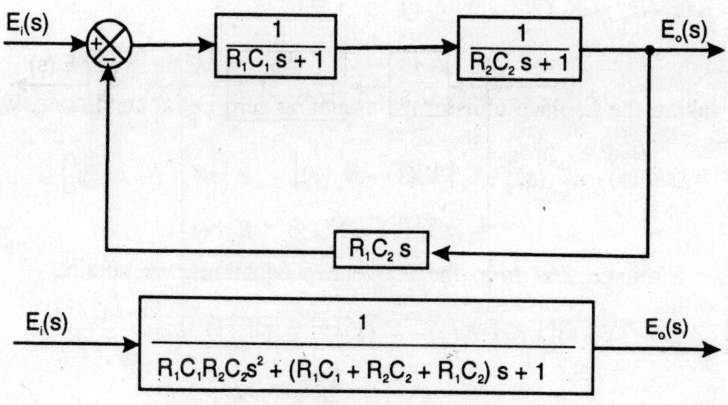

Fig. P. 5.3 (g)

Problem 5.4. Obtain the transfer function of the mechanical system shown in figure 5.4 (a). Also obtain the transfer function of the electrical system shown in figure 5.4 (b). Show that the transfer functions of the two systems are of identical form and thus they are analogous.

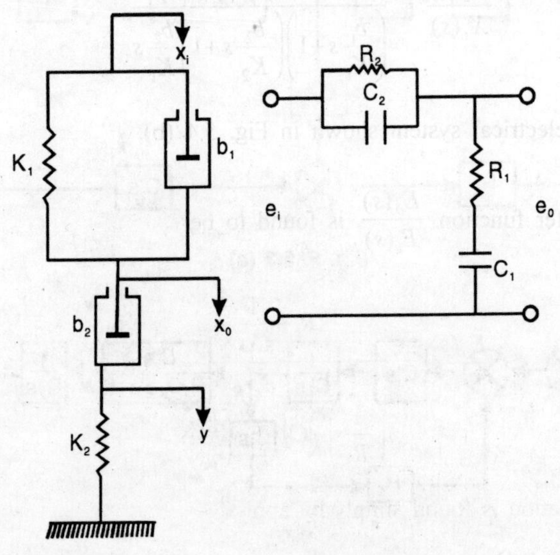

Fig. P. 5.4 (a) and (b)

Solution:

Applying equation of motion to the system

$$b_1(\dot{x}_i - \dot{x}_o) + K_1(x_i - x_o) + b_2(\dot{y} - \dot{x}_o) = 0$$

and

$$b_2(\dot{x}_o - \dot{y}) = K_2 y$$

By taking the Laplace transforms, assuming zero initial conditions, we have

$$b_1[sX_i(s) - sX_o(s)] + K_1[X_i(s) - X_o(s)] = b_2[sX_o(s) - sY(s)]$$

$$b_2[sX_o(s) - sY(s)] = K_2 Y(s)$$

If we eliminate $Y(s)$ from the above two equations, we obtain,

$$b_1[sX_i(s) - sX_o(s)] + K_1[X_i(s) - X_o(s)] = b_2 sX_o(s) - b_2 s \frac{b_2 sX_o(s)}{b_2 s + K_2}$$

$$\Rightarrow \quad (b_1 s + K_1) X_i(s) = \left(b_1 s + K_1 + b_2 s - b_2 s \frac{b_2 s}{b_2 s + K_2}\right) X_o(s)$$

Thus, the transfer function $\dfrac{X_o(s)}{X_i(s)}$ is given as

$$\frac{X_o(s)}{X_i(s)} = \frac{\left(\dfrac{b_1}{K_1}s+1\right)\left(\dfrac{b_2}{K_2}s+1\right)}{\left(\dfrac{b_1}{K_1}s+1\right)\left(\dfrac{b_2}{K_2}s+1\right)\dfrac{b_2}{K_1}s}$$

For the electrical system shown in Fig. 5.4 (b),

the transfer function $\dfrac{E_o(s)}{E_i(s)}$ is found to be

$$\frac{E_o(s)}{E_i(s)} = \frac{R_1+\dfrac{1}{C_1 s}}{\dfrac{1}{\left(\dfrac{1}{R_2}\right)+C_2 s}+R_1+\dfrac{1}{C_1 s}}$$

above relation is found simply by applying voltage division rule.

or
$$\frac{E_o(s)}{E_i(s)} = \frac{(R_1 C_1 s+1)(R_2 C_2 s+1)}{(R_1 C_1 s+1)(R_2 C_2 s+1)+R_2 C_1 s}$$

A comparison of the transfer functions show that the systems shown in Fig. 5.4 (a) to 5.4 (b) are analogous to each other.

Problem 5.5.　In the liquid-level system of given figure, assume that the outflow rate Qm³/sec through the outflow value is related to the head H m by

$$Q = K\sqrt{H} = 0.01\sqrt{H}$$

Assume also that when the inflow rate Q_i is 0.015 m³/sec the head stays constant. At $t = 0$ the inflow valve is closed and so there is no inflow for $t \geq 0$. Find the time necessary to empty the tank to half the original head. The capacitance C of the tank is $2\,\text{m}^2$.

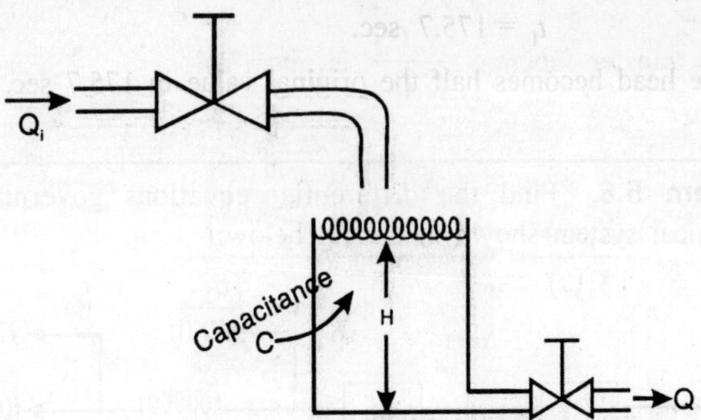

Fig. P. 5.5

Solution:

When the head is stationary, the inflow rate equals the outflow rate. Thus head H_o at $t=0$ obtained from

$$0.015 = 0.01\sqrt{H_o}$$

$$\Rightarrow \qquad H_o = 2.25 \ m$$

The equation for the system for $t > 0$ is

$$- C \ dH = Q \ dt$$

or

$$\frac{dH}{dt} = -\frac{Q}{C} = \frac{-0.01\sqrt{H}}{2}$$

Hence,

$$\frac{dH}{\sqrt{H}} = -0.005 \ dt \qquad\qquad (1)$$

Assume that, at $\qquad t = t_1$, $H = 1.125$ m i.e., $\dfrac{H}{2}$

Integrating both sides of the equation (1), we get

$$\int_{2.25}^{1.125} \frac{dH}{\sqrt{H}} = \int_{0}^{t_1}(-0.005)\,dt = -0.005\,t_1$$

$$\Rightarrow \qquad 2\sqrt{H}\Big|_{2.25}^{1.125} = -0.005\,t_1$$

or $\quad 2\sqrt{1.125} - 2\sqrt{2.25} = -0.005\,t_1$

or $\qquad t_1 = 175.7$ sec.

Thus, the head becomes half the original value in 175.7 sec.

Problem 5.6. Find the differential equations governing the mechanical system shown in the fig. below:

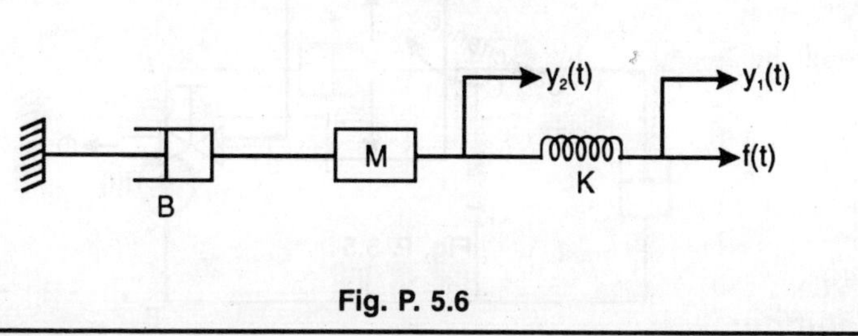

Fig. P. 5.6

Solution:

The differential equations will be as follows: applying equation of motion to the system

$$M\ddot{y}_2 + B\dot{y}_2 + K(y_2 - y_1) = 0$$

$$\Rightarrow \qquad M\ddot{y}_2 + B\dot{y}_2 = K(y_1 - y_2) \qquad (1)$$

and $\qquad K(y_1 - y_2) = f(t) \qquad (2)$

The equations (1) & (2) constitute the required diff. equations.

Problem 5.7. Find the transfer function relating displacements y and x for the mechanical system shown below:

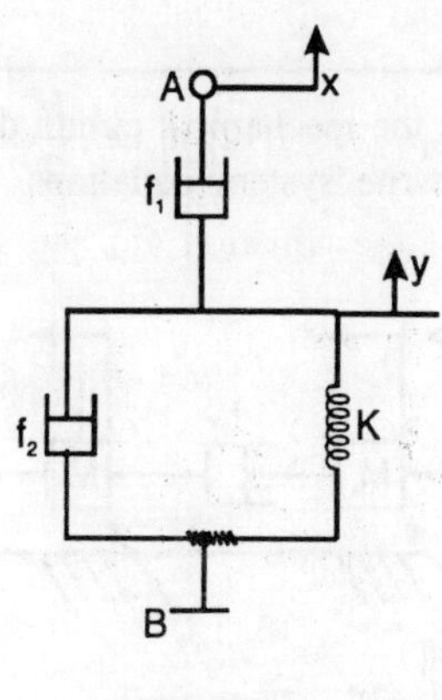

Fig. P. 5.7

Solution:

Let the force $f(t)$ be applied at point A, The mechanical circuit diagram of the system can be solved as

$$f_2\frac{dy}{dt}+Ky+f_1\frac{d}{dt}(y-x)=0 \tag{1}$$

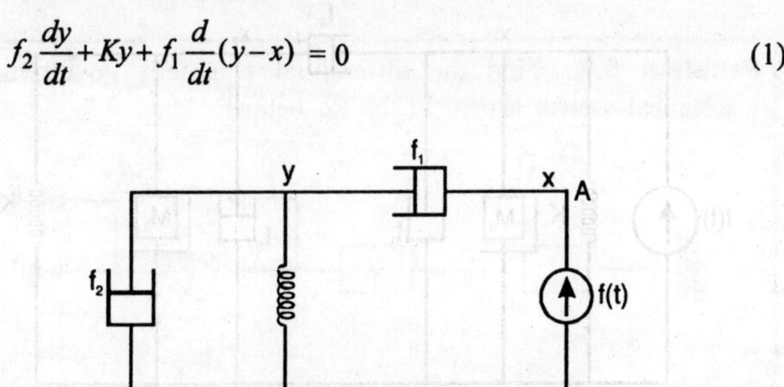

Mechanical ckt. diagram **Fig. P. 5.7 (a)**

$$f_1\frac{d}{dt}(x-y)=f(t) \tag{2}$$

Taking initial conditions as zero and Laplace transform on both sides of equation (i)

$$f_2\,sY(s)+KY(s)+f_1s\left[Y(s)-X(s)\right]=0$$

$$\Rightarrow \qquad \frac{Y(s)}{X(s)}=\frac{f_1s}{(f_1+f_2)s+K}=\frac{T_1s}{(T_2s+1)}$$

Where , $T_1=\dfrac{f_1}{K}$ and $T_2=\dfrac{f_1+f_2}{K}$

Problem 5.8.   Draw the mechanical circuit diagram for the system shown in figure and write system equations.

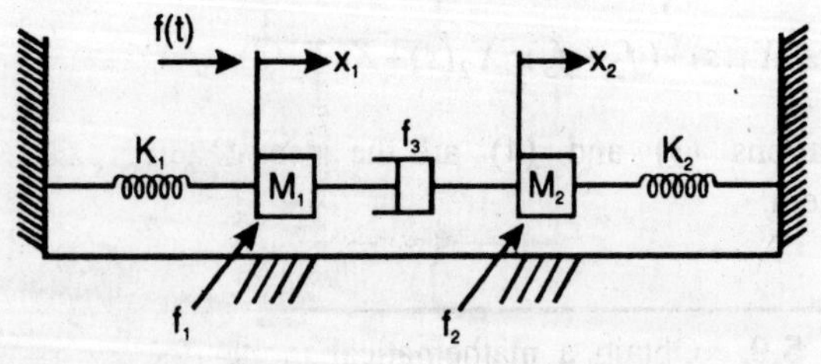

Fig. P. 5.8

Solution:

The mechanical circuit diagram is drawn below:

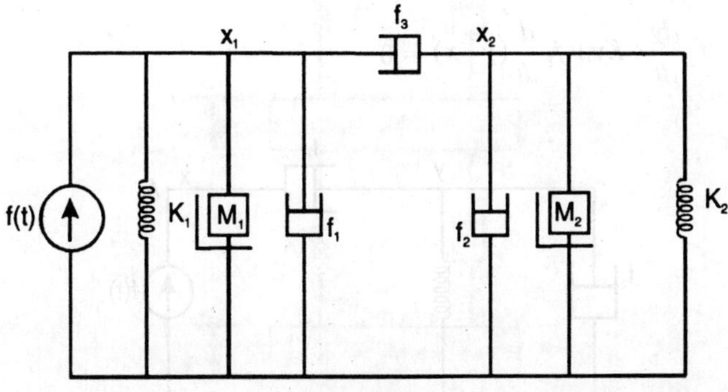

Fig. P. 5.8 (a)

The equations for the system are written below

At node (1)

$$K_1 x_1 + M_1 \frac{d^2 x_1}{dt^2} + f_1 \frac{dx_1}{dt} + f_3 \left[\frac{dx_1}{dt} - \frac{dx_2}{dt} \right] = f(t) \qquad (1)$$

At node (2)

$$k_2 x_2 + M_2 \frac{d^2 x_2}{dt^2} + f_2 \frac{dx_2}{dt} + f_3 \left[\frac{dx_2}{dt} - \frac{dx_1}{dt} \right] = 0 \qquad (2)$$

Assuming initial conditions as zero, taking Laplace transform on both sides of equations (1) and (2), we have the transformed equations for the system as below:

$$M_1 s^2 X_1(s) + (f_1 + f_3)s\, X_1(s) + K_1 X_1(s - f_3 s X_2(s) = F(s) \qquad (3)$$

and $\quad M_2 s^2 X_2(s) + (f_2 + f_3)s\, X_2(s) + K_2 X_2(s) - f_3 s X_1(s) = 0 \qquad (4)$

Thus, equations (3) and (4) are the required equations of the given system.

Problem 5.9. Obtain a mathematical model for the mechanical system shown below:

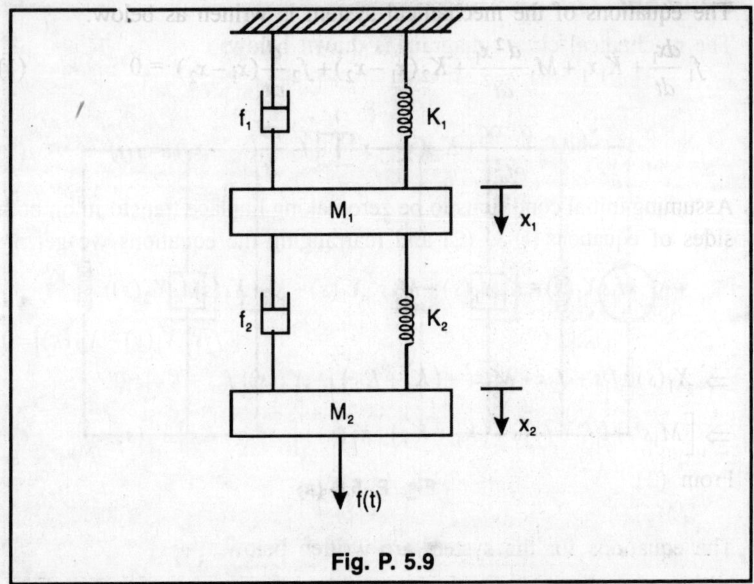

Fig. P. 5.9

Solution:

The mechanical circuit diagram for the circuit is drawn below:

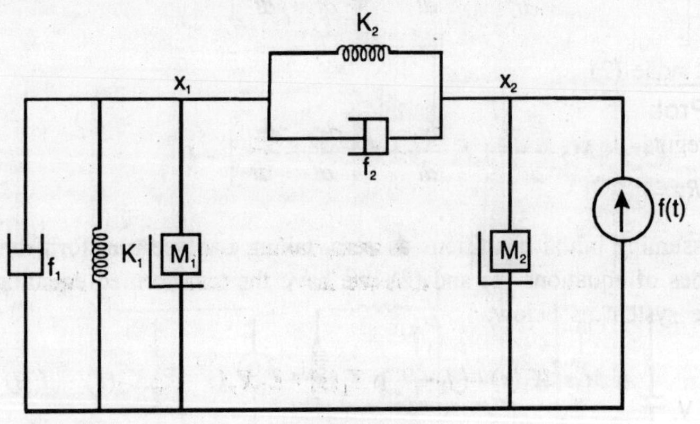

Fig. P. 5.9 (a)

We use the force—current analogy where,

1. Applied force $f(t)$ → current $i(t)$.
2. Mass is analogous to capacitance
3. Coefficient of viscous fricition f is analogous to the reciprocal of resistance $(1/R)$
4. Spring constant K is analogous to reciprocal of inductance $(1/L)$
5. Diplacement x is analogous to flux linkage.

The equations of the mechanical system is written as below:

$$f_1 \frac{dx_1}{dt} + K_1 x_1 + M_1 \frac{d^2 x_1}{dt^2} + K_2(x_1 - x_2) + f_2 \frac{d}{dt}(x_1 - x_2) = 0 \qquad (1)$$

$$M_2 \frac{d^2 x_2}{dt^2} + K_2(x_2 - x^4) + f_2 \frac{d}{dt}(x_2 - x_1) = f(t) \qquad (2)$$

Assuming initial conditions to be zero, taking Laplace transform on both sides of equations (1) & (2) and rearranging the equations we get

$$f_1 s X_1(s) + K_1 X_1(s) + M_1 s^2 X_1(s) + K_2(X_1(s) - X_2(s))$$

$$+ f_2 s \left[X_1(s) - X_2(s) \right] = 0$$

$$\Rightarrow X_1(s) \left[f_1 s + f_2 s + M_1 s^2 + (K_1 + K_2) \right] - X_2(s)[f_2 s + K_2] = 0$$

$$\Rightarrow \left[M_1 s^2 + (f_1 + f_2)s + (K_1 + K_2) \right] X_1(s) + \left[-f_2 s - K_2 \right] X_2(s) = 0 \qquad (3)$$

From (2)

$$M_2 s^2 X_2(s) + K_2 \left(X_2(s) - X_1(s) \right) + f_2 s \left(X_2(s) - X_1(s) \right) = F(s)$$

$$X_1 \left[-f_2 s - K_2 \right] + \left[M_2 s^2 + f_2 s + K_2 \right] X_2(s) = F(s) \qquad (4)$$

Rearranging in state matrix model we have,

$$\begin{bmatrix} M_1 s^2 + (f_1 + f_2)s + (K_1 + K_2) & -(f_2 s + K_2) \\ -(f_2 s + K_2) & M_2 s^2 + f_2 s + K_2 \end{bmatrix} \begin{bmatrix} X_1(s) \\ X_2(s) \end{bmatrix} = \begin{bmatrix} 0 \\ F(s) \end{bmatrix}$$

Problem 5.10. The scheme given in figure represents a voltage regulator, determine the value of the reference voltage. Given that

$$R_f = 100 \, \Omega$$

Fig. P. 5.10

Solution:

The fraction of the output voltage compared with the reference voltage is $\dfrac{400}{400+600}=\dfrac{400}{1000}=0.4$ The generator field inductance need not be considered as the system block diagram shown below:

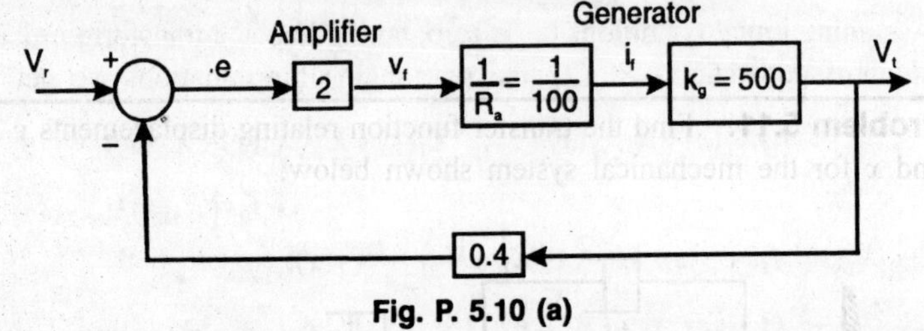

Fig. P. 5.10 (a)

The overall transfer function for steady state condition is given by

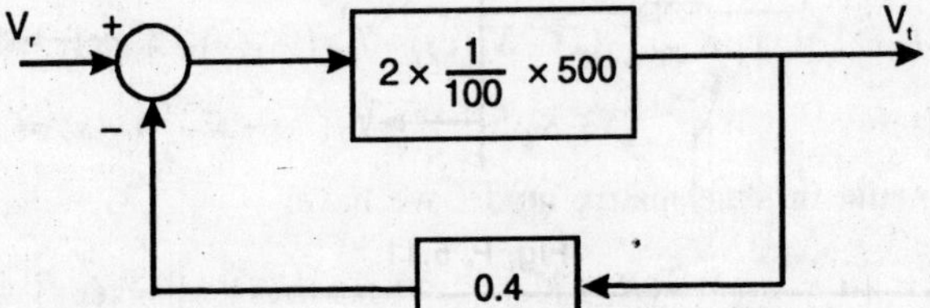

Fig. P. 5.10 (b)

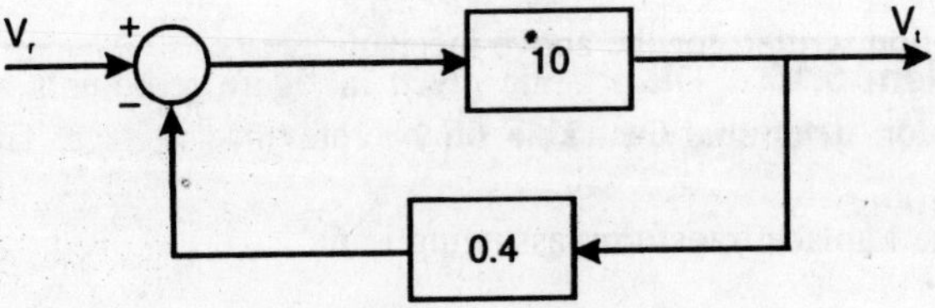

Fig. P. 5.10 (c)

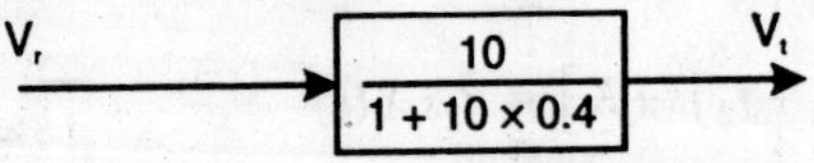

Fig. P. 5.10 (d)

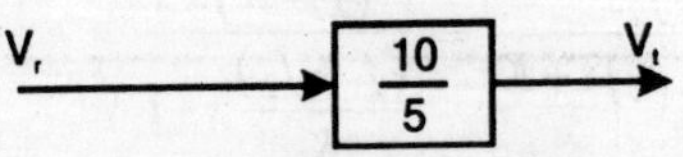

Fig. P. 5.10 (e)

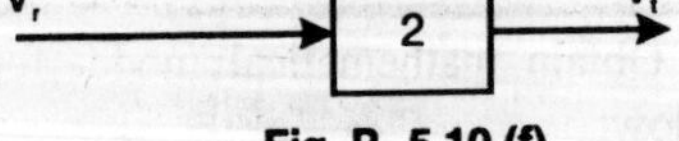

Fig. P. 5.10 (f)

Hence
$$\frac{V_t}{V_r} = 2$$

$$\Rightarrow \qquad V_r = \frac{V_t}{2} = \frac{250}{2} = 125V$$

Problem 5.11. Find the transfer function relating displacements y and x for the mechanical system shown below:

Fig. P. 5.11

Solution:

The equation written for the above mechanical system is shown below:

$$f_1\dot{y} + Ky + f_2(\dot{y} - \dot{x}) = 0$$

Taking the Laplace transform, assuming initial condition to be zero, we have,

$$f_1 s Y(s) + KY(s) + f_2 s[Y(s) - X(s)] = 0$$

$$\Rightarrow \qquad Y(s)\big[(f_1 + f_2)s + K\big] = f_2 s\, X(s)$$

$$\Rightarrow \qquad \frac{Y(s)}{X(s)} = \frac{f_2 s}{(f_1 + f_2)s + K} = \frac{(f_2/K)s}{\dfrac{(f_1 + f_2)}{K}s + 1}$$

Problem 5.12. Obtain mathematical model for the mechanical system shown below:

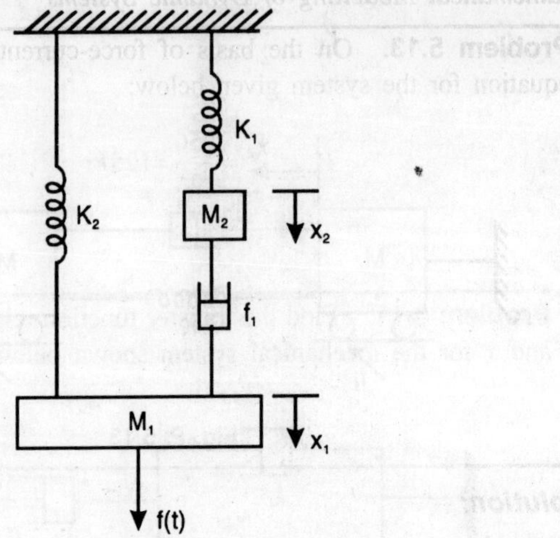

Fig. P. 5.12

Solution:

The mathematical equation for the mechanical model is written as below:

$$M_2\ddot{x}_2 + K_1 x_2 + f_1\left(\dot{x}_2 - \dot{x}_1\right) = 0 \tag{1}$$

$$M_1\ddot{x}_1 + K_2 x_1 + f_1\left(\dot{x}_1 - \dot{x}_2\right) = f(t) \tag{2}$$

Assuming initial conditions to be zero, taking the Laplace transform on both sides of the equations (1) & (2).

From (1)

$$M_2 s^2 X_2(s) + K_1 X_2(s) + f_1 s\left(X_2(s) - X_1(s)\right) = 0$$

$$-f_1 s X_1(s) + \left(M_2 s^2 + f_1 s + K_1\right) X_2(s) = 0 \tag{3}$$

From (2)

$$M_1 s^2 X_1(s) + K_2 X_1(s) + f_1 s\left(X_1(s) - X_2(s)\right) = F(s)$$

$$\left(M_1 s^2 + f_1 s + K_2\right) X_1(s) - f_1 s X_2 = F(s) \tag{4}$$

The state model can be drawn from equations (3) & (4)

$$\begin{bmatrix} \left(M_1 s^2 + f_1 s + K_2\right) & -f_1 s \\ -f_1 s & \left(M_2 s^2 + f_1 s + K_1\right) \end{bmatrix} \begin{bmatrix} X_1(s) \\ X_2(s) \end{bmatrix} = \begin{bmatrix} F(s) \\ 0 \end{bmatrix}$$

Problem 5.13. On the basis of force-current analogy write the equation for the system given below:

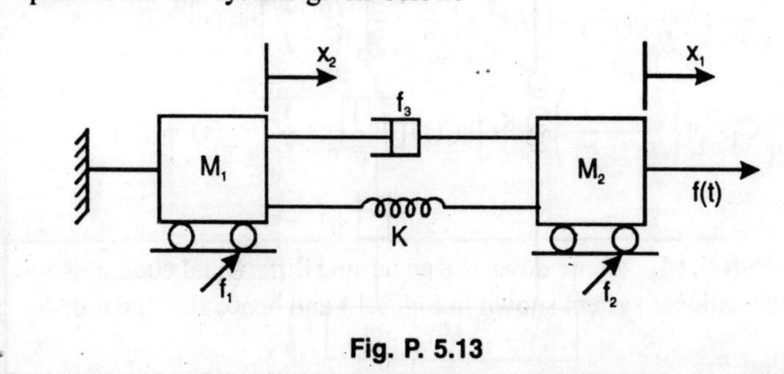

Fig. P. 5.13

Solution:

The mathematical equations of the above mechanical system is given below:

$$M_1\ddot{x}_2 + f_1\dot{x}_2 + f_3(\dot{x}_2 - \dot{x}_1) + K(x_2 - x_1) = 0 \tag{1}$$

$$M_2\ddot{x}_1 + f_3(\dot{x}_1 - \dot{x}_2) + K(x_1 - x_2) + f_2\dot{x}_1 = f(t) \tag{2}$$

Assuming initial conditions to be zero, taking the Laplace transform on both sides of equations (1) & (2)

from (1)

$$M_1s^2 X_2(s) + f_1 s\, X_2(s) + f_3 s\big(X_2(s) - X_1(s)\big) + K\big(X_2(s) - X_1(s)\big) = 0$$

$$\Rightarrow \quad -(f_3 s + K)X_1(s)\big(M_1 s^2 + (f_1 + f_3)s + K\big)X_2(s) = 0 \tag{3}$$

from (2) $M_2 s^2 X_1(s) + f_3 s[X_1(s) - X_2(s)] + K[X_1(s) - X_2(s)]$

$$+ f_2 s\, X_1(s) = F(s)$$

$$\Rightarrow \quad \big[M_2 s^2 + (f_2 + f_3)s + K\big]X_1(s) - (f_3 s + K)X_2(s) = F(s) \tag{4}$$

The force-current analogy is given as

(1) Force $f(t)$ $\leftrightarrow$ current (i)

(2) Mass (M) $\leftrightarrow$ Capacitance (C)

(3) Coeff of vicous friction (f) $\leftrightarrow$ 1/Resistance $\left(\dfrac{1}{R}\right)$

(4) Spring Constant $(k) \leftrightarrow \dfrac{1}{\text{inductance}}\left(\dfrac{1}{L}\right)$

(5) Displacement (x) $\leftrightarrow$ flux linkage (ψ)

Applying in equations (3) & (4), we get the desired Laplace Tranform:-

$$-\left(\frac{1}{R_3}s+\frac{1}{L}\right)\psi_1(s)+\left(C_1 s^2+\left(\frac{1}{R_1}+\frac{1}{R_3}\right)s+\frac{1}{L}\right)\psi_2(s)=0$$

$$\&\quad\left(C_2 s^2+\left(\frac{1}{R_2}+\frac{1}{R_3}\right)s+\frac{1}{L}\right)\psi_1(s)-\left(\frac{1}{R_3}s+\frac{1}{L}\right)\psi_2(s)=I(s)$$

Problem 5.14.   Write down the governing differential equations for the translational system shown in Fig. 5.14 and hence find the transfer function $\dfrac{f_o}{f_1}$.

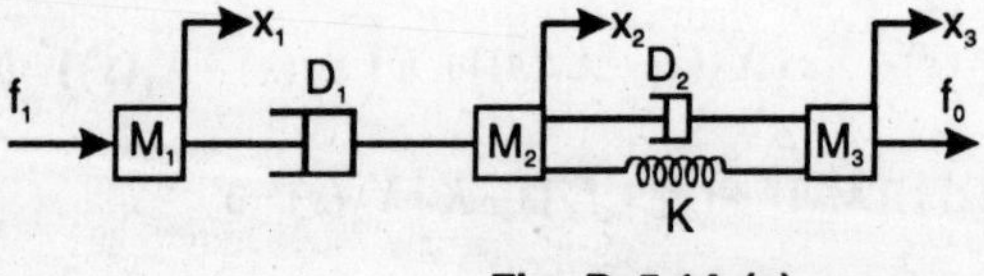

Fig. P. 5.14   (Es'2000)

Solution:

Assuming the displacements x_1, x_2 and x_3 for the masses M_1, M_2 and M_3 respectively.

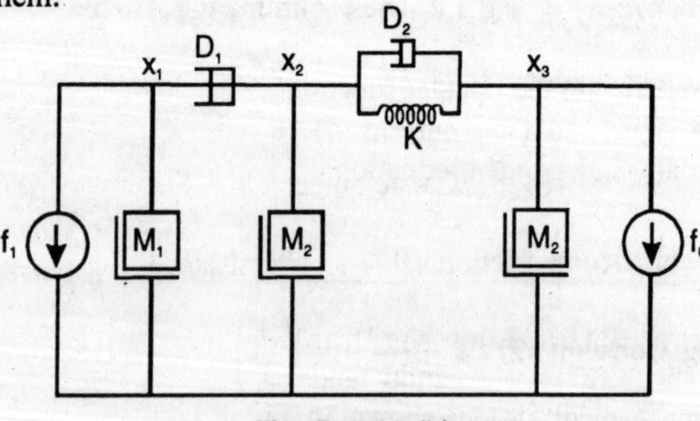

Fig. P. 5.14 (a)

Redrawing the mechanical network as our conventional method i.e., assuming x_1, x_2 and x_3 as the nodes and connecting all the elements relating them.

Fig. P. 5.14 (b)

Applying equation of motion at different nodes,

at node x_1

$$M_1\ddot{x}_1 + D_1(\dot{x}_1 - \dot{x}_2) = f_1 \tag{1}$$

at node x_2

$$M_2\ddot{x}_2 + D_1(\dot{x}_2 - \dot{x}_1) + D_2(\dot{x}_2 - \dot{x}_3) + K(x_2 - x_3) = 0 \tag{2}$$

At node x_3

$$M_3\ddot{x}_3 + D_2(\dot{x}_3 - \dot{x}_2) + K(x_3 - x_2) = -f_0 \tag{3}$$

From (1) $\qquad\qquad M_1\ddot{x}_1 + D_1\dot{x}_1 - D_1\dot{x}_2 = f_1$

applying Laplace transform to above equation,

$$\left(M_1 s^2 + D_1 s\right)X_1(s) - D_1 s\, X_2(s) = F_1(s) \tag{4}$$

$$\text{(assuming zero initial condition)}$$

from (2)

$$\left(M_2 s^2 + D_1 s + D_2 s + K\right)X_2(s) - (D_1 s)X_1(s) - \left(D_2 s + K\right)X_3(s) = 0 \tag{5}$$

from (3)

$$\left(M_3 s^2 + D_2 s + K\right)X_3(s) - \left(D_2 s + K\right)X_2(s) = -F_o(s) \tag{6}$$

Putting the value of $X_1(s)$ from (4) to (5)

$$\left(M_2 s^2 + D_1 s + D_2 s + K\right)X_2(s) - \frac{D_1 s\left[F_1(s) + D_1 s\, X_2(s)\right]}{\left(M_1 s^2 + D_1 s\right)}$$

$$-\left(D_2 s + K\right)X_3(s) = 0$$

So, we get the relation between x_3 and x_2 similarly, we can get the relation between x_3 and x_1. Thus, eliminating all the three variables x_1, x_2 & x_3, we get the transfer function $\dfrac{F_o(s)}{F_1(s)}$ by putting the respective values in any independent equation.

Problem 5.15. Find the transfer function $\dfrac{X(s)}{E(s)}$ for the electromechanical system shown in Fig. P.5.15.

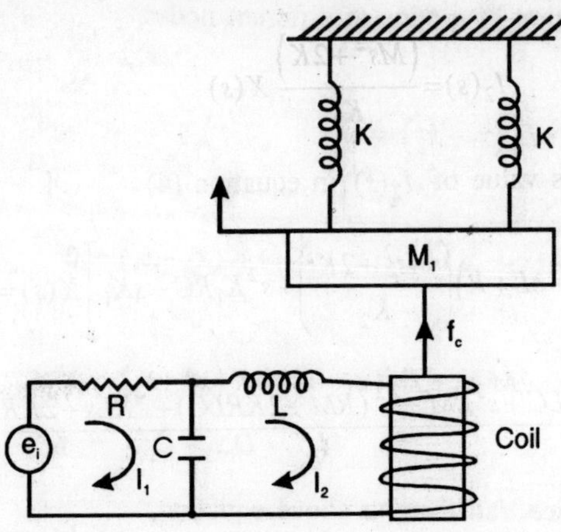

Fig. P. 5.15

Solution:

In these type of problems, we take two assumptions back emf of the coil $e_b = K_1 \dfrac{dx}{dt}$ and the coil current I_2 produces a force F_c such that $F_c = K_2 I_2$ on mass M.

Now, writing the equations of performance is Laplace form,

$$E_i(s) = RI_1(s) + \frac{1}{Cs}\left(I_1(s) - I_2(s)\right) \tag{1}$$

and

$$sLI_2(s) + \frac{1}{sC}\left[I_2(s) - I_1(s)\right] = -E_b(s) = -SK_1X(s) \tag{2}$$

Now, putting the value of $\dfrac{1}{Cs}\left(I_2(s) - I_1(s)\right)$ from (1) in (2)

$$sLI_2(s) + RI_1(s) - E_i(s) = -SK_1X(s) \tag{3}$$

Substituting for $I_1(s)$ from equation (2) in equation (1)

$$sLI_2(s) + sK_1X(s) + sRC\left[sLI_2(s) + \frac{I_2(s)}{Cs} + sK_1X(s)\right] = E_i(s)$$

or

$$\left(sL + R + s^2 RCL\right)I_2(s) + \left(s^2 K_1 RC + sK_1\right)X(s) = E_i(s) \tag{4}$$

Also,

$$F_c = k_2 I_2 = M\ddot{x} + Kx + Kx$$

or

$$K_2 I_2(s) = \left(Ms^2 + 2K\right)X(s)$$

or

$$I_2(s) = \frac{\left(Ms^2 + 2K\right)}{K_2} X(s)$$

Substituting this value of $I_2(s)$ in equation (4)

$$\left[\left(s^2 RLC + sL + R\right)\left(\frac{Ms^2 + 2K}{K_2}\right) + s^2 K_1 RC + sK_1\right] X(s) = E_i(s)$$

or $X(s)\left[\dfrac{s^4 MRLC + s^3 LM + s^2 (RM + 2KRLC) + 2KLs + 2KR}{k_2}\right.$

$$\left. + s^2 K_1 RC + sK_1\right] \doteq E_i(s)$$

or $X(s)\left[\dfrac{\begin{array}{c}s^4 MRLC + s^3 LM + s^2 (RM + 2KRLC + K_1 K_2 RC) \\ + s(2KL + K_1 K_2) + 2KR\end{array}}{K_2}\right] = E_i(S)$

Thus, the transfer function

$$\frac{X(s)}{E_i(s)} = \frac{K_2}{\begin{array}{c}s^4 MRLC + s^3 LM + s^2 (RM + 2KRLC + K_1 K_2 RC) \\ + s(2KL + K_1 K_2 + 2KR)\end{array}}$$

Problem 5.16. Find the transfer function of the system shown in the given figure.

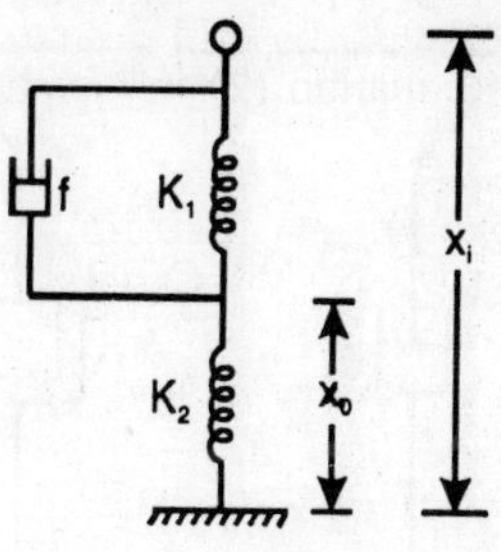

Fig. P. 5.16

Solution:

Applying equation of motion for the given system.

$$K_1(x_i - x_o) + f(\dot{x}_1 - \dot{x}_o) - K_2 x_o = 0$$

$$\Rightarrow \quad K_1(x_i) + f\dot{x}_i - K_1 x_o - f\dot{x}_o - K_2 x_o = 0$$

Applying Laplace transform in the above equation assuming the initial conditions to be zero.

$$K_1 x_i(s) + fs x_i(s) = K_1 x_o(s) + fs x_o(s) + K_2 x_o(s)$$

$$\Rightarrow \quad x_i(s)(K_1 + fs) = x_0(s)(K_1 + k_2 + fs)$$

$$\therefore \quad \text{Transfer function of the system shown} = \frac{X_o(s)}{X_i(s)} = \frac{K_i + fs}{fs + K_i + K}$$

Problem 5.17. For the mechanical system shown in fig. P. 5.17
  (a) Draw the equivalent mechanical network
  (b) Write the differential equations concerned
  (c) Draw the analogous electrical network

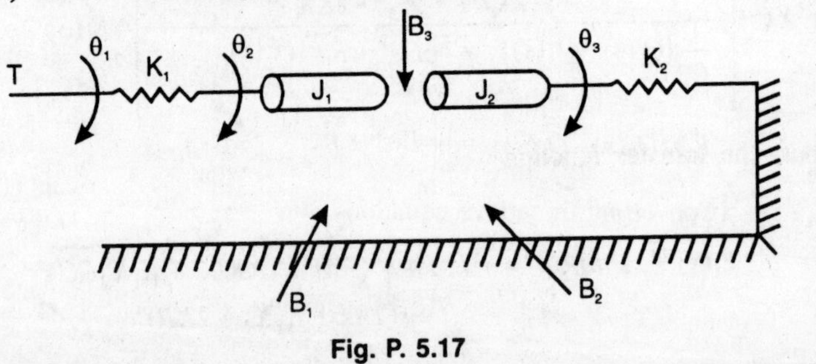

Fig. P. 5.17

Solution:

  (a) The equivalent mechanical network is shown below assuming the displacements θ_1, θ_2 & θ_3 as nodes.

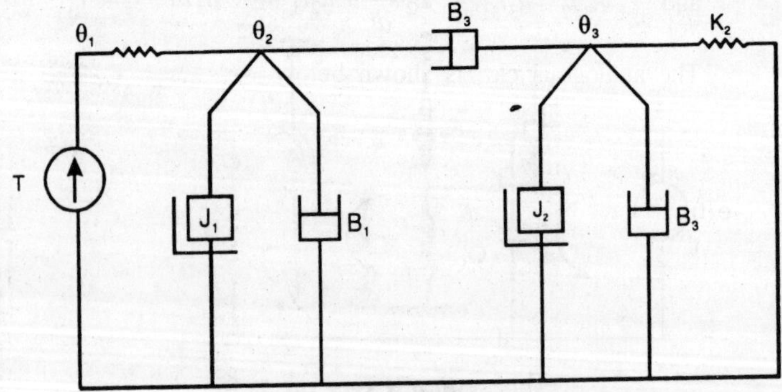

Fig. P. 5.17 (a)

(b) Applying equations of motion of Rotational dynamics

Node θ_1 $\qquad\qquad\qquad T = K_1(\theta_1 - \theta_2)$

Node θ_2 $\qquad\qquad K_1(\theta_1 - \theta_2) = J_1\ddot{\theta}_2 + B_1\dot{\theta}_2 + B_3\left(\dot{\theta}_2 - \dot{\theta}_3\right)$

and node θ_3 $\qquad B_3\left(\dot{\theta}_2 - \dot{\theta}_3\right) = J_2\ddot{\theta}_3 + B_2\dot{\theta}_3 + K_2\theta_3$

Taking Laplace transform of the above equations

$$T(s) = K_1\{\theta_1(s) - \theta_2(s)\} \tag{1}$$

$$K_1\{\theta_1(s) - \theta_2(s)\} = \left(J_1 s^2 + B_1 s\right)\theta_2(s) + B_3 s\{\theta_2(s) - \theta_3(s)\} \tag{2}$$

$$B_3 s\{\theta_2(s) - \theta_3(s)\} = \left(J_2 s^2 + B_2 s + K_2\right)\theta_3 s \tag{3}$$

(c) Based on the Force-voltage analogy we change the mechanical equations (1), (2) & (3) and finally we get the analogous electrical network:

$$E(s) = \frac{1}{C_1}\{Q_1(s) - Q_2(s)\}; \qquad Q = \text{Charge}$$

$$\frac{1}{C_1}\{Q_1(s) - Q_2(s)\} = \left(L_1 s^2 + R_1 s\right)Q_2(s) + R_3 s\{Q_2(s) - Q_3(s)\}$$

$$R_3 s\{Q_2(s) - Q_3(s)\} = \left(L_2 s^2 + R_2 s + \frac{1}{C_2}\right)Q_3(s)$$

Converting the above equations into time-domain

$$e_i(t) = \frac{1}{C_1}\int\left(i_1 - i_2\right)dt$$

$$\frac{1}{C_1}\int\left(i_1 - i_2\right)dt = L_1\frac{di_2}{dt} + R_1 i_2 + R_3 i_2 - R_3 i_3 \qquad \left(\because i = \frac{dq}{dt}\right)$$

and $\quad R_3 i_2 - R_3 i_3 = L_2\dfrac{di_3}{dt} + R_2 i_3 + \dfrac{1}{C_2}\int i_3\, dt$

The analogous ckt. is shown below

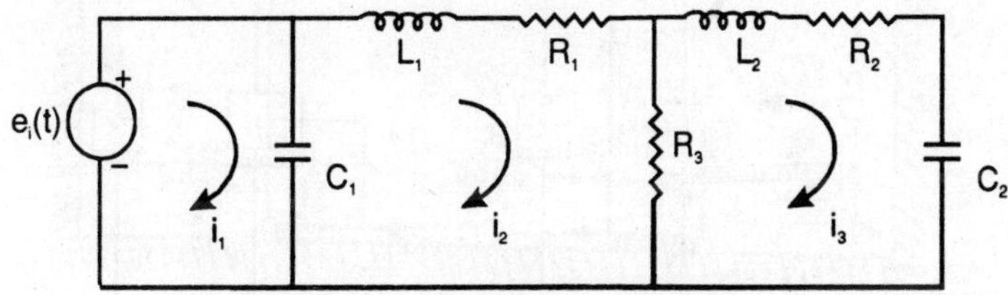

Fig. P. 5.17 (b)

Problem 5.18. Write the differential equation of the system shown in the figure.

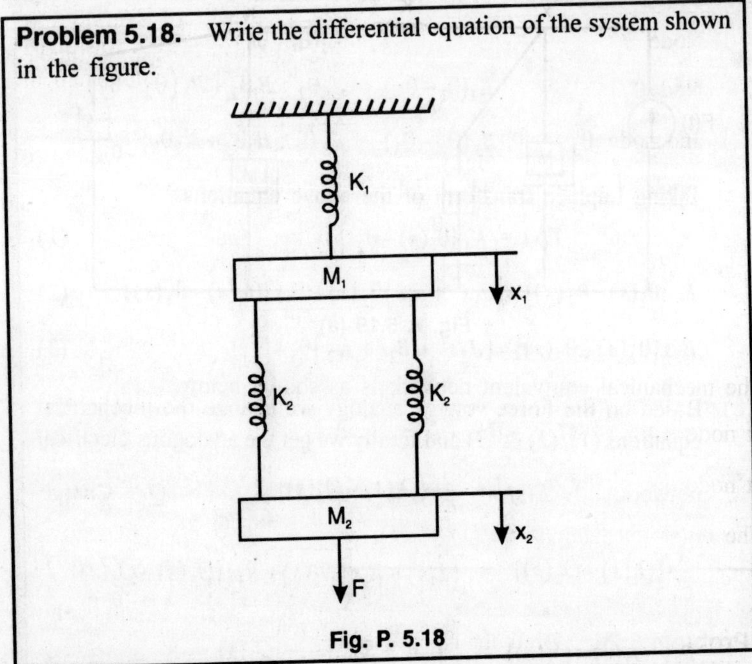

Fig. P. 5.18

Solution:

By applying directly the equation of motion to the system, we get

for M_1 $\quad M_1\ddot{x}_1 + K_1 x_1 + K_2(\bar{x}_1 - \bar{x}_2) + K_2(x_1 - x_2) = 0$

or $\qquad\qquad M_1\ddot{x}_1 + K_1 x_1 + 2K_2(x_1 - x_2) = 0$

for M_2 $\qquad\qquad M_2\ddot{x}_2 + 2K_2(x_2 - x_1) = F$

or $\qquad\qquad M_2\ddot{x}_2 + 2K_2 x_2 - 2K_2 x_1 = F$

Problem 5.19. Find nodal equations for the system shown in figure.

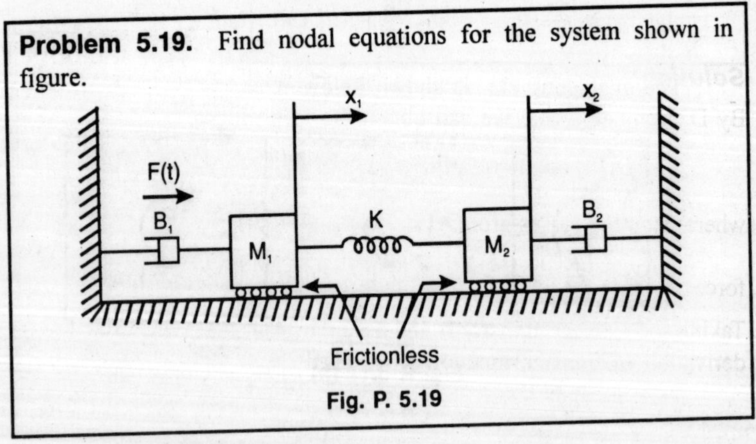

Fig. P. 5.19

Solution:

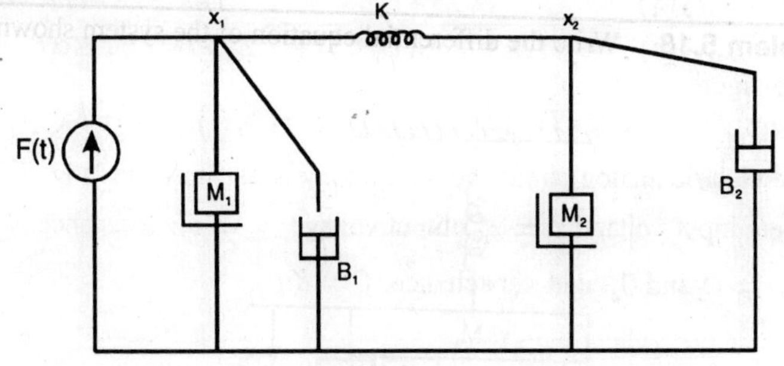

Fig. P. 5.19 (a)

The mechanical equivalent network is as shown below.

at node x_1 $\qquad M_1\ddot{x}_1 + B_1\dot{x}_1 + K(x_1 - x_2) = F(t)$ $\qquad\qquad$ (i)

at node x_2 $\qquad M_2\ddot{x}_2 + B_2\dot{x}_2 + K(x_2 - x_1) = 0$ $\qquad\qquad$ (ii)

The required equations are (1) & (2).

Problem 5.20. Draw the electric analog and derive the transfer function of a mechanical lead network as shown in Fig. 5.20 where $x_1 =$ input displacement, $x_0 =$ output displacement, $y =$ displacement of the spring; D_1, $D_2 =$ viscous damping coefficients and $K =$ compliance of the spring.

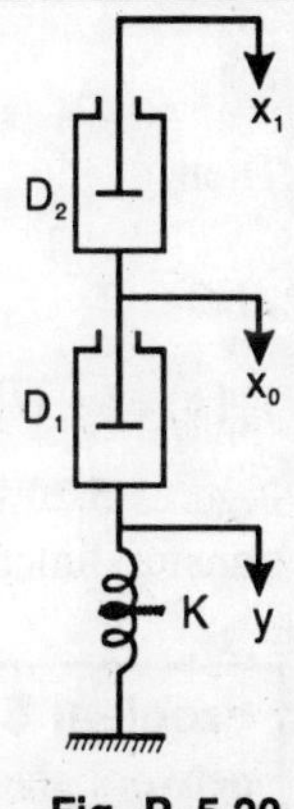

Fig. P. 5.20

Solution:

By D'Alembert's rule we can obtain the following equations:

$$D_2(\dot{x}_1 - \dot{x}_0) = D_1(\dot{x}_0 - \dot{y}) \quad \text{and} \quad D_1(\dot{x}_0 - \dot{y}) = \frac{y}{K}$$

where damping forces are $D_2(\dot{x}_1 - \dot{x}_0)$ and $D_1(\dot{x}_0 - \dot{x}_y)$ and the spring force is $\dfrac{y}{K}$.

Taking the Laplace transform and assuming zero initial conditions (for derivation of transfer function), we obtain

$$\Rightarrow \quad \frac{X_0(s)}{X_t(s)} = \frac{D_1 K s + 1}{(D_2/(D_1+D_2))(D_1 K)s+1} \times \frac{D_2}{D_1+D_2} = \frac{s+1/T}{s+1/\alpha T}$$

where $T = D_1 K, \alpha = D_2/(D_1+D_2), X_1(s) = L x_1(t) \, X_0(s) = L x_0(t)$

The electric analog circuit by *f-v* analogy is as shown in Fig. 5.20 (a) where input voltage $v_i \Rightarrow x_1$, output voltage $v_o \Rightarrow x_o$; resistances R_1 and $R_2 \Rightarrow D_1$ and D_2 and capacitance $C \Rightarrow K$

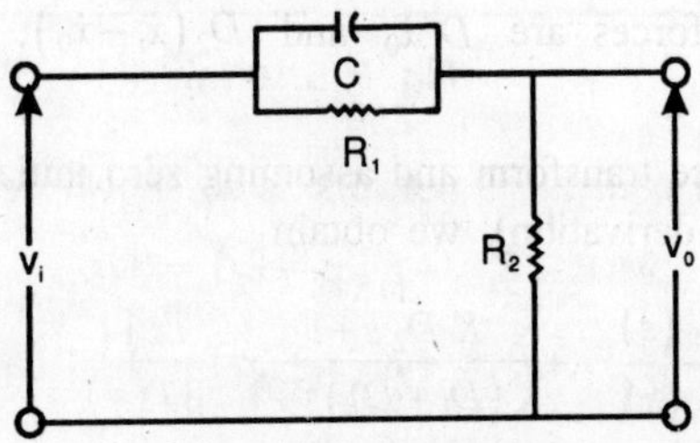

Fig. P. 5.20 (a)

Then, $\dfrac{V_o(s)}{V_i(s)} = \dfrac{Z_2}{Z_1+Z_2} = \dfrac{s+1/T}{s+1/\alpha T}$

where $Z_1 = R_1/(R_1 Cs+1), Z_2 = R_2, T = R_1 C, \alpha = R_2/(R_1+R_2),$

$V_i(s) = L v_i(t), V_o(s) = L v_o(t)$

Figures 5.20 (a) and 5.20 (b) are analogous from *f-v* analogy and their transfer functions are equivalent.

Problem 5.21. Consider the mechanical lag network shown in Fig 5.21 *(a)* where x_i = input displacement, x_0 = output displacement, D_1, D_2 = viscous Damping coefficients and K = compliance.

Fig. P. 5.21 (a)

Solution:

By D'Alembert's rule:

$$D_1 \dot{x}_0 = \frac{(x_i - x_0)}{K} + D_2(\dot{x}_i - \dot{x}_0)$$

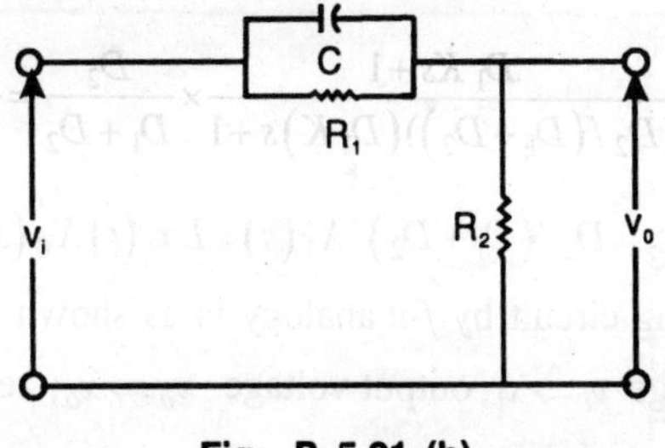

Fig. P. 5.21 (b)

where damping forces are $D_1 \dot{x}_0$ and $D_2(\dot{x}_i - \dot{x}_0)$; spring force is $(x_i - x_0)/K$.

Taking the Laplace transform and assuming zero initial conditions (for transfer function derivation), we obtain

$$\frac{X_0(s)}{X_i(s)} = \frac{K D_2 s + 1}{K(D_1 + D_2)s + 1} = \frac{Ts + 1}{\beta Ts + 1} \quad \text{where} \quad T = D_2 K$$

and

$$\beta = 1 + \frac{D_1}{D_2};$$

$$X_i(s) = L x_i(t), X_0(s) = L x_0(t)$$

The electric analog circuit by f-v analog is as shown in Fig 5.21 (b) where input voltage $v_i \Rightarrow x_i$, output voltage $v_0 \Rightarrow x_0$; resistances R_1 and $R_2 \Rightarrow D_1$ and D_2 and capacitance $C \Rightarrow$ compliance K. Then,

$$\frac{V_0(s)}{V_i(s)} = \frac{Z_2}{Z_1 + Z_2} = \frac{Ts + 1}{\beta Ts + 1}$$

where $Z_1 = R_1, Z_2 = R_2 + \dfrac{1}{Cs}, T = R_2 C, \beta = 1 + \dfrac{R_1}{R_2}$

$$V_0(s) = L v_0(t)$$

and

$$V_i(s) = L v_i(t)$$

Figures 5.21 (a) and 5.21 (b) are analogous from f-v analogy and their transfer functions are equivalent.

Problem 5.22. Draw the electric analogous circuit of the mechanical system in Fig 5.22 (a)

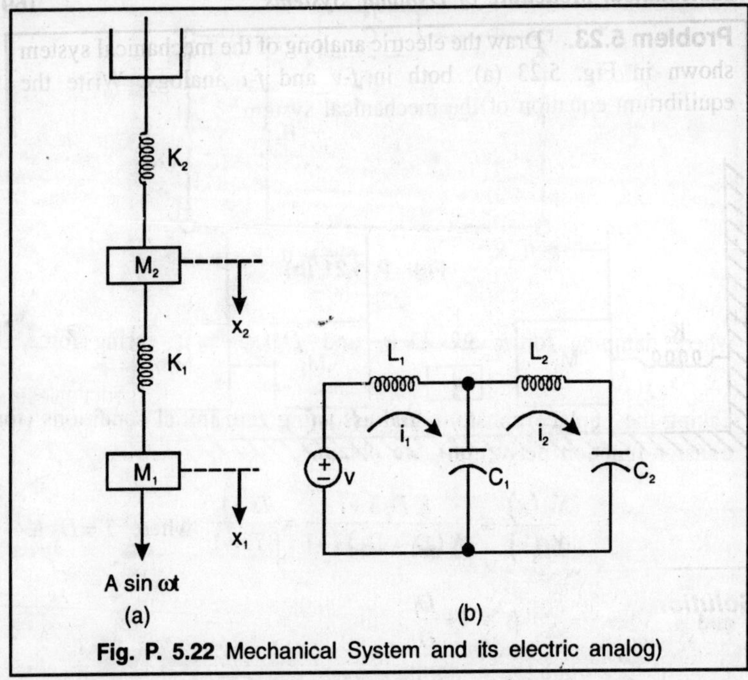

Fig. P. 5.22 Mechanical System and its electric analog)

Solution:

Let x_1 and x_2 be the displacement of masses M_1 and M_2 respectively from the initial position downward. Assume forces are positive when exerted downward. K_1 and K_2 are compliances. Spring 2 would exert a restoring force $-x_2/K_2$ on M_2. Spring 1 would be stretched to the amount $x_1 - x_2$ and would exert a force $(x_1 - x_2)/K_1$ on mass M_2 and a force- $-(x_1 - x_2)/K_1$ on M_1. The equation of motion would then be obtained by summing up the forces acting on each mass and equating to inertial force MA for that mass. The force balance equations of two masses

are
$$M_1 \ddot{x}_1 + (x_1 - x_2)/k_1 = A\sin(\omega t)$$

and
$$M_2 \ddot{x}_2 - \frac{x_1}{K_1} + x_2/(K_1 + K_2) = 0$$

The electrical analogous circuit is shown in Fig. 5.22 (b) by *f-v* analogy. The KVL equations are

$$L_1 \ddot{q}_1 + (q_1 - q_2)/C_1 = v$$

and
$$L_2 \dot{q}_2 + q_2/(C_1 + C_2) - q_1/C_1 = 0$$

where
$$\dot{q} = i, \ L = M, \ C = K, \ i = x, \ v = A\sin \omega t$$

Problem 5.23. Draw the electric analong of the mechanical system shown in Fig. 5.23 (a), both in *f-v* and *f-i* analogy. Write the equilibrium equation of the mechanical system.

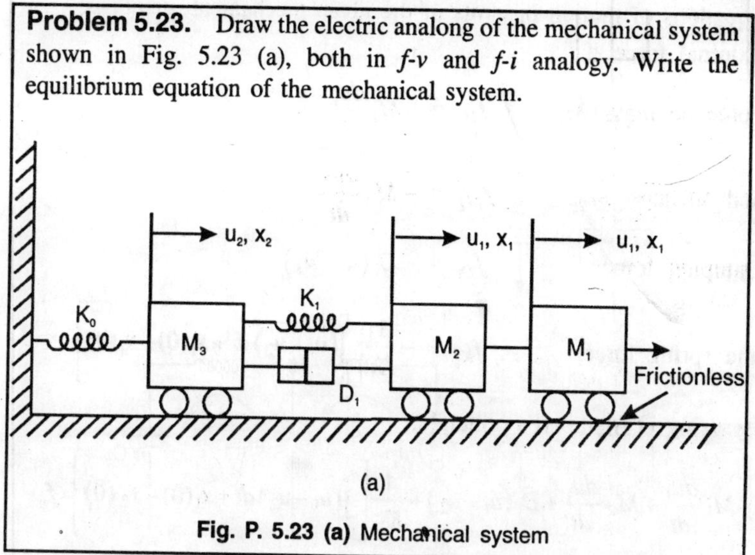

(a)

Fig. P. 5.23 (a) Mechanical system

Solution:

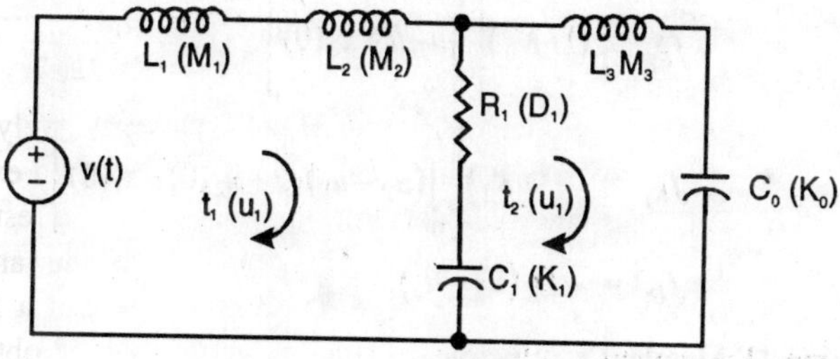

Fig. P. 5.23 (b) electrical analong circuit by *f-v* analogy

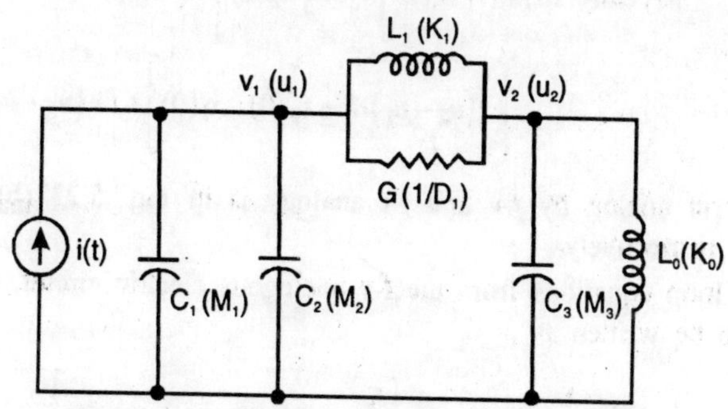

Fig. P. 5.23 (c) *f-i* analogous electric circuit

Equations of motion in terms of the given mechanical quantities:

External force $= f$

Force on mass M_1 $\qquad f_{M1} = -M_1 \dfrac{du_1}{dt}$

and on mass M_2 $\qquad f_{M2} = -M_2 \dfrac{du_2}{dt}$

Damping force, $\qquad f_{D_1} = -D_1\left(u_1 - u_2\right)$

and spring force, $\qquad f_{K_1} = -\dfrac{1}{K_1}\left[\int_0^t (u_1 - u_2)\,dt + x_1(0) - x_1(0)\right]$

By using D'Alembert's principle,

$$M_1 \frac{du_1}{dt} + M_2 \frac{du_1}{dt} + D_1\left(u_1 - u_2\right) + \frac{1}{K_1}\left[\int_0^t (u_1 - u_2)\,dt + x_1(0) - x_2(0)\right] = f$$

For mass M_3 $\qquad f_{M3} = M_3\left(du_2 / dt\right);$

$$f_{K_0} = \left(1/K_0\right)\left[\int_0^t u_2\,dt + x_2(0)\right]$$

$$f_{K_1} = -\left(1/K_1\right)\left[\int_0^t (u_2 - u_1)\,dt + x_2(0) - x_1(0)\right]$$

$$f_{D_1} = -D_1\left(u_2 - u_1\right)$$

Then, using D'Alembert's principle,

$$M_3\left(du_2 / dt\right) + \left(1/K_0\right)\left[\int_0^t u_2\,dt + x_2(0)\right] + \left(1/K_1\right)$$

$$\left[\int_0^t (u_2 - u_1)\,dt + x_2(0) - x_1(0)\right] + D_1\left(u_2 - u_1\right) = 0$$

The electric analog by f-v and f-i analogy is in Fig. 5.23 (b) and 5.23 (c), respectively.

The two loop equations from the f-v analogous electric circuit, using KVL, can be written as

$$\left(L_1 + L_2\right)\frac{di_1}{dt} + R_1\left(i_1 - i_2\right) + \frac{1}{C_1}\left[\int_0^t (i_1 - i_2)\,dt + q_1(0) - q_2(0)\right] = v$$

and
$$L_3 \frac{di_2}{dt} + \frac{1}{C_o}\left[\int_0^t i_2 dt - q_2(0)\right] + R_1(i_2 - i_1)$$

$$+ \frac{1}{C_1}\left[\int_0^t (i_2 - i_1)dt + q_2(0) - q_1(0)\right] = 0$$

The two node equations from the *f-i* analogous electric circuit, using KCL, can be written as

$$(C_1 + C_2)\frac{dv_1}{dt} + G_1(v_1 - v_2) + \frac{1}{L_1}\left[\int_0^t (v_1 - v_2)dt + \phi_1(0) - \phi_2(0)\right] = i$$

and
$$C_3 \frac{dv_1}{dt} + \frac{1}{L_0}\left[\int_0^t v_2 dt - \phi_2(0)\right] + G_1(v_2 - v_1) +$$

$$+ \frac{1}{L_1}\left[\int_0^t (v_2 - v_1)dt + \phi_2(0) - \phi_1(0)\right] = 0$$

Problem 5.24. Draw the electric analog, by *f-v* and *f-i* analogy, of the mechanical system in Fig. P. 5.24 (a). Write the equilibrium equations of the mechanical system.

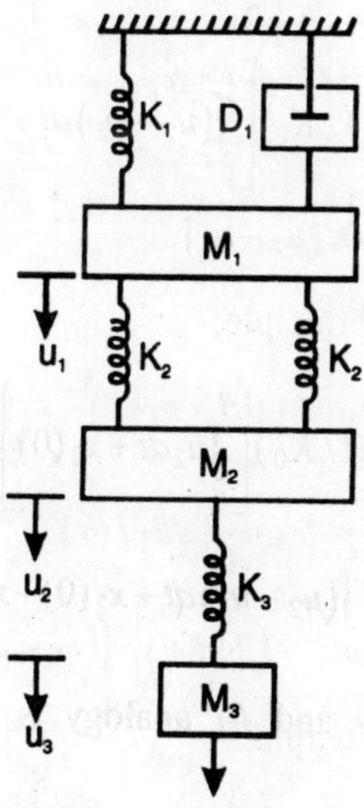

Fig. P. 5.24 (a) Mechanical system

Solution:

Equations of motion in terms of given mechanical quantities: For mass M_3; external force, $f = F \sin \omega t$

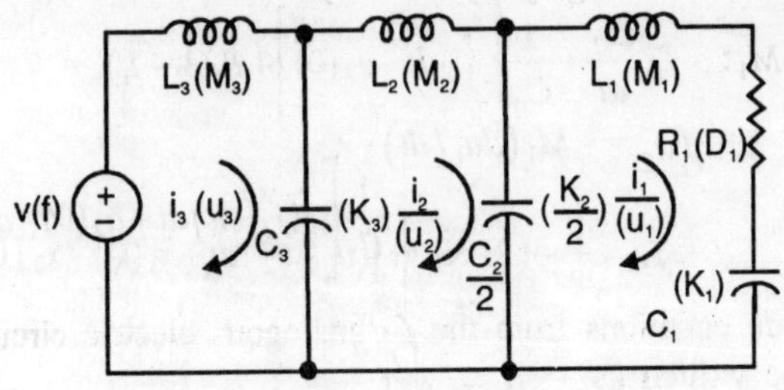

Fig. P. 5.24 (b) Electric analog circuit by *f-v* analogy

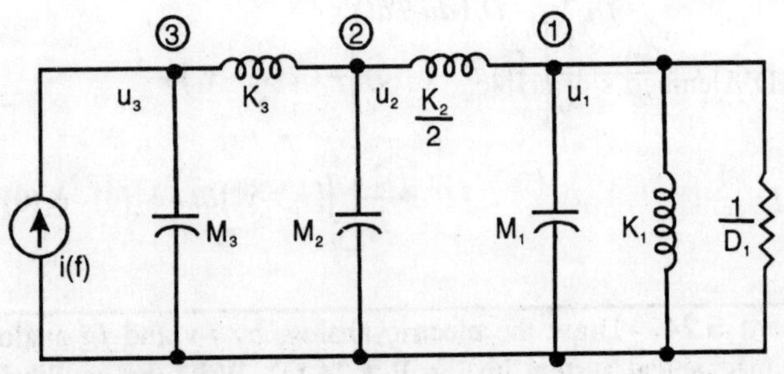

Fig. P. 5.24 (c) Electric analog circuit by *f-i* analogy

$$M_3 = -M_3 \left(du_3 / dt \right)$$

and

$$f_{k3} = -\left(1/K_3\right)\left[\int_0^t \left(u_3 - u_2\right) dt + x_3(0) - x_2(0)\right]$$

Using D'Alembert's principle,

$$M_2 \left(du_3 / dt\right) + \left(1/K_3\right)\left[\int_0^t \left(u_3 - u_2\right) dt + x_3(0) - x_2(0)\right] = f = F\sin\omega t$$

For mass $\quad M_2: \quad f M_2 = -M_2 \left(du_2 / dt\right)$

$$f_{K2} = -\left(2/K_2\right)\left[\int_0^t \left(u_2 - u_2\right) dt + x_2(0) - x_1(0)\right]$$

Then, by D'Alembert's principle,

$$M_2 \left(du_2 / dt\right) + \left(2/K_2\right)\left[\int_0^t \left(u_2 - u_1\right) dt + x_2(0) - x_1(0)\right]$$

$$+ \left(1/K_3\right)\left[\int_0^t \left(u_2 - u_1\right) dt + x_2(0) - x_1(0)\right] = 0$$

For mass M_1:

$$f_{M_2} = -M_1 \left(du_1 / dt \right)$$

$$f_{K_2} = -\left(2/K_2 \right) \left[\int_0^t \left(u_1 - u_2 \right) dt + x_1(0) - x_2(0) \right]$$

$$f_{K_1} = -\left(1/K_1 \right) \left[\int_0^t u_1 \, dt + x_1(0) \right]$$

$$f_{D_1} = -D_1 \left(du_1 / dt \right)$$

Using D'Alembert's principle,

$$M_1 \left(du_1 / dt \right) + \left(2/K_2 \right) \left[\int_0^t \left(u_1 - u_2 \right) dt + x_1(0) - x_2(0) \right]$$

$$+ \left(1/K_1 \right) \left[\int_0^t u_1 dt + x_1(0) \right] + D_1 \left(du_1 / dt \right) = 0$$

The electric analog circuits by f–v and f–i analogy are shown in Figs. 5.24 (b) and 5.24 (c), respectively.

The three loop equations from the f–v analogous electric circuit, using KVL, can be written as

$$L_3 \frac{di_3}{dt} + \frac{1}{C_3} \left[\int_0^t \left(i_3 - i_2 \right) dt + q_3(0) - q_2(0) \right] = v$$

$$L_2 \frac{di_2}{dt} + \frac{2}{C_2} \left[\int_0^t \left(i_2 - i_1 \right) dt + q_2(0) - q_1(0) \right]$$

$$+ \frac{1}{C_3} \left[\int_0^t \left(i_2 - i_3 \right) dt + q_2(0) - q_3(0) \right] = 0$$

$$\text{and} \quad L_1 \frac{di_1}{dt} + R_1 i_1 \frac{1}{C_1} \left[\int_0^t i_1 \, dt + q_1(0) \right] + \frac{2}{C_2} \left[\int_0^t \left(i_1 - i_2 \right) dt + q_1(0) - q_2(0) \right] = 0$$

The three node equations from the f–i analogous electric circuit, using KCL, can be written as

$$C_3 \frac{dv_3}{dt} + \frac{1}{L_3} \left[\int_0^t \left(v_3 - v_2 \right) dt + \phi_3(0) - \phi_2(0) \right] = i$$

$$C_2 \frac{dv_2}{dt} + \frac{1}{L_3}\left[\int_0^t (v_2 - v_3)\,dt + \phi_2(0) - \phi_3(0)\right]$$

$$+ \frac{2}{L_2}\left[\int_0^t (v_2 - v_1)\,dt + \phi_2(0) - \phi_1(0)\right] = 0$$

and $\quad \dfrac{2}{L_2}\left[\displaystyle\int_0^t (v_1 - v_2)\,dt + \phi_1(0) - \phi_2(0)\right] + C_1 \dfrac{dv_1}{dt} + G_1 v_1$

$$+ \frac{1}{L_1}\left[\int_0^t v_1\,dt + \phi_1(0)\right] = 0$$

Problem 5.25.   Draw the electric analog of the mechanical system in Fig. 5.25 (a), by *f–v* and *f–i* analogy. Write the equilibrium equation of the mechanical system.

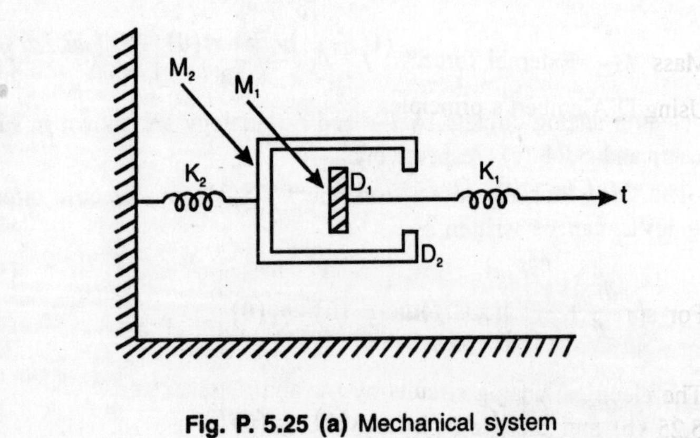

Fig. P. 5.25 (a) Mechanical system

Solution:

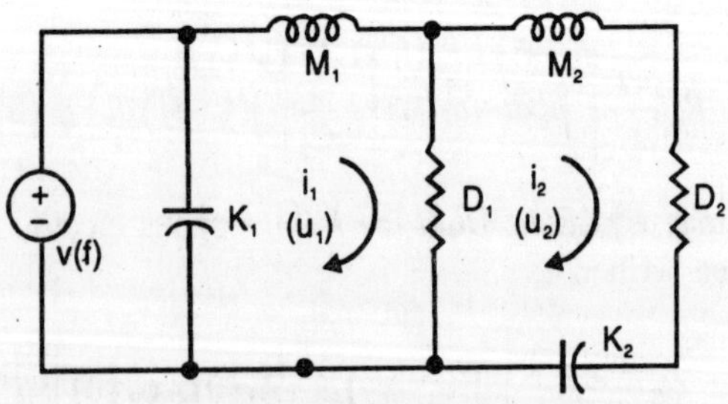

Fig. P. 5.25 (b) Electric analog circuit by *f–v* analogy

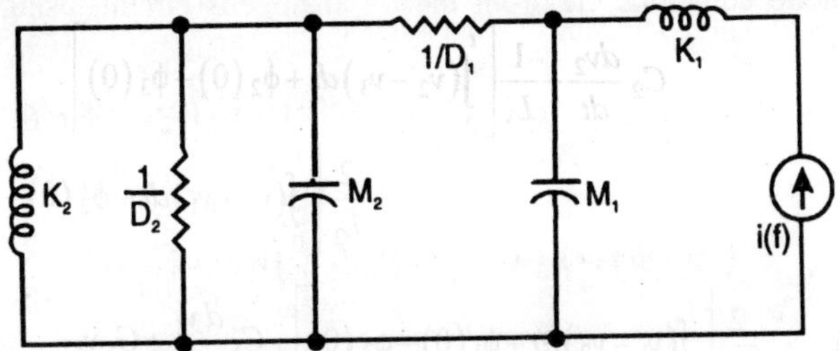

Fig. P. 5.25 (c) *f–i* analogous electric circuit

Equation of motion in terms of given mechanical quantities:

Mass M_1: External force, $f = 0$

Using D'Alembert's principle,

$$M_1\left(du_1/dt\right)+D_1\left(u_1-u_2\right)+\frac{1}{K_1}\left[\int_0^t u_1\,dt+x_1(0)\right]=0$$

Mass M_2: External force, $f = 0$

Using D'Alembert's principle,

$$M_2\left(du_2/dt\right)+D_2\,u_2+D_1\left(u_2-u_1\right)+\frac{1}{K_2}\left[\int_0^t u_2\,dt+x_2(0)\right]=0$$

For spring K: External force $= f$; Then, $\dfrac{1}{K_1}\left[\int_0^t u_1\,dt+x_1(0)\right]=f$

The electrical analog circuits by *f–v* and *f–i* analogy are shown in Fig. 5.25 (b) and 5.25 (c), respectively.

The loop equations from the *f–v* analogous electric circuit, using KVL can be written as

$$L_1\left(di_1/dt\right)+R_1\left(i_1-i_2\right)+\left(1/C_1\right)\left[\int_0^t i_1\,dt+q_1(0)\right]=0$$

$$L_2\left(di_2/dt\right)+R_2\,i_2+R_1\left(i_2-i_1\right)+\left(1/C_2\right)\left[\int_0^t i_2\,dt+q_2(0)\right]=0$$

and $\qquad\left(1/C_1\right)\left[\int_0^t i_1\,dt+q_1(0)\right]=v$

The node equations are from the *f–i* analogous circuit, using KCL

$$C_2\left(dv_2/dt\right)+G_2\,v_2+\left(1/L_2\right)\left[\int_0^t v_2\,dt+\phi_2\left(0\right)\right]+G_1\left(v_2-v_1\right)=0$$

and

$$C_1\left(dv_1/dt\right)+G_1\left(v_1-v_2\right)+\left(1/L_1\right)\left[\int_0^t v_1\,dt+\phi_1\left(0\right)\right]=0$$

$$\left(1/L_1\right)\left[\int_0^t v_1\,dt+\phi_1\left(0\right)\right]=i$$

Problem 5.26. Draw the electric analog circuit by *f-v* and *f-i* analogy and write the equilibrium equation of the mechanical system of Fig. 5.26 (a).

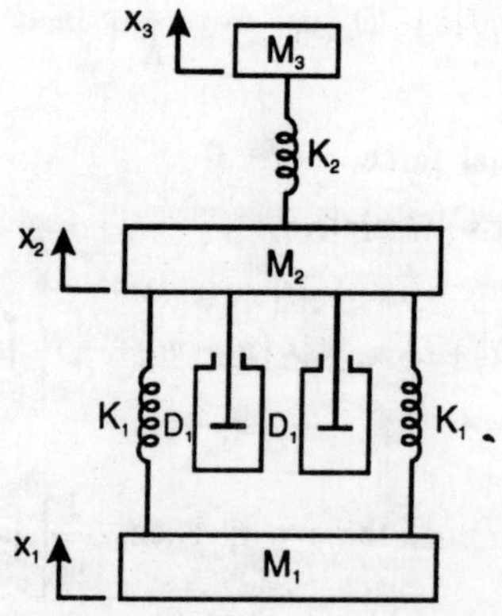

Fig. P. 5.26 (a) Mechanical system

Solution:

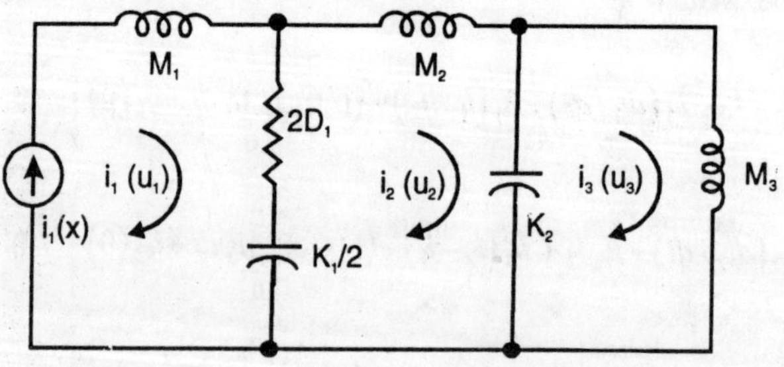

Fig. P. 5.26 (b) Electric analog by *f–v* analogy

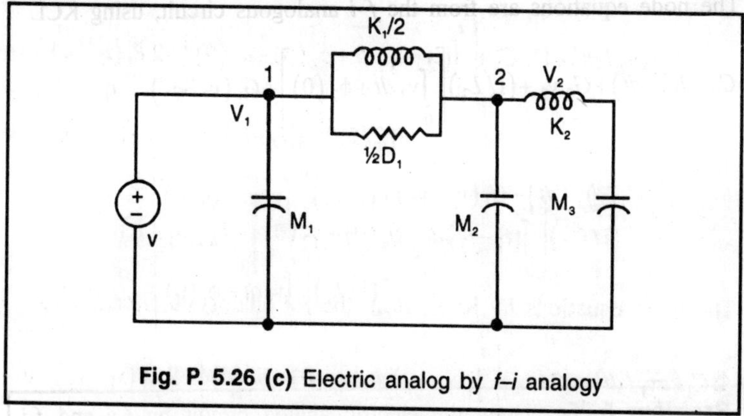

Fig. P. 5.26 (c) Electric analog by *f–i* analogy

Equation of motion in terms of mechanical quantities:

For mass M_1:

$$M_1 \frac{du_1}{dt} + \frac{2}{K_1}\left[\int_0^t (u_1 - u_2)\,dt + x_1(0) - x_2(0)\right] + 2D_1(u_1 - u_2) = 0$$

Force applied on mass M_1 is zero, i.e., in *f–v* electric analog circuit, the first mesh is shorted; i.e. $v\,[f] = 0$. Only the mesh current $i_1\,[u_1]$ is flowing in mesh 1.

For mass M_2:

$$M_2 \frac{du_2}{dt} + \frac{1}{K_2}\left[\int_0^t (u_2 - u_3)\,dt + x_2(0) - x_3(0)\right]$$

$$+ \frac{2}{K_1}\left[\int_0^t (u_2 - u_1)\,dt + x_2(0) - x_1(0)\right] + 2D_1(u_2 - u_1) = 0$$

For mass M_3: $M_3 \frac{du_3}{dt} + \frac{1}{k_2}\left[\int_0^t u_3\,dt + x_3(0)\right] = 0$

The loop equations from the *f–v* analogous electric circuit, using KVL, can be written as

$$L_1(di_1/dt) + (2/C_1)\left[\int_0^t (i_1 - i_2)\,dt + q_1(0) - q_2(0)\right] 2R_1(i_1 - i_2) = 0$$

$$L_2\left(di_2/dt\right)+\left(1/C_2\right)\left[\int_0^t\left(i_2-i_3\right)dt+q_2\left(0\right)-q_3\left(0\right)\right]+2R_1\left(i_2-i_1\right)$$

$$+\left(2/C_1\right)\left[\int_0^t\left(i_2-i_1\right)dt+q_2\left(0\right)-q_1\left(0\right)\right]=0$$

and $\quad \left(1/C_2\right)\left[\int_0^t\left(i_3-i_2\right)dt+q_3\left(0\right)-q_2\left(0\right)\right]+L_3\left(di_3/dt\right)=0$

The node equations by KCL, from the f–i analogous circuit, are:

$$C_1\left(dv_1/dt\right)+\left(2/L_1\right)\left[\int_0^t\left(v_1-v_2\right)dt+\phi_1\left(0\right)-\phi_2\left(0\right)\right]+2G_1\left(v_1-v_2\right)=0$$

$$C_2\left(dv_2/dt\right)+\left(1/L_1\right)\left[\int_0^t\left(v_2-v_3\right)dt+\phi_2\left(0\right)-\phi_3\left(0\right)\right]$$

$$+\left(2/L_1\right)\left[\int_0^t\left(v_2-v_1\right)dt+\phi_2\left(0\right)-\phi_1\left(0\right)\right]+2G_1\left(v_2-v_1\right)=0$$

and $\quad C_3\left(\dfrac{dv_3}{dt}\right)+\dfrac{1}{L_2}\left[\int_0^t\left(v_3-v_2\right)dt+\phi_3\left(0\right)-\phi_2\left(0\right)\right]=0$

The electric analog circuits by f–v and f–i analogy are shown in Figs 5.26 (b) and 5.26 (c), respectively.

Problem 5.27. Draw the electric analog circuit of the mechanical system by f–v and f–i analogy write the equilibrium equation of the mechanical system of Fig. 5.27 (a).

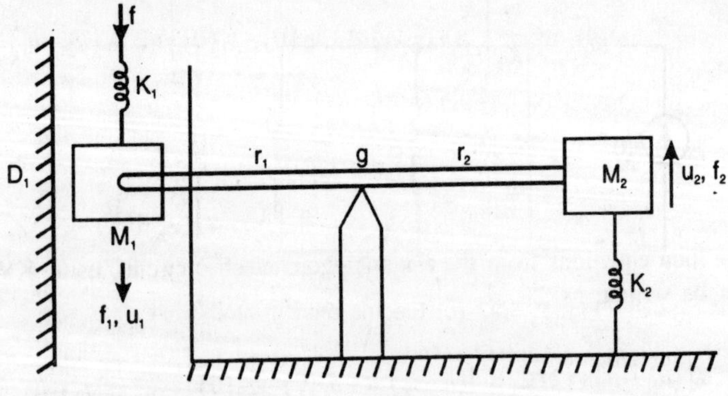

Fig. P. 5.27   (a) Mechanical system

Solution:

Taking moments about g, we get $\dfrac{f_1}{f_2}=\dfrac{r_2}{r_1}$ and $\dfrac{u_1}{u_2}=\dfrac{r_1}{r_2}$

Equation of motion:

For spring:
$$\frac{1}{K_1}\left[\int_0^t (u_1)\,dt + x_1(0)\right] = f$$

For mass M_1:
$$M_1\frac{d u_1}{dt} + D_1 u_1 + \frac{1}{K_1}\left[\int_0^t u_1\,dt + x_1(0)\right] + f_1 = 0$$

For mass M_2:
$$M_2\frac{d u_2}{dt} + \frac{1}{K_2}\left[\int_0^t u_2\,dt + x_2(0)\right] + f_2 = 0$$

The electric analog circuits are shown in Figs. 5.27 (b) and 5.27 (c).

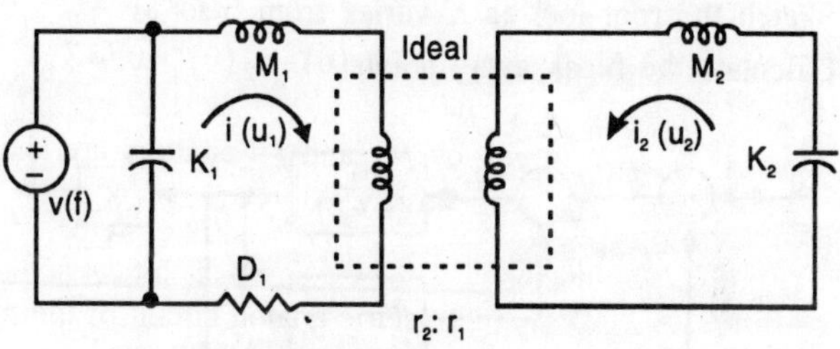

Fig. P. 5.27 (b) Electric analog by *f–v* analogy

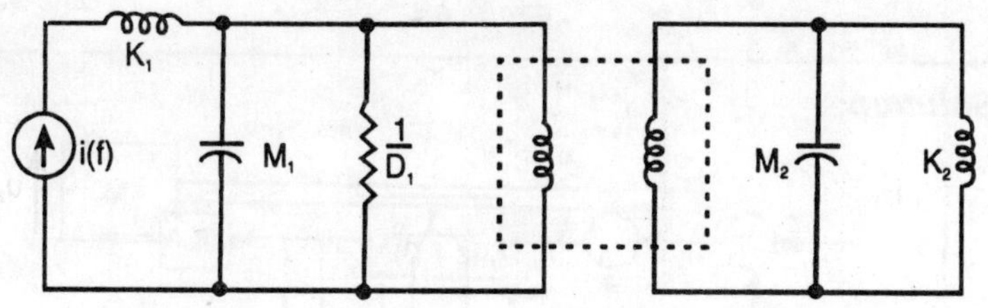

Fig. P. 5.27 (c) Electric analog circuit by *f-i* analogy

CHAPTER 6

Root Locus Technique

Problem 6.1. (a) Find the overall transfer function of the system shown below.

(b) Sketch the root loci as K varies from 0 to ∞

(c) Calculate the break away points

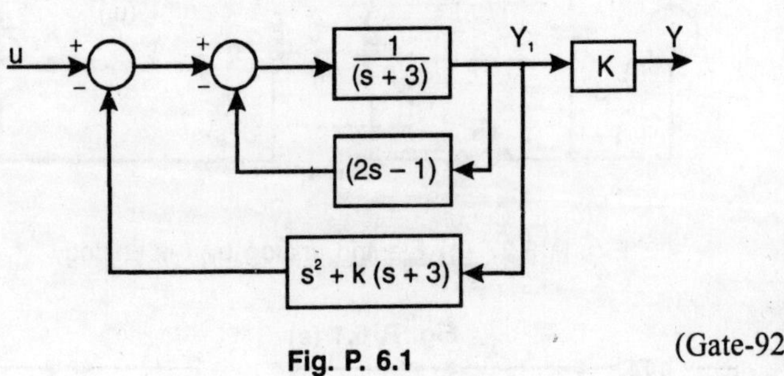

Fig. P. 6.1

(Gate-92)

Solution:

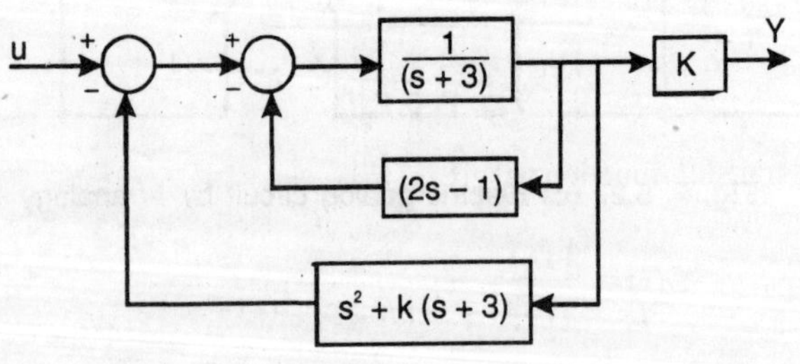

Fig. P. 6.1 (a)

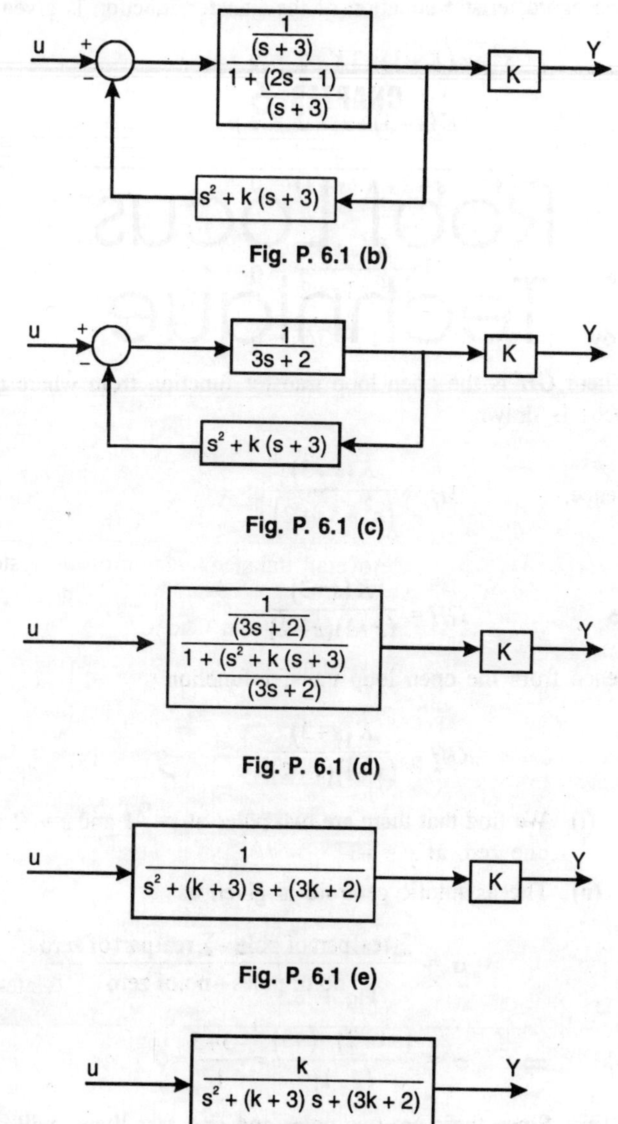

Fig. P. 6.1 (b)

Fig. P. 6.1 (c)

Fig. P. 6.1 (d)

Fig. P. 6.1 (e)

Fig. P. 6.1 (f)

Therefore, transfer function is given from the block diagram as

$$G(s) = \frac{Y(s)}{u(s)} = \frac{K}{s^2 + (K+3)s + (3K+2)}$$

(b) The characteristic equation of the transfer function is given as.

$$s^2 + s(K+3) + 3K + 2 = 0$$

$$\Rightarrow \qquad s^2 + K(s+3) + 3s + 2 = 0$$

$$\Rightarrow \qquad s^2 + 3s + 2 + K(s+3) = 0$$

$$\Rightarrow \qquad 1 + \frac{K(s+3)}{s^2+3s+2} = 0$$

Now $\qquad\qquad 1 + GH. = 0$

Where GH is the open loop transfer function from where root locus is drawn.

Hence, $\qquad GH = \dfrac{K(s+3)}{\left(s^2+3s+2\right)}$

$$\Rightarrow \qquad GH = \frac{K(s+3)}{(s+1)(s+2)}$$

Hence from the open loop transfer function

$$GH = \frac{K(s+3)}{(s+1)(s+2)}$$

(i) We find that there are two poles at $s = -1$ and $s = -2$ and one zero at $s = -3$

(ii) The asymtotic centroid is given as

$$\sigma = \frac{\sum \text{real part of pole} - \sum \text{real part of zero}}{\text{no. of poles} - \text{no. of zero}}$$

$$\Rightarrow \qquad \sigma = \frac{(-1-2)-(-3)}{(2-1)} = \frac{-3+3}{1} = 0$$

(iii) Since there are two poles and one zero there will one pole terminating to infinity.

(iv) To determine the breakeven point, $\dfrac{dK}{ds} = 0$

$$K = \frac{-(s+1)(s+2)}{(s+3)} \qquad \{\text{from } 1 + GH = 0\}$$

$$\Rightarrow \qquad \frac{dK}{ds} = \frac{-d}{ds}\left[\frac{s^2+3s+2}{(s+3)}\right]$$

$$\Rightarrow \qquad \frac{dK}{ds} = \left\{\frac{(s+3)(2s+3)-(s^2+3s+2)\cdot 1}{(s+3)^2}\right\}$$

$$\Rightarrow \qquad 0 = \left\{\frac{2s^2+6s+3s+9-s^2-3s-2}{(s+3)^2}\right\}$$

$$\Rightarrow \qquad s^2+6s+7 = 0$$

$$\Rightarrow \qquad s = \frac{-6\pm\sqrt{36-28}}{2}$$

$$\Rightarrow \qquad s = \frac{-6\pm 2.828}{2} = -4.414, -1.586$$

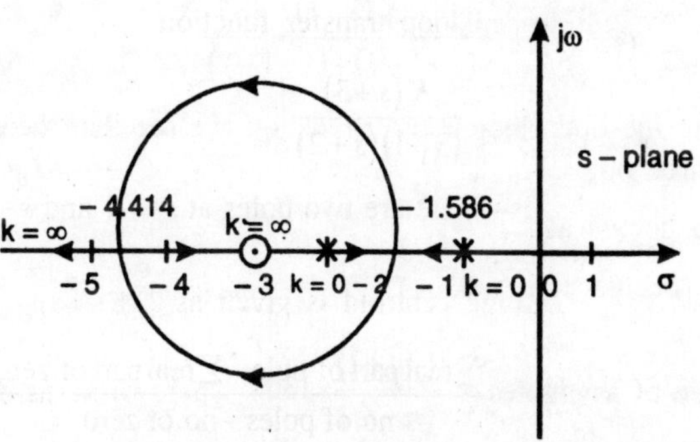

Fig. P. 6.1 (g) Root locus sketch

Problem 6.2. Consider the closed loop control system shown in Fig. P.6.2. Sketch the root loci diagram of the system for $\infty \geq K \geq 0$. Show on the sketch the following:

(a) asymtotes of root loci as

(b) intersection of the asymtotes.

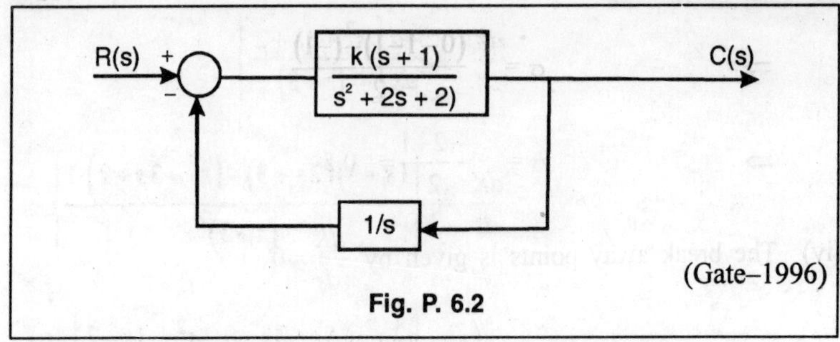

Fig. P. 6.2

Solution:

(a)The open loop transfer function is *GH*.

i.e.
$$GH(s) = \frac{K(s+1)}{\left(s^2+2s+2\right)} \times \frac{1}{s.}$$

$$\Rightarrow \qquad G(s)H(s) = \frac{K(s+1)}{s.\left[(s+1)^2+1\right]}$$

$$\Rightarrow \qquad G(s)H(s) = \frac{K(s+1)}{s(s+1+j)+(s+1-j)}$$

Hence, from the open loop transfer function we see that there are one zero and three poles.

(i)   Zero occurs at $s = -1$.

Pole occurs at $s = 0,\ -1+j, -1-j$

(ii)   Angle of asymtotes $= \dfrac{(2q+1)}{(n-m)} \times 180°$ where $(n-m)$ is difference

between no. of poles and zero & $q = 0, 1, 2$ -------- $(n-m-1)$

Here $(n - m) = 3 - 1 = 2$

∴ $q = 0, 1$

$$\theta = \frac{(2q+1)}{2} \times 180° = 90°,\ 270°$$

(iii)   Controid of asymtotes will be,

$$\sigma = \frac{\sum \text{real part of pole} - \sum \text{real part of zero}}{(n-m)}$$

$$\Rightarrow \qquad \sigma = \frac{(0-1-1)-(-1)}{2}$$

$$\Rightarrow \qquad \sigma = \frac{-2+1}{2} = -0.5$$

(iv) The break away points is given by $\dfrac{dK}{ds}=0$

$$K = \frac{-s\left(s^2+2s+2\right)}{(s+1)}$$

$$\Rightarrow \qquad \frac{dK}{ds} = \frac{-d}{ds}\left[\frac{s^3+2s^2+2s}{s+1}\right]$$

$$\Rightarrow \qquad \frac{dK}{ds} = -\left[\frac{(s+1)+\left(3s^2+4s+2\right)-\left(s^3+2s^2+2s\right)}{(s+1)^2}\right]$$

$$\Rightarrow \qquad 0 = \left[\frac{3s^3+3s^2+4s^2+4s+2s+2-s^3-2s^2-2s}{(s+1)^2}\right]$$

$$\Rightarrow \qquad 2s^3+4s^2+4s+2 = 0$$

$$\Rightarrow \qquad s^3+2s^2+2s+1 = 0$$

$$\Rightarrow \qquad s^2(s+1)+s(s+1)+(s+1) = 0$$

$$\Rightarrow \qquad (s+1)+\left(s^2+s+1\right) = 0$$

$$\Rightarrow \qquad s=-1 \,\&\, s = \frac{-1\pm\sqrt{1-4}}{2}$$

$$= \frac{-1\pm j\sqrt{3}}{2} = \frac{-1}{2}\pm j\frac{\sqrt{3}}{2}$$

(v) Apply Routh criterion to get the point where the root locus cuts the imaginary ($j\omega$) axis.

$$1+GH=0$$

$$\Rightarrow \qquad 1+\frac{K(s+1)}{s\left(s^2+2s+2\right)} = 0$$

$$\Rightarrow \quad s^3 + 2s^2 + 2s + Ks + K = 0$$

$$\Rightarrow \quad s^3 + 2s^2 + (2+K)s + K = 0$$

s^3	1	$2+K$
s^2	2	K
s^1	$\dfrac{4+2K-K}{2}$	0
s^0	K	0

For stability, $K > 0$ & $\dfrac{4+K}{2} > 0 \Rightarrow K > -4$

Hence the range of K is $0 < K < \infty$

(vi)   Angle of departure

$$\theta_1 = 180° - 135° - 90° + 90° - 90°$$
$$\theta_1 = -45°$$
$$\theta_2 = +45° \text{ (Similarly)}$$

The root locus is drawn using the above data, since the OLTF has one zero and three poles the number of root locus terminating to infinity will be $(n - m)$ i.e. $3 - 1 = 2$

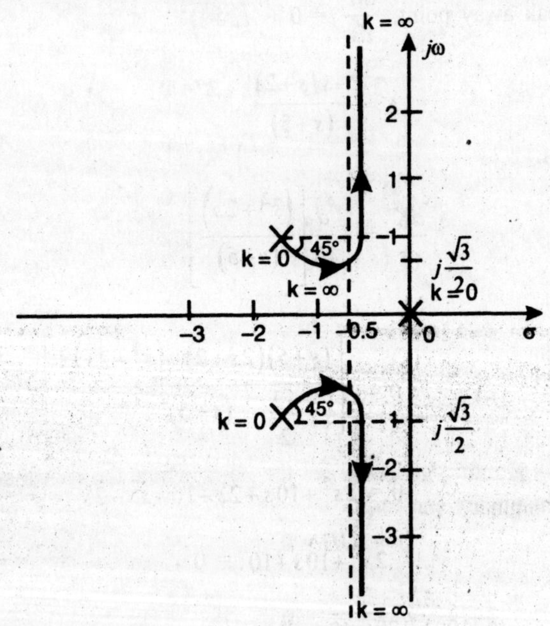

Fig. P. 6.2 (a)

Problem 6.3. A unity feedback system has the open loop transfer function

$$G(s) = \frac{K(s+5)}{s(s+2)}; \quad K \geq 0$$

(a) Draw a rough sketch of the root locus plot, given that the complex roots of the characteristic equation move along a circle.

(b) As K increases does the system become less stable? Justify your answer.

(c) Find the value of K (if it exists) so that the damping of $`\xi`$ the complex closed loop poles is 0.3 (Gate – 2000)

Solution:

$$G(S) = \frac{K(s+5)}{s(s+2)}; \quad K \geq 0$$

(i) There are two poles and one zero poles at $s = 0, -2$. Zero at $s = -5$

(ii) Break away point, $\dfrac{dK}{ds} = 0$

$$K = \frac{-s(s+2)}{(s+5)}$$

$$\Rightarrow \qquad \frac{dK}{ds} = -\frac{d}{ds}\left[\frac{(s^2+2s)}{(s+5)}\right]$$

$$\Rightarrow \qquad 0 = -\left[\frac{(s+5)(2s+2)-(s^2+2s).1}{(s+5)^2}\right]$$

$$\Rightarrow \qquad 0 = 2s^2 + 10s + 2s + 10 - s^2 - 2s$$

$$\Rightarrow \qquad 2s^2 + 10s + 10 = 0$$

$$\Rightarrow \quad s = \frac{-10 \pm \sqrt{100-40}}{2} = \frac{-10 \pm \sqrt{60}}{2} = -8.873, -1.127$$

(iii) Root locus terminating to infinity $= n - m = 2 - 1 = 1.$

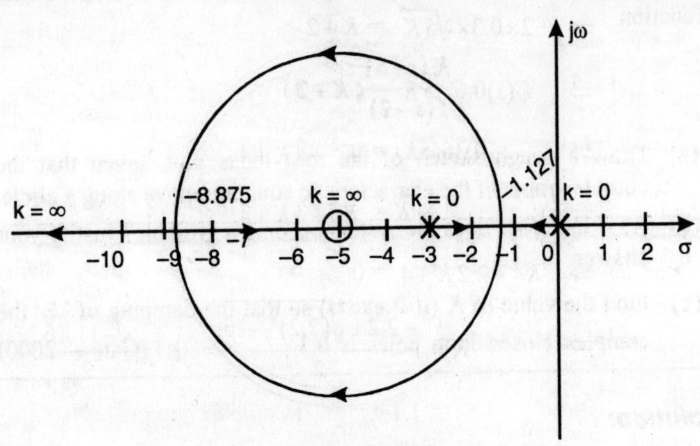

Fig. P. 6.3

(b) Apply Routh criterion for stability

$$1 + G(s) = 0$$

$$\Rightarrow \quad 1 + \frac{K(s+5)}{s(s+2)} = 0$$

$$\Rightarrow \quad s^2 + 2s + Ks + 5K = 0$$

$$\Rightarrow \quad s^2 + (K+2)s + 5K = 0$$

for stability

s^2	1	$5K$	$K > 0$
s^1	$K+2$	0	$K+2 > 0$
s^0	$5K$		$K > -2$

i.e. $0 < K < \infty$

Hence, as the value of K is increased the system becomes more stable as the closed loop poles moves away from origin in left half part.

(c) For $\xi = 0.3$, the value of K comes out to be a complex number, so it does not exist.

$$1 + G(s) = 0$$

$$\Rightarrow \quad s^2 + (K+2)s + 5K = 0$$

Now $\qquad 2\xi\omega_n = K + 2, \quad \omega_n = \sqrt{5K}$

$$\Rightarrow \quad 2\times 0.3\times\sqrt{5K} = K+2$$

$$\Rightarrow \quad 0.6\sqrt{5K} = (K+2)$$

$$\Rightarrow \quad 0.36(5K) = K^2 +4K+4$$

$$\Rightarrow \quad 1.8\,K = K^2 +4K+4$$

$$\Rightarrow \quad K^2 +2.2K+4 = 0$$

$$\Rightarrow \quad K = \frac{-2.2\pm\sqrt{(2.2)^2 -16}}{2} = -1.1\pm\frac{\sqrt{-11.16}}{2}$$

$$= 1.1\pm j1.67 \;(\text{Complex number})$$

Problem 6.4. (a) Predict the number of branches of the root locus terminating at infinity for the following transfer function.

$$G(s) = \frac{10\left(s^3 +2s^2 +5s+1\right)}{(s+1)(s+6)}$$

(b) Find the no. of separate root loci of the characteristic equation of unity feedback control system with an open loop transfer function of

$$G(s) = \frac{K(s+1)\,(s+3)(s+5)}{s(s+2)} \qquad \text{(ES' 2003)}$$

Solution:

(a) As we know that in the root locus technique, each locus starts at an open loop pole and ends either at an open loop zero or infinity.

Here, there are two poles and three zeroes

$$\because \; Z > P,$$

Therefore root locus does not need to terminate at infinity and hence, number required = 0

(b) The number of separte root loci

$$= P, \text{ if } P > Z$$

and $\qquad\quad = Z, \text{ if } Z > P$

Thus, for the given case, number of separate root loci = $Z = 3$

> **Problem 6.5.**   Sketch the root locus for the characteristic equation
>
> $$s(s+1)(s+2)+K(s+1.5)= 0$$

Solution:

The standard form is given as below;

$$s(s+1)(s+2)+K(s+1.5) = 0$$

$$\Rightarrow \qquad 1+\frac{K(s+1.5)}{s(s+1)(s+2)} = 0$$

$$\Rightarrow \qquad 1 + GH = 0$$

The open loop transfer function is given as

$$GH = \frac{K(s+1.5)}{s(s+1)(s+2)}$$

(i)   Poles are at $s = 0, -1, -2$

Zero is at $s = -1.5$

No. of Zero at infinity $= 3 - 1 = 2$

(ii)   Centroid $\quad \sigma = \dfrac{\sum \sigma_p - \sum \sigma_z}{(n-m)} = \dfrac{\left[0-1+(-2)-(-1.5)\right]}{2}$

$$= \frac{-3+1.5}{2} = \frac{-1.5}{2} = -0.75$$

(iii)   Angle of asymtote $= \left(\dfrac{2q+1}{n-m}\right) \times 180°$

$$= \left(\frac{2q+1}{2}\right) \times 180° = (2q+1)90 = 90°, 270°$$

(iv)   Breakway points, $\dfrac{dK}{ds} = 0$

$$K = \frac{-s(s+1)(s+2)}{(s+1.5)}$$

$$\Rightarrow \qquad \frac{dK}{ds} = \frac{-d}{ds}\left[\frac{s^3+3s^2+2s}{(s+1.5)}\right]$$

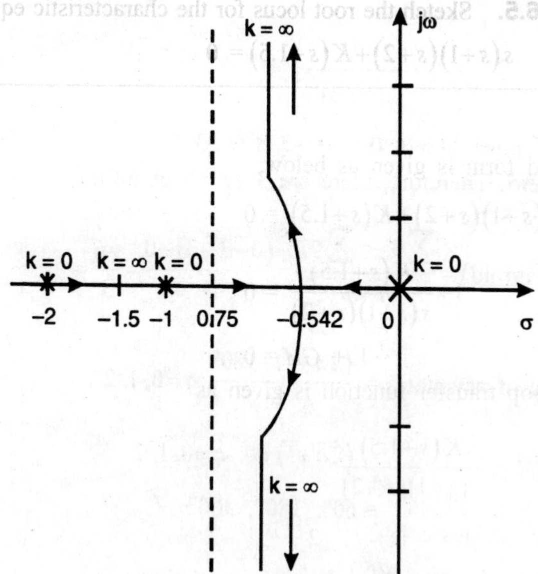

Fig. P. 6.5

$$\Rightarrow \quad \frac{dK}{ds} = -\left\{ \frac{(s+1.5)\left(3s^2+6s+2\right)-\left(s^3+3s^2+2s\right)\cdot 1}{(s+1.5)^2} \right\}$$

$$\Rightarrow \quad 0 = -\left\{ \frac{3s^3+4.5s^2+6s^2+9s+2s+3-s^3-3s^2-2s}{(s+1.5)^2} \right\}$$

$$\Rightarrow \quad 2S^3+7.5s^2+9s+3 = 0$$

$$\Rightarrow \quad s = -0.542$$

Problem 6.6. The loop transfer function of a unity feedback control system by

$$G(s) = \frac{K}{s(s+3)^2}$$

Sketch the root locus plot of the closed loop system for positive values of K and therefrom determine the value of K that would make the system work at a damping factor of 0.5

Solution:

$$G(s) = \frac{K}{s(s+3)^2}$$

(i) Three poles at $s = 0$, $s = -3$, $s = -3$

No zero. Therefore, there are 3 zeroes at infinity

(ii) $\sigma(\text{Centroid}) = \dfrac{\sum \sigma_p - \sum \sigma_z}{n-m} = \dfrac{(0-3-3)-0}{3} = \dfrac{-6}{3} = -2$

(iii) Angle of asymtotes $= \dfrac{(2q+1)-180°}{3}; q=0, 1, 2$

$$= (2q+1)\, 60; \quad q=0, 1, 2$$

$$= 60°, \ 180°, \ 300°$$

(iv) Breakaway point, $\dfrac{dK}{ds} = 0$

$$K = \frac{-s(s+3)^2}{1}$$

$\Rightarrow \qquad \dfrac{dK}{ds} = \dfrac{-d}{ds}\left[s^3 + 6s^2 + 9s\right]$

$\Rightarrow \qquad \dfrac{dK}{ds} = -\left(3s^2 + 12s + 9\right)$

$\Rightarrow \qquad 0 = s^2 + 4s + 3$

$\Rightarrow \qquad s^2 + 3s + s + 3 = 0$

$\Rightarrow \qquad (s+3)(s+1) = 0$

$\Rightarrow \qquad s = -1, -3$

(v) Characteristic Equation $1 + G(s) = 0$

$\Rightarrow \qquad 1 + \dfrac{K}{s(s^2 + 6s + 9)} = 0$

$\Rightarrow \qquad s^3 + 6s^2 + 9s + K = 0$

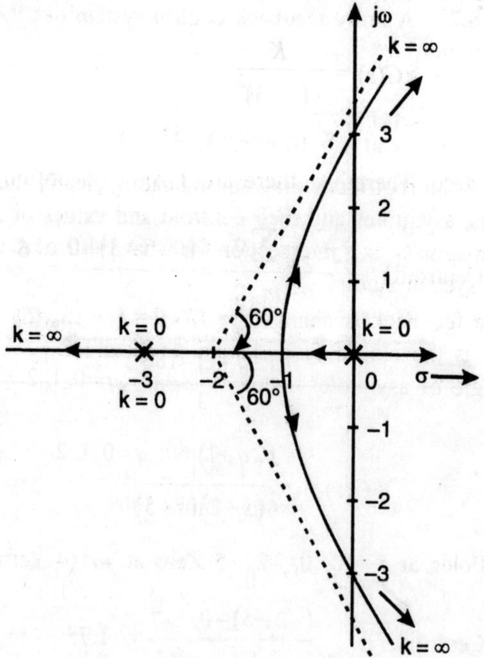

Fig. P. 6.6

Apply Routh criterion for stability

			For stability
s^3	1	9	
s^2	6	K	$K > 0$
s^1	$\dfrac{54-K}{6}$	0	$54 - K > 0$ $K < 54$
s^0	K	0	$0 < K < 54$

The root locus will the imaginary axis at

$$6s^2 + K = 0$$

$$\Rightarrow \quad 6s^2 + 54 = 0$$

$$\Rightarrow \quad s^2 + 9 = 0$$

$$\Rightarrow \quad s = \pm j3$$

Problem 6.7. A unity feedback control system has the open loop transfer function

$$G(s) = \frac{K}{s^2 + (s+2)(s+5)}$$

(a) Sketch the root locus diagram indicating clearly the breakaway points, asymtotes and their centroid and values of K and ω at the imaginary axis intersection for what value of K is the closed loop system stable?

(b) If the feedback is changed to $H(s) = 1 + 2s$, for what values of K is the closed loop system now stable?

Solution:

(a)
$$G(s) = \frac{K}{s^2(s+2)(s+5)}$$

(i) Poles at $s = 0, 0, -2, -5$ Zero at ∞ (4 Zeroes)

(ii) Centroid, $\sigma = \dfrac{(-2-5)-0}{4} = \dfrac{-7}{4} = -1.75$

(iii) Angle of asymtote,

$$\theta = \frac{(2q+1)180°}{4}, \quad q = 0, 1, 2, 3$$

$$= (2q+1)45° = 45, \ 135°, \ 225°, \ 315°$$

$$= \pm 45°, \pm 135°$$

(iv) Breakaway point $\Rightarrow \dfrac{dK}{ds} = 0$

$$\Rightarrow \qquad \frac{dK}{ds} = \frac{-d}{ds}\left[s^2(s+2)(s+5) \right]$$

$$\Rightarrow \qquad \frac{dK}{ds} = \frac{-d}{ds}\left[s^2(s^2+7s+10) \right]$$

$$\Rightarrow \qquad 0 = -\left[4s^3 + 21s^2 + 20s \right]$$

$$\Rightarrow \qquad s + \left(4s^2 + 21s + 20 \right) = 0$$

$$\Rightarrow \quad s = \frac{-21 \pm \sqrt{(21)^2 - 320}}{8}, 0$$

$$\Rightarrow \quad s = \frac{-21 \pm 11}{8} = \frac{-32}{8}, \frac{-10}{8}, 0$$

$$\Rightarrow \quad s = -4, -1.25, 0$$

(v) Characteristic Equation,

$$1 + G(s) = 0$$

$$\Rightarrow \quad s^2\left(s^2 + 7s + 10\right) + K = 0$$

$$\Rightarrow \quad s^4 + 7s^3 + 10s^2 + K = 0$$

Applying Routh stability criterion,

$$
\begin{array}{ccccl}
s^4 & 1 & 10 & K & \\[4pt]
s^3 & 7 & 0 & 0 & K \geq 0 \\[4pt]
s^2 & 10 & K & 0 & \dfrac{-7K}{10} \geq 0 \\[4pt]
s^1 & \dfrac{-7K}{10} & 0 & 0 & K \geq 0 \\[4pt]
s^0 & K & 0 & 0 & \text{Range}\, 0 \leq K \leq \infty
\end{array}
$$

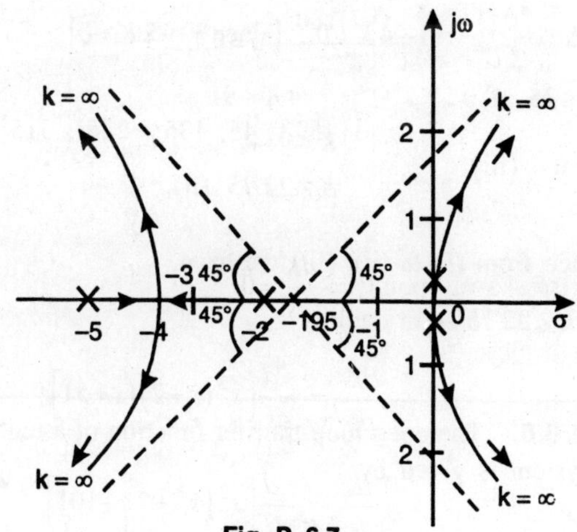

Fig. P. 6.7

(b)
$$G(s)\,H(s) = \frac{K(1 \pm 25)}{s^2(s+2)(s+5)}$$

The characteristic equation is

$$1 + G(s)\,H(s) = 0$$

$$\Rightarrow \quad s^4 + 7s^3 + 10s^2 + K(1+2s) = 0$$

$$\Rightarrow \quad s^4 + 7s^3 + 10s^2 + 2Ks + K = 0$$

Applying Routh stability criterion

$$
\begin{array}{c|ccc}
s^4 & 1 & 10 & K \\
s^3 & 7 & 2K & 0 \\
s^2 & \dfrac{70-2K}{7} & K & 0 \\
s^1 & \dfrac{K(4K-91)}{2K-70} & 0 & 0 \\
s^0 & K & &
\end{array}
$$

The condition for stability is given by

$$70 - 2K > 0 \qquad\qquad \frac{91-4K}{70-2K} > 0$$

$$K < \frac{70}{2} \qquad 91 - 4K > 0 \quad [\text{since } 70 - 2K > 0]$$

$$K < 35 \quad \text{(i)} \qquad\qquad 4K < 91$$

$$\qquad\qquad\qquad K < 91/4$$

$$K > 0 \quad \text{(iii)} \qquad\qquad K < 22.75 \quad \text{(ii)}$$

Hence from (i) & (ii), (iii) we have,

$$0 \le K < 22.75 \rightarrow \text{for stability}$$

Problem 6.8. The open loop transfer function of a unity feedback control system is given by

$$G(s) = \frac{K}{s(s+4)(s+6)}$$

Draw the root locus of the system. On the root locus mark the breakaway points and the imaginary axis crossing points on it.

Solution:

$$G(s) = \frac{K}{s(s+4)(s+6)}$$

(i)   Poles at $s = 0,\ -4,\ -6,\ 3$ Zero at ∞ (infinity)

(ii) Centroid, $\sigma = \dfrac{-4-6}{3} = \dfrac{-10}{3} = -3.33$

Asymtode angle, $\theta = \dfrac{(2q+1)\times 180°}{3}$, $q = 0, 1\ 2$

$$= 60,\ 180°,\ 300°$$

(iii) Breakaway point $\Rightarrow \dfrac{dK}{ds} = 0$

$\Rightarrow \qquad K = -(s+4)(s+6)s$

$\Rightarrow \qquad K = -\left(s^3 + 10s^2 + 24s\right)$

$\Rightarrow \qquad \dfrac{dK}{ds} = \dfrac{-d}{ds}\left(s^3 + 10s^2 + 24s\right)$

$\Rightarrow \qquad 0 = 3s^3 + 20s + 24$

$\Rightarrow \qquad s = \dfrac{-20 \pm \sqrt{400 - 12\times 24}}{2\times 3}$

$\Rightarrow \qquad s = \dfrac{-20 \pm \sqrt{400 - 288}}{6}$

$\Rightarrow \qquad s = \dfrac{-20 \pm \sqrt{112}}{6} = -5.09, -1.57$

For root crossing imaginary axis,

$$1 + G(s) = 0$$

$\Rightarrow \qquad K + s^3 + 10s^2 + 245 = 0$

$\Rightarrow \qquad s^3 + 10s^2 + 245 + K = 0$

Apply Routh criterion for stability

$$
\begin{array}{ccc}
s^3 & 1 & 24 \\
s^2 & 10 & K \\
s^1 & \dfrac{240-K}{10} & 0 \\
s^0 & K & 0
\end{array}
$$

$$K>0 \qquad (i)$$
$$240-K>0$$
$$K<240 \qquad (ii)$$
$$0<K<240$$

Critical value $= 20$

$10s^2 + K = 0$ (when $K=2$ the system osillates with frequency ω)

$10s^2 + 240 = 0$

$$s = \pm j\sqrt{24} \quad \omega = \sqrt{24} = 4.9\,\text{rad/sec}$$

The root locus cuts the imaginary $(j\omega)$ axis at $\pm j\,4.9$. The diagram is shown below:

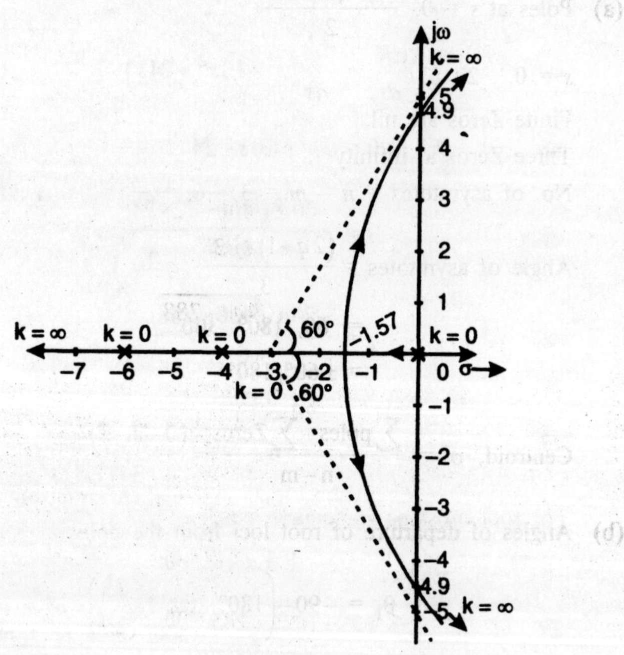

Fig. P. 6.8

Problem 6.9. A unity feedback system has an open loop transfer function,

$$G(s) = \frac{K}{s\left(s^2 + 4s + 13\right)}$$

Make a rough sketch of the root locus plot by determining the following:

(a) Controid, member and angles of asymtotes
(b) Angle of departure of root loci from the poles
(c) Breakaway points, if any
(d) The point of root loci intersecting the imaginary line. Determine the value of K at which the system exhibits oscillatory behaviour.

Solution:

$$G(s) = \frac{K}{s\left(s^2 + 4s + 13\right)}$$

(a) Poles at $s = 0$, $\dfrac{-4 \pm \sqrt{16-52}}{2}$

$s = 0,\ -2 \pm j3$

Finite Zeros are nil.

Three Zeros at infinity

No. of asymtotes $= n - m = 3 - 0 = 3$

$$\text{Angle of asymtotes} = \frac{(2q+1)\times 180°}{3},\ q=0, 1, 2$$

$$= 60,\ 180°,\ 300°$$

$$= \pm 60°,\ 180°$$

$$\text{Centroid, } \sigma = \frac{\sum \text{poles} - \sum \text{Zeros}}{n-m} = \frac{(0-2-2)}{3} - 0 = \frac{-4}{3} = -1.33$$

(b) Angles of departure of root loci from the poles

$$\theta_1 = -90 - \left(180° - \tan^{-1}\left(\frac{3}{2}\right)\right) + 180°$$

$$\theta_1 = -33.69°$$

Similarly,
$$\theta_2 = +\ 33.69$$

$$\theta_3 = 180°$$

(c) Breakaway point, $\dfrac{dK}{ds} = 0$

$$K = -\left(s^3 + 4s^2 + 13s\right)$$

$$\Rightarrow \qquad \frac{dK}{ds} = \frac{-d}{ds}\left(s^3 + 4s^2 + 13s\right)$$

$$\Rightarrow \qquad 0 = 3s^3 + 8s + 13$$

$$\Rightarrow \qquad s = \frac{-8 \pm \sqrt{64 - 156}}{6}$$

$$\Rightarrow \qquad s = \frac{-8 \pm j9.6}{6}$$

$$\Rightarrow \qquad s = -1.33 \pm j1.6$$

(d) The characteristic equation
$$1 + G(s) = 0$$

$$\Rightarrow \qquad 1 + \frac{K}{s\left(s^2 + 4s + 13\right)} = 0$$

$$\Rightarrow \qquad s^3 + 4s^2 + 13s + K = 0$$

Routh Array is

			Condition of stability
s^3	1	13	
s^2	4	K	$K \geq 0$
s^1	$\dfrac{52 - K}{4}$	0	$52 - K > 0$
			$K \leq 52$
s^0	K	0	$0 \leq K \leq 52$

For $K = 52$, the system oscillates with frequency,

$$4s^2 + 52 = 0$$

$$\Rightarrow \qquad s^2 + 13 = 0$$

$$\Rightarrow \qquad s = \pm j\sqrt{13}$$

$$\omega = \sqrt{13} \ \text{rad/sec} = 3.61 \ \text{rad/sec}$$

This is point of intersection with $j\omega$ axis.

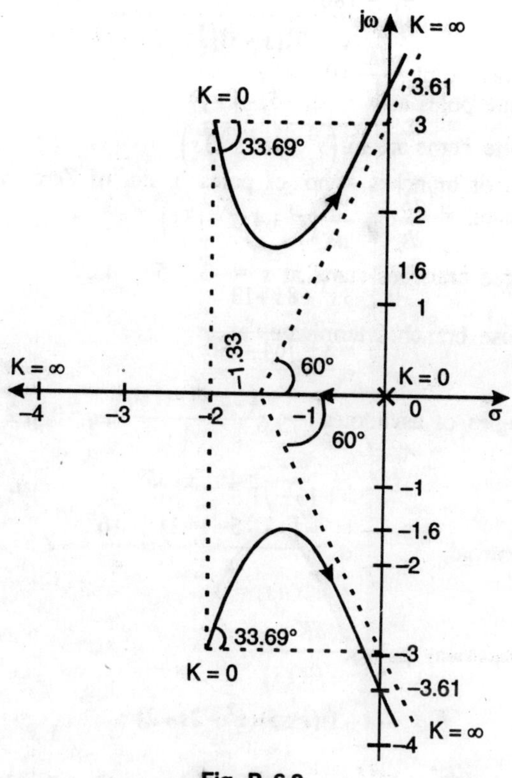

Fig. P. 6.9

Problem 6.10. Consider the open loop transfer function of a unity feedback system.

$$G(s) = \frac{K}{(s+3)(s+5)\left(s^2+2s+2\right)}$$

In the root locus diagram determine

(a) the number of branches of root loci

(b) at what location these branches start

(c) where these branches terminate

(d) Angles of asymtotes and at which they cuts the real axis.

(e) Breakaway point

(f) points at which root loci cuts $j\omega$ axis

Solution:

$$G(s) = \frac{K}{(s+3)(s+5)(s^2+2s+2)}$$

(a) Finite poles at $s = -3, -5, -1 \pm j1$

Finite Zeros are nil

No. of branches = no. of poles or no. of Zeros whichever is greater. = 4.

(b) These branches starts at $s = -3, -5, -1 \pm j1$

(c) These branches terminates at ∞

(d) Angles of asymtotes $= \dfrac{(2q+1)\times 180^\circ}{4}$, $q = 0, 1, 2, 3$

$$= \pm 45^\circ \; \pm 135^\circ$$

Centroid, $\sigma = \dfrac{(-3-5-1-1)}{4} = \dfrac{-10}{4} = -2.5$

(e) Breakaway points, $\dfrac{dK}{ds} = 0$

$$K = -(s+3)(s+5)(s^2+2s+2)$$

$$\Rightarrow \quad \frac{dK}{ds} = \frac{-d}{ds}\left[(s+3)(s^3+5s^2+2s^2+10s+2s+10)\right]$$

$$\Rightarrow \quad \frac{dK}{ds} = 4s^3+30s^2+66s+46$$

$$\Rightarrow \quad 0 = 2s^3+15s^2+33s+23$$

$$\Rightarrow \quad (s+2)(2s^2+11s+11)=0$$

$$\Rightarrow \quad s = -2, -3.5, -2.04.$$

(f) Characteristic equation is given as.

$$1 + G(s) = 0$$

$$\Rightarrow \quad s^4 + 10s^3 + 33s^2 + 46s + (K+30) = 0$$

Apply Routh Array for stability;

$$
\begin{array}{llll}
s^4 & 1 & 33 & K+30 \\
s^3 & 10 & 46 & 0 \\
s^2 & 28.4 & 30+K & 0 \\
s^1 & K_1 & 0 & 0 \\
s^0 & 30+K & 0 & 0
\end{array}
\qquad \text{where } K_1 \dfrac{46(28.4)-(30+K)10}{28.4}
$$

when $K_1 = 0$, $K = 100.6$

The auxiliary equation

$$28.4\,s^2 + (30+K) = 0$$

$$\Rightarrow \qquad 28.4\,s^2 = -136.6$$

$$\Rightarrow \qquad s = \pm j2.15$$

$$\omega = 2.15 \text{ rad/sec. (frequency of oscillation)}$$

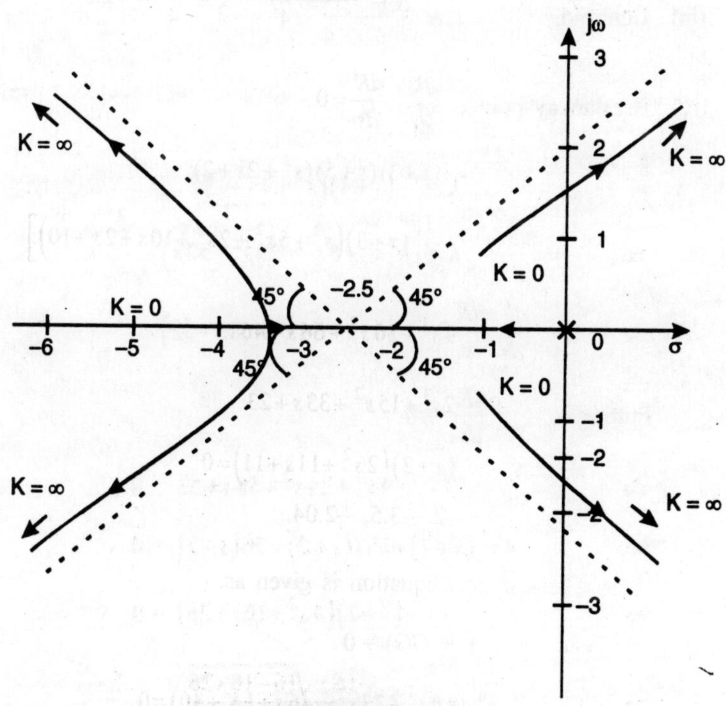

Fig. P. 6.10

Problem 6.11.   Sketch the root locus plot for the system when open loop transfer function is given by

$$G(s)\,H(s) = \frac{K}{s(s+4)\left(s^2+4s+13\right)}$$

Solution:

(i) $$G(s)\,H(s) = \frac{K}{s(s+4)\left(s^2+4s+13\right)}$$

Finite poles at $s = 0,\ -4,\ -2\pm j3$

Finite Zeros are nil

(ii) Angles of asymtotes $= \dfrac{(2q+1)}{4}\times 180, q=0, 1, 2, 3.$

$$= \pm 45° + 135°$$

(iii) Centroid, $\sigma = \dfrac{(-4-2-2)}{4} = -2$

(iv) Breakaway points, $\dfrac{dK}{ds} = 0$

$$K = -(s+4)\left(s^2+4s+13\right)$$

$$\Rightarrow \qquad K = -\left(s^4+8s^3+29s^2+52s\right)$$

$$\Rightarrow \qquad \frac{dK}{ds} = -\left(4s^3+24s^2+58s+52\right)$$

Putting $\dfrac{dK}{ds} = 0$

$$\Rightarrow \qquad 4s^3+24s^2+58s+52 = 0$$

$$\Rightarrow \qquad 4s^2(s+2)+16s(s+2)+26(s+2) = 0$$

$$\Rightarrow \qquad (s+2)\left(4s^2+16s+26\right) = 0$$

$$\Rightarrow \qquad s = -2,\ \frac{-16\pm\sqrt{16-16\times26}}{8}$$

$$\Rightarrow \quad s = -2, \; -2 \pm j1.58$$

(v) For intersection of root locus with $j\omega$ axis,

$$1 + G(s)\,H(s) = 0$$

$$s^4 + 8s^3 + 29s^2 + 52s + K = 0$$

s^4	1	29	K
s^3	8	52	0
s^2	22.5	K	0
s^1	$52 - 0.35K$	0	0
s^0	K	0	0

For stability

$$K > 0$$

$$52 - 0.35K \geq 0$$

$$K \leq 148.6$$

$$22.5s^2 + K = 0$$

$$22.5s^2 + 148.6 = 0$$

$$s^2 = -\frac{148.6}{22.5}, \quad s = \pm j2.56$$

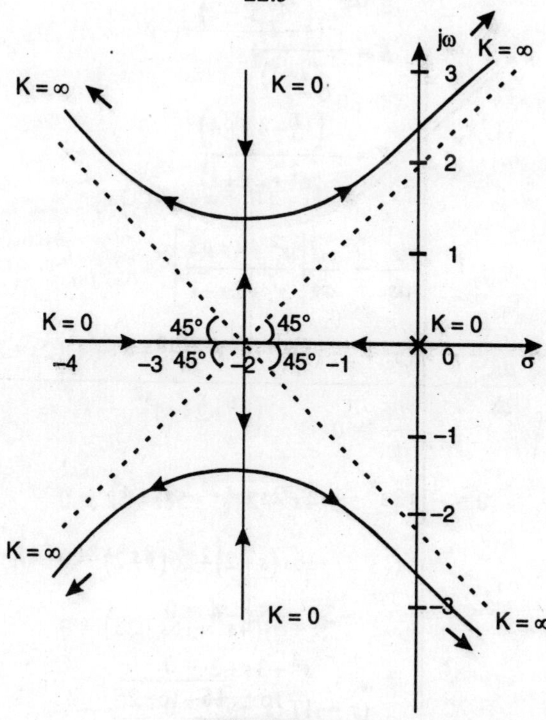

Fig. P. 6.11

Problem 6.12. The open loop transfer function of a control system is given as.

$$G(s)\,H(s) = \frac{K(s+1)^2}{(s+2)^2}$$

Draw the root locus.

Solution:

$$G(s)\,H(s) = \frac{K(s+1)^2}{(s+2)^2}$$

(i) Two poles at $s = -2, -2$

(ii) Two zeros at $s = -1, -1.$

(iii) Since number of poles & zeros are same hence there is no asymtotes.

(iv) Breakaway points, $\dfrac{dK}{ds}$

$$K = -\frac{(s+2)^2}{(s+1)^2}$$

$$\Rightarrow \qquad K = -\frac{\left(s^2+4s+4\right)}{\left(s^2+2s+1\right)}$$

$$\Rightarrow \qquad \frac{dK}{ds} = \frac{-d}{ds}\left[\frac{s^2+4s+4}{s^2+2s+1}\right]$$

$$\Rightarrow \qquad \frac{dK}{ds} = -\frac{\left[\left(s^2+2s+1\right)(2s+4)-\left(s^2+4s+4\right)(2s+2)\right]}{\left(s^2+2s+1\right)^2}$$

$$\Rightarrow \qquad 0 = -\Big[\left(2s^3+4s^2+2s+4s^2+8s+4\right)$$
$$-\left(2s^3+8s^2+8s+2s^2+8s+8\right)\Big]$$

$$\Rightarrow \qquad -2s^2-6s-4 = 0$$

$$\Rightarrow \qquad s^2+3s+2 = 0$$

$$\Rightarrow \qquad (s+1)\,(s+2) = 0$$

$$\Rightarrow \qquad s = -1,\, -2$$

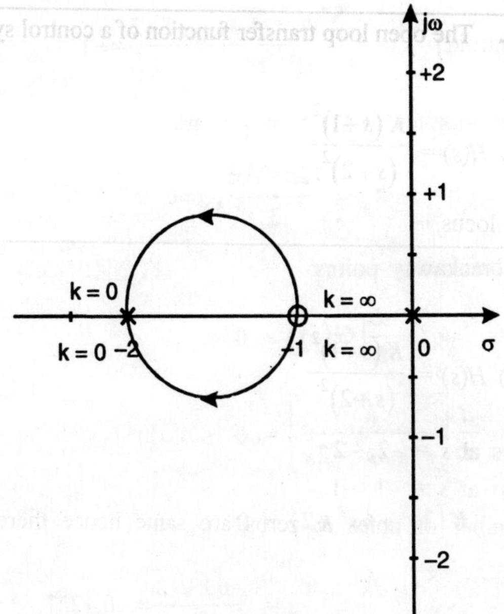

Fig. P. 6.12

Problem 6.13. A unity feedback system has the open-loop transfer function.

$$G(s) = \frac{K}{s(s+1)(s+2)}$$

(i) Calculate the breakaway point and the imaginary axis crossing points.

(ii) Draw the root loci as K varies from 0 to ∞

(iii) Give the procedure for finding closed-loop transfer function directly from the root locus diagram for a given damping ratio of dominant root pair.

Solution:

$$G(s) = \frac{K}{s(s+1)(s+2)}$$

(a) Finite poles at $\quad s = 0,\ -1,\ -2$

$$p = 3$$

Finite zeros at, $\quad z = 0.\ s$

(b) $\sigma(\text{centroid})$ $= \dfrac{\sum P_j - \sum z_k}{P - z} = \dfrac{-1-2}{3} = -1$

The asymptotes of the root loci are

$$\theta_A = \frac{(2n+1)\pi}{3}; \; n=0, 1, 2 = 60°, \; 180°, \; -60°$$

(i) For breakaway points

$$\frac{d}{ds}\big[G(s)\big] = 0$$

or $\dfrac{d}{ds}\left[\dfrac{K}{s^3 + 3s^2 + 2s}\right] = 0$

or $-K\big(3s^2 + 6s^2 + 2\big) = 0$

or $s = \dfrac{-6 \pm \sqrt{12}}{6} = -0.4227, -1.5773$

To obtain the imaginary axis crossing, construct the Routh array. The characteristic equation is

$$1 + G(s) = 0$$

or $s^3 + 3s^3 + 2s + K = 0$

Constructing Routh array, we have

$$
\begin{array}{c|ccc}
s^3 & 1 & 2 & 0 \\
s^2 & 3 & K & 0 \\
s^1 & \dfrac{6-K}{3} & 0 & 0 \\
s^0 & K & 0 & 0
\end{array}
\qquad \dfrac{6-K}{3} = 0 \text{ or } K = 0
$$

Putting $K = 6$, the auxiliary equation is

or $3s^2 + 6 = 0$

or $s^2 = -2$

$s = j\omega$

or $\omega = \pm\sqrt{2}$

(ii) The root loci is drawn below:

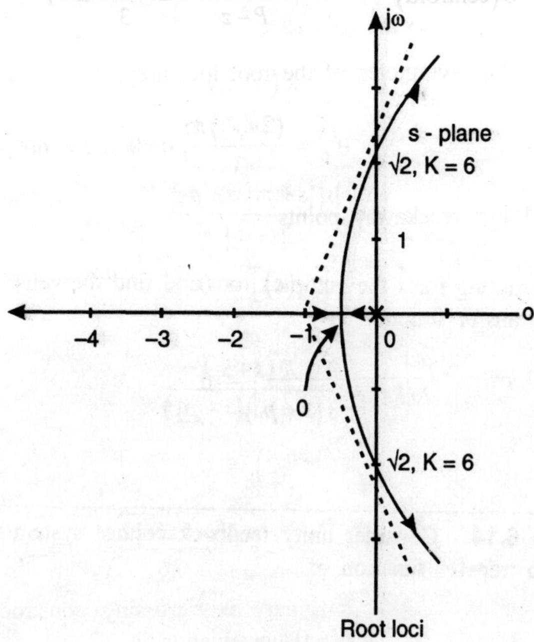

Fig. P. 6.13

(iii) (a) From the root locus diagram, the open-loop poles and zeros
are determined. If the number of zeros are not equal the number
of poles, the missing zeros are considered to be infinity.
Therefore the transfer function is

$$H(s) = \frac{K(s+z_1)}{s(s+p_1)(s+p_2)}$$

(b) When $K = 0$, the roots, poles and zeros, are as given in root locus
diagram corresponding to open loop transfer function on the real
axis $\zeta > 1$. When K is increased the roots move along the real axis
until they meet. At this point the roots are equal $\zeta = 1$. As the gain
is increased the roots break away from the real axis and become
complex conjugate pairs.

(c) Distance of the root from the origin corresponds to the natural
frequency, ω_n. Since damping ratio is given, we can find

$$\omega_n = \frac{p_1}{\omega_n} = \cos\theta = \xi,$$

where ω_n crosses the root locus and will give the complex root.

At any point on the root locus

$$|G(s)| = \frac{K(s+z_1)}{s(s+p_1)(s+p_2)} = 1$$

or

$$K = \frac{|s||s+p_1||s+p_2|}{|s+z_1|}$$

Substituting for s the complex root and find the value of K. So the transfer function is

$$H(s) = \frac{K(s+z_1)}{s(s+p_1)(s+p_2)}$$

Problem 6.14. Consider unity feedback control system with an open loop transfer function of

$$G(s) = \frac{K(s+1)(s+2)}{(s+0.1)(s-1)}$$

Draw the root loci of the system with gain K as a variable. As an aid to plotting determine: asymptotes, centroid, breakaway point, the gain at which the root locus crosses j-axis.
Find from the root locus plot the value of gain K for which a closed loop system is critically damped.

Solution:

System poles: $- 0.1, 1$

         Zeros: $- 1, - 2$

These are marked in the figure

* The plot has no asymptotes

* To determine the gain at which the root locus crosses the $j\omega$ axis, consider it's characteristic equation.

$$(s+0.1)(s-1) + K(s+1)(s+2) = 0$$

or, $(K+1)s^2 + (3K-0.9)s + (2K-0.1) = 0$

Apply Routh Hurwitz criterion.

$$s^2(K+1) \quad (2K-0.1)$$
$$s^1(K-0.3) \quad\quad 0$$
$$s^0(2K-0.1) \quad\quad 0$$

Hence for stability, $K > 0.3$

The root locus crosses the $j\omega$-axis at, $K = 0.3$

at a frequency of, $\quad\quad 1.4s^2 + 0.5 = 0$

or, $\quad\quad s = \pm J\sqrt{0.5}/1.4 = 0.6$ rad/s

or, Breakaway points

$$B(s) = (s+1)(s+2) = s^2 + 3s + 2$$

$$A(s) = (s+0.1)(s-1) = s^2 - 0.9s - 0.1$$

From equation, $\quad\quad A(s)B'(s) - A'(s)B(s) = 0$

$$\left(s^2 - 0.9s - 0.1\right)(2s+3) - (2s-0.9)\left(s^2 + 3s + 2\right) = 0$$

or, $\quad\quad 3.9s^2 + 3.8s - 2.1 = 0$

$$s = -\,1.368,\ 0.394$$

The value of K at the breakaway point is calculated below,

$$K_1 = \left|\frac{(s+0.1)(s-1)}{(s+1)(s+2)}\right|_{s=-1.368} = 12.9$$

$$K_2 = \left|\frac{(s+0.1)(s-1)}{(s+1)(s+2)}\right|_{s=0.394} = 0.09$$

Gain for critical damping

The above information is confirmed by the root locus plot of the figure obtained by means of Method Control tool.

$\varsigma = 1$ is achieved when the two roots are equal and negative (real). This happens at the breakaway point in the left half S-plane. The conrresponding value of gain is. $K = 12.9$

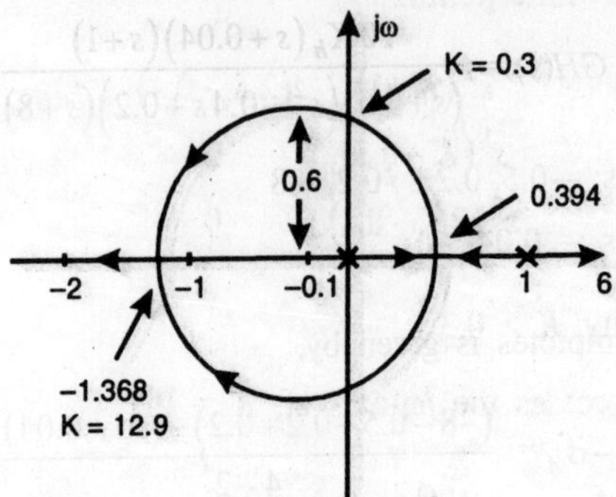

Fig. P. 6.14

Problem 6.15. In order to overcome the inherent instability of a helicopter (let us say wrt pitch attitude), a stabilizing feedback loop is introduced in it's autopilot system as shown in the Fig 6.15(a). The forward transfer function of the helicopter is estimated to be

$$G(s) = \frac{10(s+0.04)}{(s+0.5)\left(s^2 - 0.4s + 0.2\right)}$$

The feedback path transfer function is shown in the figure where gain of the transfer function (K_h) is adjustable.

Plot the root locus of the system with variable K_h and find it's value for the closed loop system to have a dominant pole damping of $1/\sqrt{2}$

With this value of K_h, find the steady state error $(e = r - c)$ caused by a wind gust disturbance of $T_d(s) = 1/s$ as shown in the block diagram.

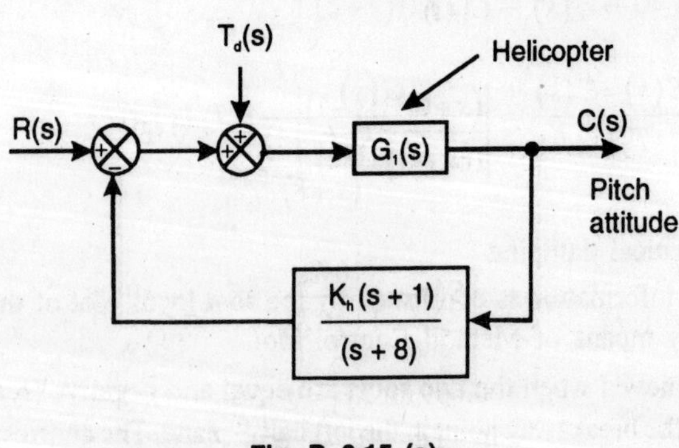

Fig. P. 6.15 (a)

Solution:

$$GH(s) = \frac{10 K_h (s+0.04)(s+1)}{(s+0.5)(s^2 - 0.4s + 0.2)(s+8)}$$

Open loop poles: $-0.5,\ 0.2 \pm j0.2,\ -8$

Open loop zeros :$- 0.04,\ -1$

Centroid of asymptotes is given by,

$$-\sigma_A = \frac{(-8-0.5+0.2+0.2)-(-1-0.04)}{4-2} = -3.53$$

The root locus plot is drawn in fig, with $K = 10 K_h$ as variable. We can now find the value of K at which the root locus crosses over to negative half of the s-plane. The system's characteristic equation is

$$(s+0.5)(s^2 - 0.4s + 0.2)(s+8) + K(s+0.04)(s+1) = 0$$

or, $\quad s^4 + 8.1 s^3 (K+0.8) s^2 + (1.04 K - 1.43)s + (0.04 K + 0.8) = 0$

Applying the Routh Criterion we get the following result.

The root locus crosses the $j\omega$ axis (to left) at $K = 2.385$ at $\pm j\omega = 0.575$ rad/s.

The root locus is plotted to scale in fig. The $\zeta = 1/\sqrt{2}$ line intersects the root locus at two points with K as,

$$K_1 = 1.555 \quad \text{or} \quad K_{n1} = 0.156$$

$$K_2 = 0.785 \quad \text{or} \quad K_{n2} = 0.079$$

Steady state error for unit wind gust disturbance.

$$R(s) = 0 \rightarrow E(s) = C \ (I)$$

$$\frac{E(s) = C(s)}{T_d(s)} = \frac{G(s)}{1 + G(s)\dfrac{K_n(s+1)}{(s+8)}}, \quad T_d(s) = 1/s$$

$$e_{ss} = \underset{s \to 0}{sE(s)} = \frac{10 K_n}{1 + 10 K_n}; \text{after substituting for } G(s)$$

This gives, $\qquad e_{ss} = 0.609 \text{ for } Kh_1 = 0.156$

$$= 0.44 \text{ for } Kh_2 = 0.079$$

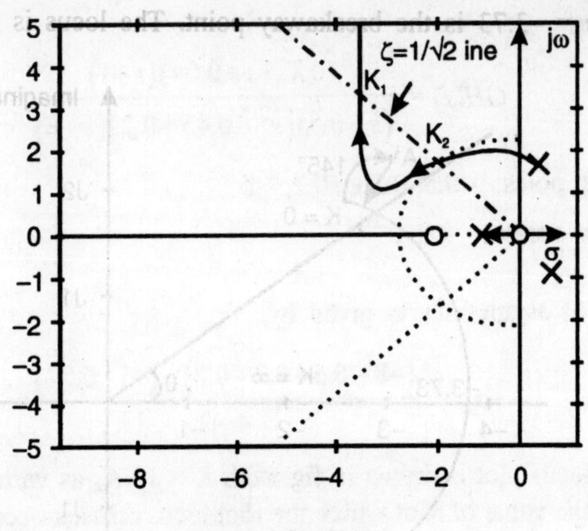

Fig. P. 6.15 (b)

Problem 6.16. Sketch the root locus plot and determine the approximate damping ratio for a value of $K = 1.33$ for a control system having a forward transfer function.

$$G(s) = \frac{k(s+2)}{s^2 + 2s + 3}$$

Solution:

$$\text{Poles} = -1 \pm j\sqrt{2}$$

$$\text{Zero} = -2$$

The characteristic equation is

$$s^2 + (2+K)s + 2K + 3 = 0$$

Since, the order of the characteristic equation is two, the number of root loci are two. Each root locus starts from one of the two complex conjugate poles and breaks in the part of the negative real axis between −2 and −∞.

$$\phi_1 = \pm\frac{180}{2-1} = \pm 180°$$

There is only one asymptote coinciding with the negative real axis.

Angle of Departure $= 180° - 90° - \left(-\tan\sqrt{2}\right) = 145°$

Breakaway point

$$\frac{dG(s)}{ds} = s^2 + 4s + 1 = 0$$

or, $s = -3.73$ is the breakaway point. The locus is shown in the fig.

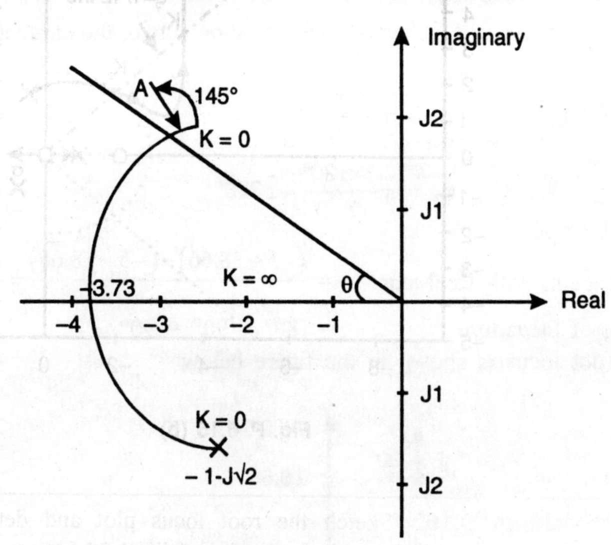

Fig. P. 6.16

Point on locus where $K = 1.33$

$$\frac{K(s+2)}{s^2 + 2s + 3} = -1$$

or, $$\frac{1.33(s+2)}{s^2 + 2s + 3} = -1$$

or, $$s^2 + 3.33s + 5.66 = 0$$

i.e, $$s = -1.7 \pm j1.7$$

$$s = 2.4\angle \pm 45°$$ (This point is shown as point A)

Problem 6.17. Sketch the root locus of

$$G(s) = \frac{K}{s^2 + 10s + 100}$$

Solution:

Poles $s = -5 \pm j8.66$

The characteristic equation is, $s^2 + 10s + 100 = 0$

Since the order of the characteristic equation is two, the number of root locus are two.

Asymptotes

$$\phi_1 = \pm \frac{180°}{2-0} = \pm 90°$$

$$\phi_3 = \pm \frac{3 \times 180°}{2-0} = \pm 270°$$

Intersection with Real-axis $\quad s = \dfrac{(-5 + j8.66) + (-5 - j8.66)}{2} = -5$

Angle of Departure $\qquad = 180° - 90° = 90°$

The root locus is shown in the figure below

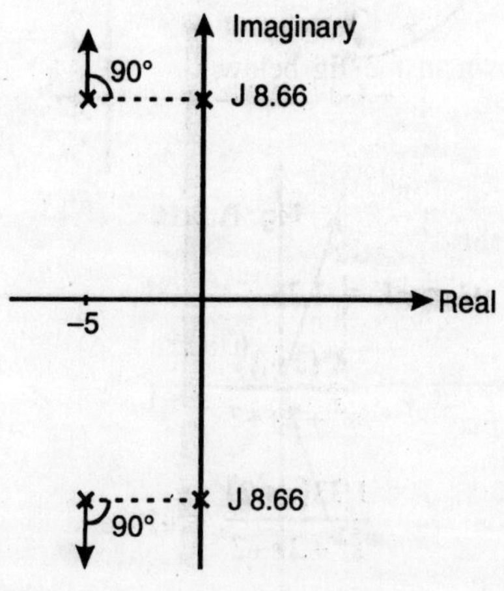

Fig. P. 6.17

Problem 6.18.   Sketch the root locus of

$$G(s) = \frac{K(s+1)}{s^2(s+2)}$$

Solution:

$$\text{Poles} = 0,\ 0,\ -2,\quad \text{Zeros} = -1$$

The characteristic equation is,

$$s^3 + 2s = 0$$

Since, the order of the characteristic equation is three, the number of
root loci are three.

Asymptotes $\qquad \phi_1 = \pm\dfrac{180°}{3-1} = \pm 90°$

$$\phi_3 = \pm\dfrac{3\times180°}{3-1} = \pm 270°$$

Intersection of the root-locus is

$$s = \frac{-2-(-1)}{3-1} = -0.5$$

The root locus is shown in the fig below,

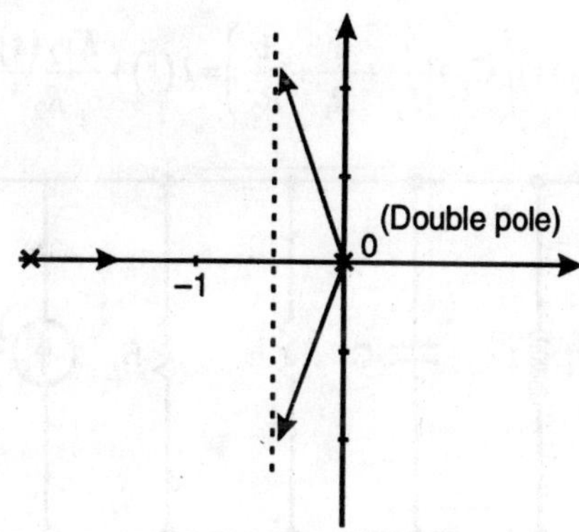

Fig. P. 6.18

Problem 6.19.   For illustration, consider the active circuit shown
in fig 6.19. Let us find the plot of root loci for variable gain parameter
K, for a network function $V_2(s)/I(s)$. Given

$$R_1 = 1/2 \text{ ohm}$$

$$R_2 = 1 \text{ ohm}$$

$$L = 1/2 = 1/2 \text{ henry}$$

$$C = 1 \text{ farad}$$

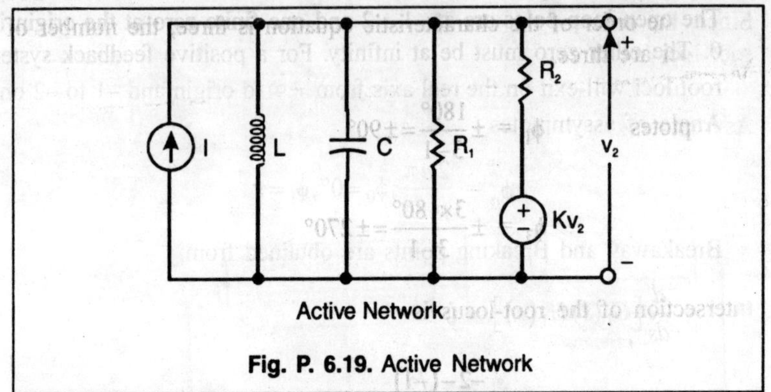

Fig. P. 6.19. Active Network

Solution:

The equivalent circuit is drawn as shown in the fig 6.19 (a), by Norton's equivalence.

Applying KCL,

$$V_2(s)\left(C_s+\frac{1}{L_s}+\frac{1}{R_1}+\frac{1}{R_2}\right)=I(s)+\frac{Kv_2(s)}{R_2}$$

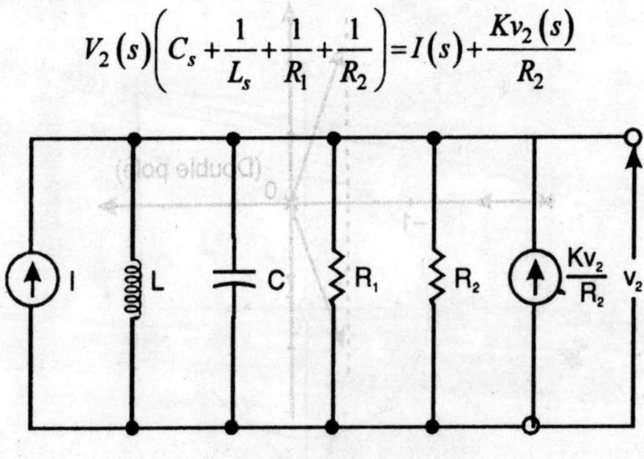

Fig. P. 6.19 (a)

Substituting values,

$$\frac{V_2(s)}{I(s)}=\frac{s}{s^2+(3-K)s+2}=\frac{s/(s^2+3s+2)}{1-\dfrac{jK}{s^2+3s+2}}$$

Comparing with the characteristic equation $F(s)=\pm\,jG(s)\cdot H(s)=0$ of the positive feedback closed loop system, we get the open loop transfer function as—

$$G(s)\,H(s)=\frac{sK}{s^2+3s+2}=\frac{sK}{(s+1)(s+2)}$$

The open loop poles are at -1, -2 and one finite zero at the origin i.e. 0. The other zero must be at infinity. For a positive feedback system, root loci will exit on the real axis from $+\infty$ to origin and -1 to -2 only.

Angle of assymptotes

$$\phi_n = \frac{2n\pi}{p-z}, \phi_0 = 0°, \phi_1 = \pi$$

Breakaway and Breaking points are obtained from,

$$\frac{d}{ds}\left[G_1(s)H_1(s)\right] = 0$$

$$s_1 = +\sqrt{2},$$

$$s_2 = -\sqrt{2}$$

i.e, $+\sqrt{2}$ is the breakaway point, and $-\sqrt{2}$ is the breakin point.

The complete root loci is drawn in fig 6.19 (b),

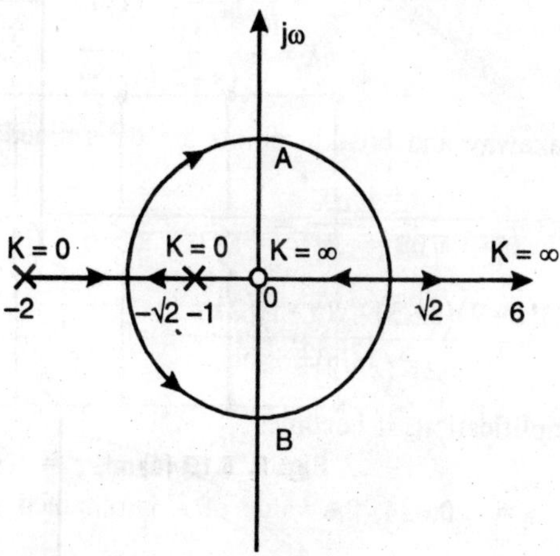

Fig. P. 6.19 (b)

Note that the points of intersection of the root loci with the imaginary axis, are at joints A and B i.e, $\pm j\sqrt{2}$. The value of gain can be determined by Routh's criterion, which is, say, $K_1 (= 3)$. Then, the stable region of gain is $0 < K < K_1$

In this range of gain, the roots of the closed loop poles are in the negative half of the s-plane. For the range of gain in $K_1 < K < \infty$, the closed loop poles lie in the positive half of the s-plane, the system is unstable.

Problem 6.20. Sketch the root locus plot of a negative feedback system, whose open loop transfer function is

$$\frac{K(s+2)(s+3)}{s(s+1)}$$

Solution:

1. Plot the open loop poles and zeros on the s-plane. Root loci exist on the negative real axis between 0 and -1 and -2 and -3.

2. The number of finite poles and zeros are the same. This means there is no asymptote in the complex region of the s-plane.

3. Determine the break away and breakin points. The characteristic equation of the system is

$$1+\frac{K(s+2)(s+3)}{s(s+1)} = 0$$

or,
$$K = -\frac{s(s+1)}{(s+2)(s+3)}$$

The breakaway and breakin points are determined as

$$\frac{dK}{ds} =$$

$$-\frac{2(s+1)(s+2)(s+3)-s(s+1)(2s+5)}{(s+2)^2(s+3)^2}=0$$

after simplification, it becomes
$$s = -\,0.634 \text{ and, } s = -2.366$$

At point $s = -\,0.634$, the value of K is obtained from equation

$$K = \frac{-s(s+1)}{(s+2)(s+3)} \text{ as } 0.0718.$$

Similary, at $s = -\,2.366$, $K = 14$

Note that $s = -0.634$ and $s = -2.366$ are actual break-away and break-in points.

Because $s = -\,0.634$ lies between two poles, it is a break away point.

4. Determine a sufficient number of points that satisfy the angle condition. The root locus is a circle with centre at -1.5, that passes through the break away and break in points.

The root locus is shown in figure below.

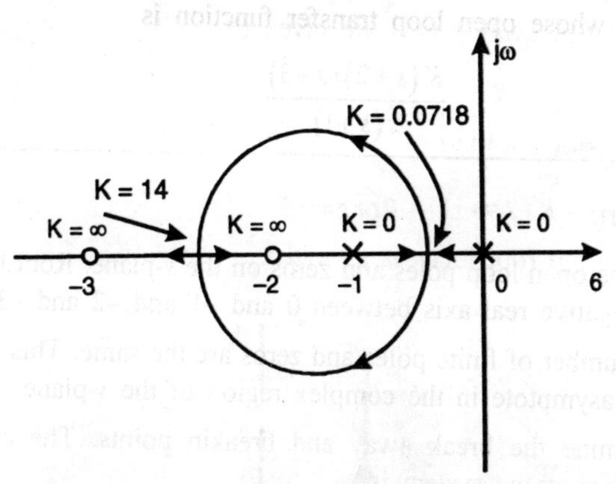

Fig. P. 6.20

Note that this system is stable for any positive value of K since all the root loci lie in the left half of the s plane.

Problem 21. Draw the root loci for the unity feedback control system with a forward gain $G(s)$ is given by,

(i) $G_1(s) = \dfrac{K}{s(s+1)}$

(ii) $G_2(s) = \dfrac{K}{s(s+1)(s+2)}$

Indicate the change in root locus if an equalizer is given by

$$G_e(s) = \left(\frac{1+0.2s}{1+0.04s}\right) \text{ is placed in cascade with } G_2(s)$$

Solution:

(i) Here, $G_1(s) = \dfrac{K}{s(s+1)}$

Poles are at, 0 and −1.

Zeros are at infinity (or $-\infty, \infty$)

The centroid (meeting points of asymptotes) is at,

$$\sigma = -\frac{1}{2} = -0.5$$

The angle of asymptotes,
$$\phi = \pm 90°$$

The break away points are given by

$$\frac{d}{ds}\left(\frac{K}{s(s+1)}\right) = 0$$

or, $\quad -K(2s+1) = 0 \text{ or } s = 0.5$

Hence the root locus is as under:

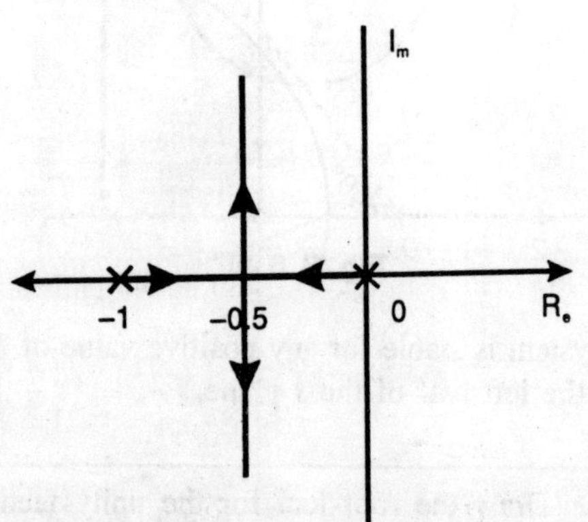

Fig. P. 6.21

(ii) Given $G_2(s) = \dfrac{K}{s(s+1)(s+2)}$

Finite poles one at, $s = 0, -1, -2$

There is no finite zeros.

The Centroid is given by $\sigma = \dfrac{-1-2}{3} = -1$

The angle of asymptotes are given by, $\pm 60°, 180°$

So, the breakaway points are given by

$$\frac{d}{ds}\left[\frac{K}{s(s+1)(s+2)}\right] = 0$$

$$\frac{d}{ds}\left[\frac{K}{s^3+3s^2+2s}\right] = 0$$

i.e., $\qquad -K\left[3s^2+6s+2\right]=0$

i.e., $\qquad s=\dfrac{-6\pm\sqrt{36.24}}{6}=\dfrac{-6\pm\sqrt{12}}{6}=-0.42,\ -1.57$

Using these given data root locus is constructed

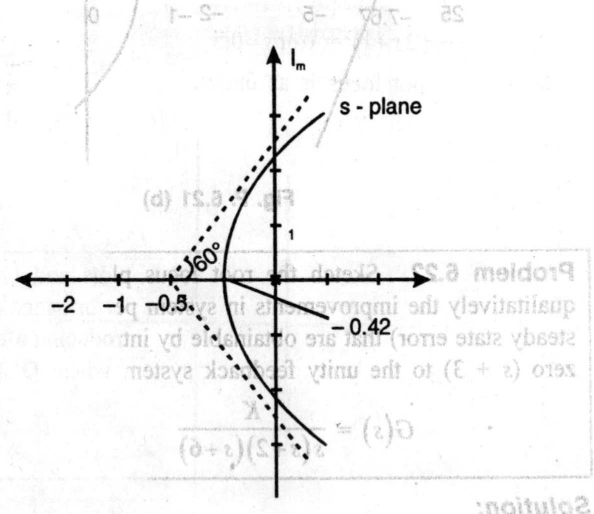

Fig. P. 6.21 (a)

With equalizer

The modified or Equalized gain is given by,

$$GH=\dfrac{K}{s(s+1)(s+2)}\dfrac{(1+0.2s)}{(1+0.04s)}$$

$$=\dfrac{5K(s+5)}{s(s+1)(s+2)(s+25)}$$

Finite poles are at, $\quad s=0,\ -1,\ -2,\ -25$

Finite zeroes are at, $s=-5$

The centroid $\qquad C=\dfrac{1-2-25+5}{3}=\dfrac{-23}{3}=-7.67$

The angle of asymptotes are given by $\pm60°,\ 180°$. Using these data, root locus will be given by,

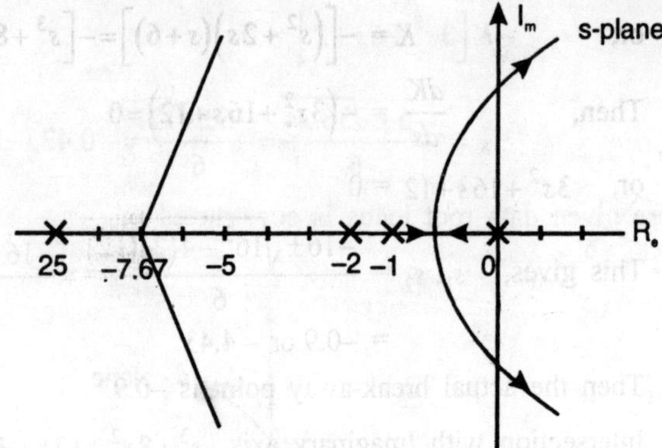

Fig. P. 6.21 (b)

Problem 6.22. Sketch the root locus plots and briefly explain qualitatively the improvements in system performance (stability and steady state error) that are obtainable by introducing a compensating zero $(s + 3)$ to the unity feedback system where OLTF is,

$$G(s) = \frac{K}{s(s+2)(s+6)}$$

Solution:

Given, $G(s).H(s) = \dfrac{K}{s(s+2)(s+6)}$

(a) The open loop pole are at: $s = 0, -2, -6$
  There is no finite zeros.
  Then the no. of branches = 3 and all terminating at ∞

(b) The real axis segments between 0 and -2 and between -6 and ∞ lie on the root locus.

(c) $\phi = \left(\dfrac{2q+1}{n-m}\right); n=3, m=0$

  $q = 0, 1, 2,\ldots$

  $\phi_0 = 60°, \phi_1 = 180°, \phi_2 = -60°$
  There are 3 asymptotes.

(d) The centroid, $-\sigma_A = \dfrac{-2-6}{3} = \dfrac{-8}{3} = -2.67$

(e) The break-away points.

$$K = -s(s+2)(s+6)$$

$$\text{or,} \qquad K = -\left[\left(s^2+2s\right)(s+6)\right] = -\left[s^3+8s^2+12s\right]$$

$$\text{Then,} \qquad \frac{dK}{ds} = -\left(3s^2+16s+12\right)=0$$

$$\text{or,} \quad 3s^2+16s+12 = 0$$

$$\text{This gives,} \quad s_1, s_2 = \frac{-16\pm\sqrt{16^2-4(3)(12)}}{6} = \frac{-16\pm10.58}{6}$$

$$= -0.9 \text{ or} -4.43$$

Then the actual break-away point is -0.9

(f) Intersection with Imaginary axis, $s^3+8s^2+12s+K=0$

$$
\begin{array}{c|cc}
s^3 & 1 & 12 \\
s^2 & 8 & K \\
s^1 & \dfrac{96-K}{8} & \\
s^0 & K &
\end{array}
\qquad
\begin{array}{l}
\dfrac{96-K}{8}>0 \\[2mm]
\therefore K<96 \\[1mm]
\text{and } K>0
\end{array}
$$

$$\text{For} \qquad K = 90°, \quad 8s^2+96 = 0$$

$$s^2 = -12, \quad s = \pm j2\sqrt{3}$$

Hence, the root locus branches meet the imaginary axis at $\pm j2\sqrt{3}$.

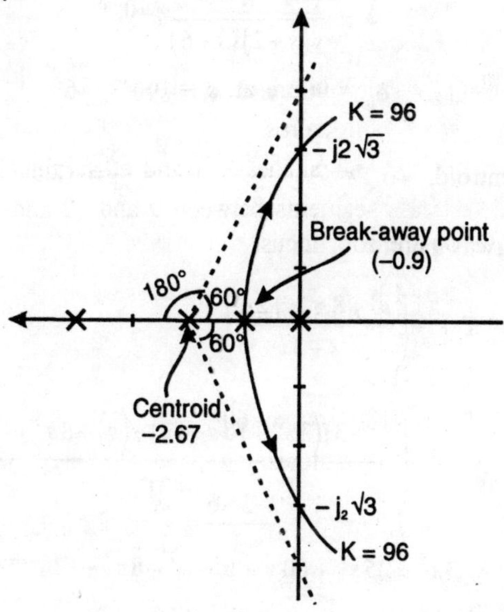

Fig. P. 6.22

Now if $H(s) = (s + 3)$

$$G(s).H(s) = \frac{K(s+3)}{s(s+2)(s+6)}$$

Now we have two branches.

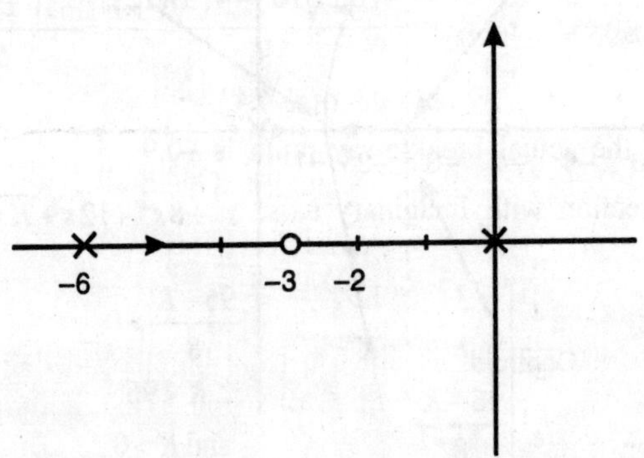

Fig. P. 6.22 (a)

The region between 0 and –2; –3 and –6 lie on root locus.
Now asymptotic angle,

$$\phi = \frac{(2q+1)180°}{3-1};q=0, 1$$

$$\phi_0 = 90° \text{ and } \phi_1 = -90°$$

The centroid, $-\sigma_A = \dfrac{(-2-6)-(-3)}{3-1} = \dfrac{-5}{2} = -2.5$

Break-away points:

$$K = -\left(\frac{s^3+8s^2+12s}{s+3}\right)$$

$$\frac{dK}{ds} = -\left(\frac{(s+3)\left(3s^2+16s+12\right)-\left(s^3+8s^2+12s\right)}{(s+3)^2}\right)$$

$$3s^3+25s^2+60s+36-s^3-8s^2-12s = 0$$

or, $2s^3+17s^2+48s+36 = 0$

Whose one value is $s = -1.17$.

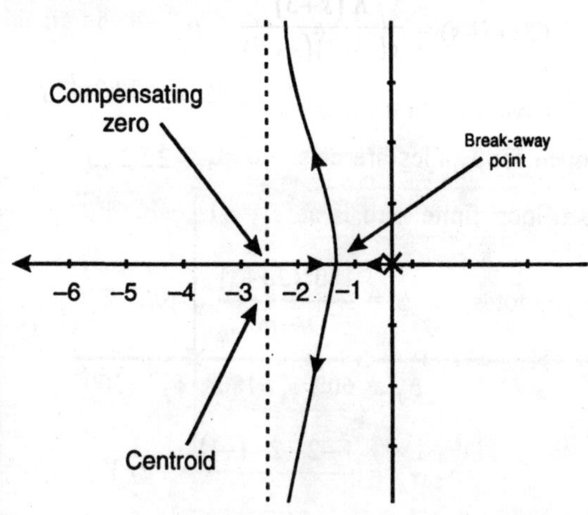

Fig. P. 6.22 (b)

The addition of zero presents results in

(1) All branches of root locus now lie completely in left half of
   s. plane. So K can be adjusted to any +ve value without causing
   instability.

(2) The complex root branches are now bent away from $j\omega$ axis

(3) By suitable adjustment of zero (compensating) along −ve real
   axis, the steady state error is within specified limits.

Problem 6.23.   The characteristic equation of a feedback control
system is

$$s^4 + 3s^3 + 12s^2 + (K-16)s + K = 0$$

Sketch the root locus plot for $0 \le K < \infty$ and show that the system
is conditionally stable (stable for only a range of gain. K) Determine
the range of gain for which the system is stable.

Solution:

The given characteristic equation is,

$$s^4 + 3s^3 + 12s^2 + (K-16)s + K = 0$$

or $\qquad 1 + \dfrac{K(s+1)}{s\left(s^3 + 3s^2 + 12s - 16\right)} = 0$

Now
$$\left(s^3 + 3s^2 + 12s - 16\right) = 0$$

$$(s-1)\left(s^2 + 4s + 16\right) = 0$$

$$s = 1, \ -2 \pm 2\sqrt{3}\,j$$

Hence, the open loop poles are at s = 0, 1, $-2 \pm 2\sqrt{3}$

And the open loop finite zero is at, $s = -1$.

Angle of asymptotes, $\quad \phi = \dfrac{180°(2q+1)}{n-m}, q = 0, 1, 2$

$$\phi_0 = 60°, \ \phi_1 = 180°, \ \phi_2, = 300°$$

Centroid, $\qquad -\sigma_A = \dfrac{1-2-2-(-1)}{3} = -2/3$

For the range of gain for the stability of the system:
Constructing Routh's Array

s^4	1	12	K
s^3	3	$K-16$	0
s^2	$\dfrac{52-K}{3}$	K	0
s^1	$\left[(K-16)\left(\dfrac{52-K}{3}\right)-3K\right]\Big/\left(\dfrac{52-K}{3}\right)$		
s^0	K		

For the system to be stable,

$$\frac{\left[\dfrac{(K-16)(52-K)}{3}-3K\right]}{\left(\dfrac{52-K}{3}\right)} > 0$$

$$K^2 - 59K + 832 < 0 \qquad \text{and} \quad 52 - K > 0$$

$$\Rightarrow \qquad (K-35.68)(K-23.31) < 0 \qquad K < 52$$

Hence for stability,
$$23.31 < K < 35.68$$

The Root-loci is drawn as follows:

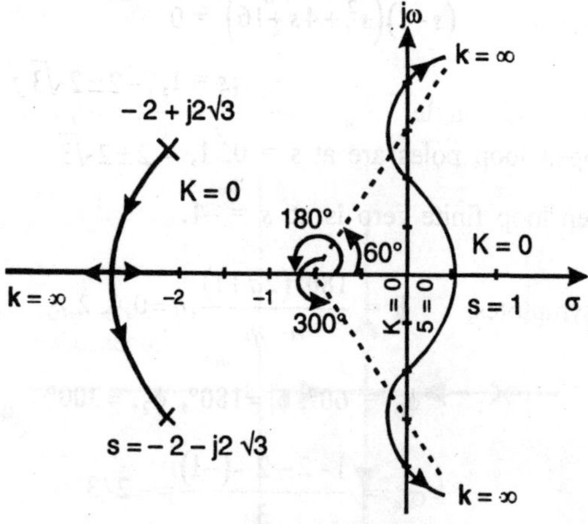

Fig. P. 6.23

Problem 6.24. The block diagram of Fig. P. 6.24 shows a feedback system using a proportional plus integral controller.
(a) Draw the root contours of the system for $a = 0$, 1, 2, and 4.
(b) Find the range of values of a for which the system is stable.
(c) Determine the damping ratio of the dominant roots for various values of a ($K_A = 5$) and show that the system is very sensitive to the value of a.

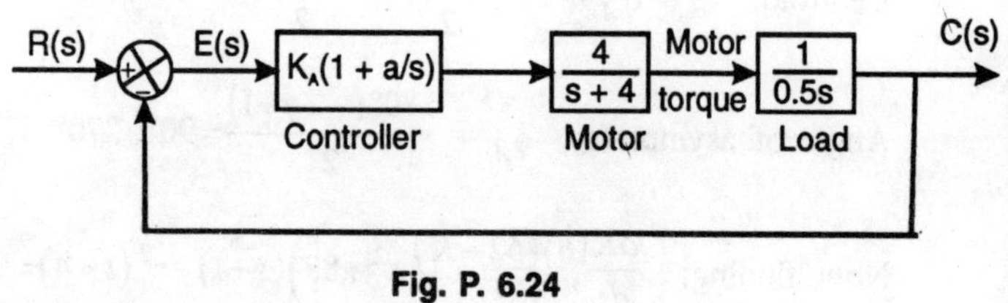

Fig. P. 6.24

Solution:

From the given block diagram;

(a)
$$G(s) = \frac{K_A\left(1 + a/s\right) \times 4 \times 1}{(s + 4) \times 0.5\,s}$$

For $a = 0$,
$$G(s) = \frac{8K_A}{s(s + 4)}$$

The open loop poles are at $s = 0, -4$. This is no finite zero.

Centroid, $\quad -\sigma_A = \dfrac{-4}{2} = -2$

Asymptotic angle, $\quad \phi_A = \dfrac{180°(2q+1)}{2}; q=0,1\, q = 90°,\, 270°$

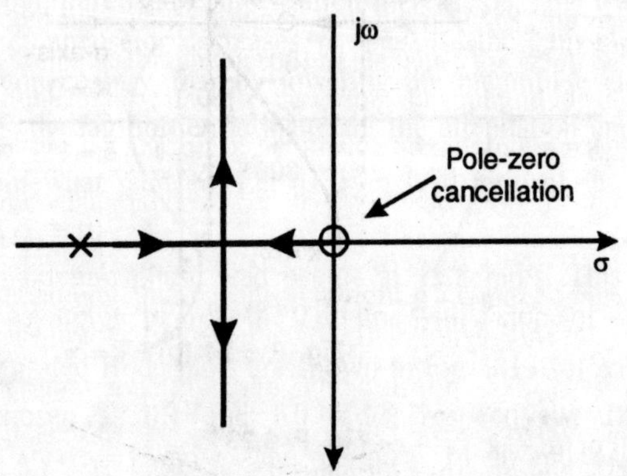

Fig. P. 6.24 (a)

For $a = 1$,

$$G(s) = \dfrac{8K_A(s+1)}{s^2(s+4)}$$

Centroid, $\quad -\sigma_A = \dfrac{-4+0-(-1)}{2} = \dfrac{-3}{2}$

Angle of asymptotes, $\quad \phi_A = \dfrac{180°(-2q+1)}{2} = 90°,\, 270°$

Now, finding: $\dfrac{dK}{ds}$, $\quad \dfrac{dK}{ds} = \left(3s^2+8s\right)(s+1) - s^2(s+4) = 0$

or, $\quad 2s^3 + 7s^2 + 8s = 0$

Then $\quad s = 0, \left(\dfrac{-7 \pm j\sqrt{31}}{4}\right)$

Since $\left(\dfrac{-7 \pm j\sqrt{31}}{4}\right)$ does not satisfy the angle criterion hence, it is not the break-away point.

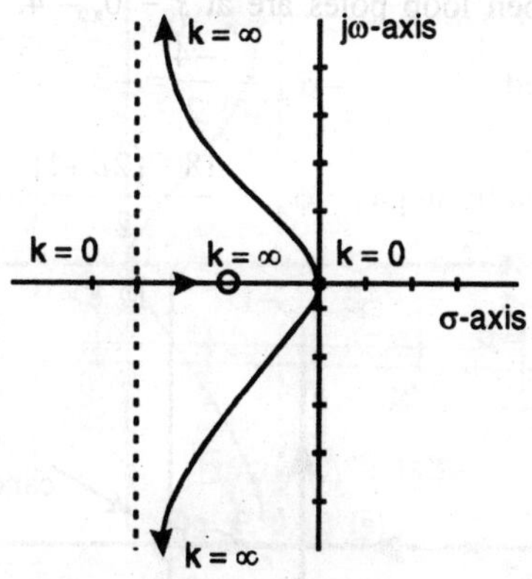

Fig. P. 6.24 (b)

For $a = 2$

The open loop T.F. $G(s) = \dfrac{8K_A(s+2)}{s^2(s+4)}$

Open loop poles are at, $\qquad s = 0,\, 0,\, -4$

and open loop zero is at $\qquad s = -2$ The Centroid,

$$-\sigma_A = \frac{-4+2}{2} = -1$$

The angle of asymptotes, $\qquad \phi_A = \dfrac{180°(2q+1)}{2};\ q = 0, 1$

$$= 90°,\ 270°$$

Now, for break-away points; $\qquad \dfrac{dK}{ds} = 0$

which gives $\qquad 2s^3 + 10s^2 + 16s = 0$

$$s = 0,\ \text{and}\ s = -2.5 \pm j1.32$$

Therefore, $s = 0$ is the break-away point, as the other points do not lie on the root locus, hence they are not the break-away point The required-root locus: For $a = 4$,

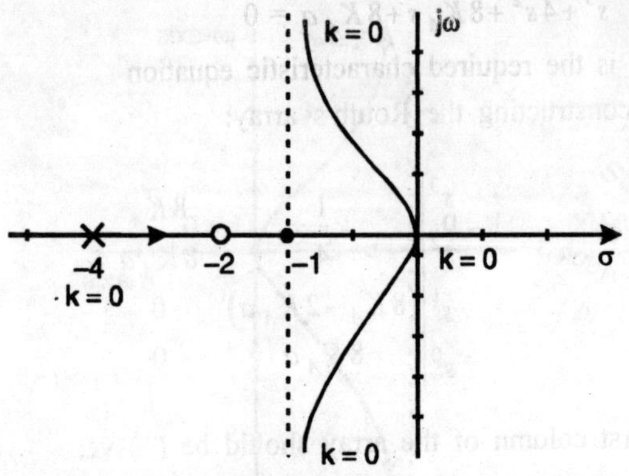

Fig. P. 6.24 (c)

$$G(s) = \frac{8K_A}{s^2}$$

The open loop poles are at,　　　　$s = 0, 0$
There is no finite zeros.

Then, angle of asymptotes,　　　$\phi_A = 90°, 270°$
Centroid,　　　　　　　　　　　$-\sigma_A = 0$
Then the required root locus is,

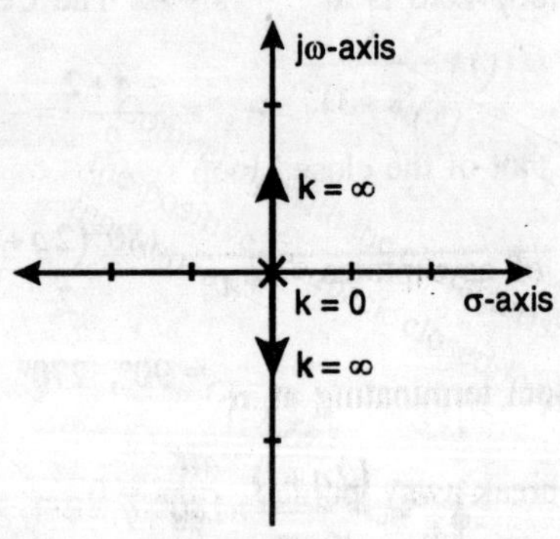

Fig. P. 6.24 (d)

The characteristic equation,

$$1 + G(s) = 0$$

$$1 + \frac{8K_A(s+a)}{s^2(s+4)} = 0$$

or, $\quad s^3 + 4s^2 + 8K_A s + 8K_A a = 0$

which is the required characteristic equation

Now constructing the Routh's array:

$$
\begin{array}{c|cc}
s^3 & 1 & 8K_A \\
s^2 & 4 & 8K_A a \\
s^1 & (8K_A - 2K_A a) & 0 \\
s^0 & 8K_A a & 0
\end{array}
$$

The first column of the array should be (+) ve,

Then, $K_A > 0$, $a > 0$ and $8K_A - 2K_A a > 0$ i.e. $a < 4$

Therefore for stability, $0 < a < 4$ **Ans.**

(c) $\qquad K_a = \underset{s \to 0}{\text{Lim}}\ s^2 G(s) = 2K_A a$

Then $\qquad\qquad e_{ss} = \left(\dfrac{1}{2K_A a} \right)$ **Ans.**

Problem 6.25. The open loop T.F. of a unity feedback control system is given by $G(s) = \dfrac{K}{s(s+3)^2}$

Draw the root locus plot of the closed loop system for (+) ve value of K.

Solution:

There are three root loci terminating at α

Angle of asymptotes, $\qquad \phi_A = \dfrac{(2q+1)}{n-m} \times 180°; q = 0, 1, 2$

$$= 60°,\ 180°,\ 300°$$

$$\sigma_A\ (\text{Centroid}) = \frac{-3-3}{3} = -2$$

For breakaway points $\qquad \dfrac{dK}{ds} = 0$

$$\frac{-d}{ds}\left(s^3+6s^2+9s\right) = 0$$

$$-\left(3s^2+12s+9\right) = 0$$

or $$s^2+4s+3 = 0$$

or $$(s+1)(s+3) = 0$$

or $$s = -1, \ s = -3$$

The characteristic equation is

$$1 + G(s) = 0$$

or $$s^3+6s^2+9s+K = 0$$

Constructing the Routh array for the characteristic equation, we get.

$$
\begin{array}{c|cc}
s^3 & 1 & 9 \\
s^2 & 6 & K \\
s^1 & \dfrac{54-K}{6} & 0 \\
s^0 & K &
\end{array}
$$

For stability $0 < K < 54$.

Critical condition $K = 54$

To find the frequency of oscillation at this condition we form the auxiliary

equation. $$6s^2+54 = 0$$

or $$-6\omega^2+54 = 0$$

or $$\omega = 3 \ \text{rad/sec.}$$

The root locus plot for the system is given in Fig. P.6.25.

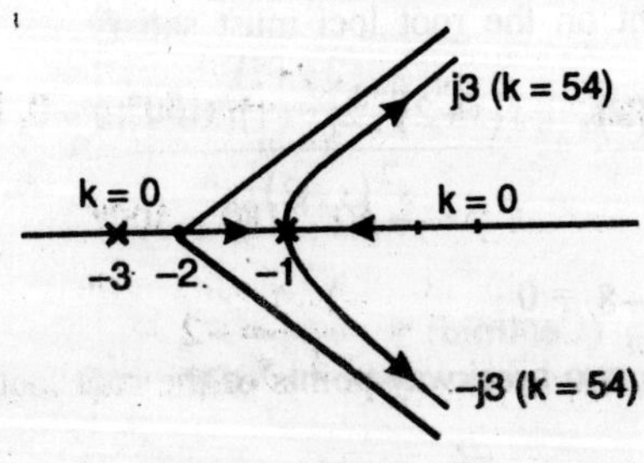

Fig. P. 6.25

Problem 6.26. Consider the second-order equation

$$s(s+2)+K(s+4)=0$$

Draw the root locus.

Solution:

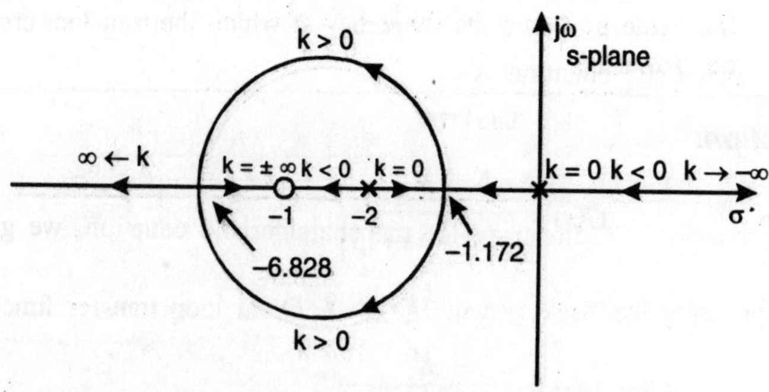

Fig. P. 6.26. Root Loci of s (s + 2) + K (s + 4) = 0

Given
$$s(s+2)+K(s+4)=0$$

The root loci of above equation are drawn as shown in the figure above for $-\infty < K < \infty$

It can be proved the complex portion of the root loci is a circle. The two breakaway points are all on the real axis, one between 0 and −2 and the other between − 4 and − ∞.

Now
$$G_1(s)H_1(s) = \frac{s+4}{s(s+2)}$$

The breakaway point on the root loci must satisfy

$$\frac{dG_1(s)H_1(s)}{ds} = \frac{s(s+2)-2(s+1)(s+4)}{s^2(s+2)^2}=0$$

or
$$s^2+8s+8 = 0$$

Solving we get, the two breakway points of the root loci at $s = -1.172$ and −6.828.

The figure above shows the two breakaway points are all on the RL (Root Locus)

Problem 6.27. A unity feedback control system has an open-loop transfer function

$$G(s) = \frac{K}{s\left(s^2+4s+13\right)}$$

Draw the root locus plot of the system by determining the following:
 (i)   Centroid, number and angle of asymptotes.
 (ii) Angle of departure of root loci from the poles.
 (iii) Breakaway point if any
 (iv) The value of K and the frequency at which the root loci cross the $j\omega$-axis.

Solution:

Given $\qquad G(s) = \dfrac{K}{s\left(s^2+4s+13\right)}$

For the unity feedback system $H(s) = 1$. Open loop transfer function

is, $\qquad G(s)\ H(s) = \dfrac{K}{s\left(s^2+4s+13\right)}$

There are no zeroes. Only three poles at $s = 0$ and $s = -2 \pm 3j$

Now, $-\sigma_A$ (Centroid) $= \dfrac{\sum \text{real part of poles} - \sum \text{real part of zeros}}{\text{number of poles} - \text{number of zero}}$

$$= \frac{-2-2}{3} = \frac{-4}{3}$$

Angle of asymptotes, $\qquad \phi_A = \dfrac{(2q+1)180°}{\text{number of poles} - \text{number of zeros}}$

where $q = 0, 1, 2 \therefore \qquad \phi_A = 60°,\ 180°,\ 300°$

characteristic equation is,

$$1 + G(s)\ H(s) = 0$$

or, $\qquad s^3 + 4s^2 + 13s + K = 0$

$$\frac{dK}{ds} = 0 = 3s^2 + 8s + 13$$

Solving quadratic equation, we get

$$s = -1.3 \pm j1.6$$

Since both the points does not satisfy angle criterion. So there is no breakway point.

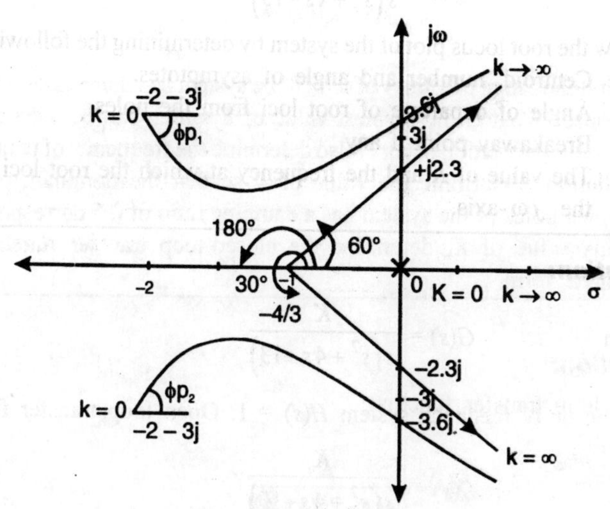

Fig. P. 6.27

At $\qquad\qquad\qquad s = -2 + 3j$

Angle of departure, $\quad \phi P_1 = 180° - (180° - 56.3°) - 90° = -33.7°$

At $\qquad\qquad\qquad s = -2 - 3j$

Angle of departure, $\quad \phi P_2 = +33.7°$

From the characteristic equation, constructing Routh Array,

$$
\begin{array}{c|cc}
s^3 & 1 & 13 \\
s^2 & 4 & K \\
s^1 & 52-K & 0 \\
s^0 & K & 0
\end{array}
$$

The root locus crosses the $j\omega$ axis for $52 - K = 0$, i.e. $K = 52$ at frequency given by auxiliary equation

$$4S^2 + 52 = 0$$

$$\omega = \pm j3.6$$

Problem 6.28.   Draw the root locus plot of a unity feedback system with an open-loop transfer function.

$$G(s) = \frac{K}{s(s+2)(s+4)}$$

Find the range of values of which the system has damped oscillatory response. What is the greatest value of K which can be used before continous oscillation occur? Also determine the frequency of continous oscillation. Determine the value of K so that the dominant pair of complex poles of the system has a damping ratio of 0.5 corresponding to this value of K, determine the closed-loop transfer function in factored form.

Solution:

Open loop transfer function,

$$G(s) = \frac{K}{s(s+2)(s+4)}$$

$$-\sigma_A \,(\text{Centroid}) = \frac{-2-4}{3} = -2,$$

$$\phi_A \,(\text{angle of asymptotes}) = \frac{-(2q+1)180°}{3} = 60°, 180°, 300°$$

From the characteristic equation

$$1 + G(s) = 0$$

$$1 + \frac{K}{s(s+2)(s+4)} = 0$$

or
$$K = -\left(s^3 + 6s^2 + 8s\right)$$

$$\frac{dK}{ds} = 0 = 3s^2 + 12s + 8$$

given
$$s = 3.15, -0.84$$

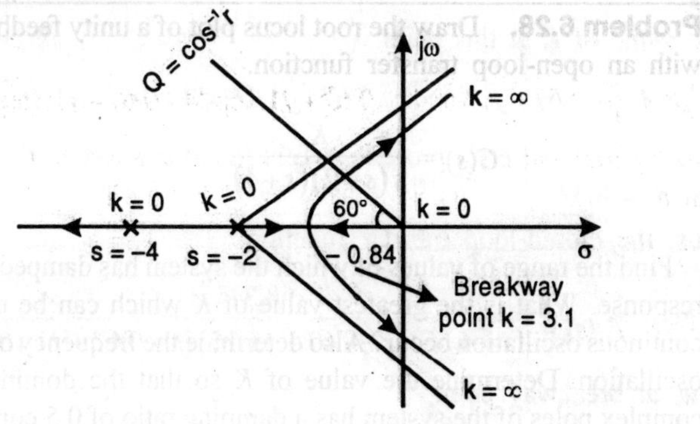

Fig. P. 6.28. Root locus plot of $\left[1+\dfrac{K}{s(s+2)(s+4)}\right]$

Now, as on the segment between 0 and –2, the two real roots branches are oppositely directed, there should be breakaway point on this segment.

Hence, $s = -0.84$ is the actual breakaway point. Now the root-locus sketched is drawn in the above figure.

Constructing Routh Array from characteristic equation.

$$
\begin{array}{c|cc}
s^3 & 1 & 8 \\
s^2 & 6 & K \\
s^1 & \dfrac{48-K}{6} & 0 \\
s^0 & K & 0
\end{array}
$$

The root locus branches will intersect the imaginary axis at a value of K given by $\quad 48 - K = 0$

or $\qquad\qquad\qquad K = 48$

The auxiliary equation

$$6s^2 + 48 = 0$$

or $\qquad\qquad s = \pm\sqrt{8} = \pm j2.83$

Gives frequency of continued oscillation at

$$\omega = 2.83 \text{ rad/sec.}$$

Now, a damping line making an angle $\theta = \cos^{-1}$ with negative real axis.

By trial and error procedure, we find that dominant roots are at

$$s_{12} = -0.67 \pm j1.16$$

The value of K at this point is

$$K = \{|-0.67+j1.16| \times |2-0.67+j1.16| \times |4-0.67+j1.16|\} = 8.34$$

Also, by trial and error procedure, it is found at $s = 4.56$ the open-loop gain $K = 8.34$

Thus, the closed-loop transfer function,

$$T\,(s) = \frac{8.31}{(s+4.56)(s+0.67+j1.16)(s+0.67-j1.16)}$$

Now, at breakway point,

$$s = -\,0.84$$

The value of $\qquad K = |-0.84| \times |1.16| \times |3.16| = 3.1$

Thus, range of values of K for the system having oscillatory response is $3.1 < K < 48$

Problem 6.29.   A unity feedback system has an open-loop transfer function

$$G(s) = K(s+1)/s(s-1)$$

Sketch the root locus plot with K as a variable parameter and show that the loci of complex roots are part of a circle with $(-1, 0)$ as centre and radius $= \sqrt{2}$.

   Is the system stable for all values of K? If not, determine the range of K for stable system operation. Find also the marginal value of K which causes sustained oscillations and the frequency of these oscillations.

From the root locus plot, determine the value of K such that the resulting system has a settling time of 4 sec. What are the corresponding values of the roots?

Solution:

Given $\qquad\qquad\qquad G(s) = \dfrac{K(s+1)}{s(s-1)}$ $\qquad\qquad\qquad\qquad$ (1)

Poles are at $s = 0, 1$ and a zero at $s = -1$.

Characteristic equation is,

$$1 + G(s)\ H(s) = 0$$

$$1 + \frac{K(s+1)}{s(s-1)} = 0$$

$$s(s-1) + K(s+1) = 0$$

$$s^2 + s(K-1) + K = 0 \qquad (2)$$

From equation (2),

$$K = -\frac{s^2 - s}{(s+1)} = \frac{-s(s-1)}{(s+1)}$$

$$\frac{dK}{ds} = 0 = \frac{-\left[(s+1)(2s-1) - (s^2 - s)(1)\right]}{(s+1)^2}$$

$$= \frac{-\left[2s^2 - s + 2s - 1 - s^2 + s\right]}{(s+1)^2}$$

$$= \frac{-\left[s^2 + 2s - 1\right]}{(s+1)^2}$$

which gives, $\qquad s = \sqrt{2} - 1, \ \sqrt{2} - 1$

The required root locus is plotted below . It gives the radius of circle as $\sqrt{2}$ and centre $(-1, 0)$

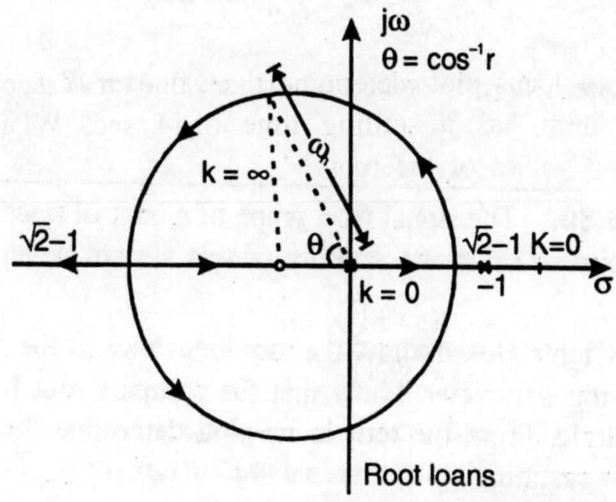

Fig. P. 6.29

Constructing Routh Array from characteristic equation (2)

$$
\begin{array}{c|cc}
s^2 & 1 & K \\
s^1 & K-1 & 0 \\
s^0 & K & 0
\end{array}
$$

for stability all elements in the first column should be positive i.e.

$K - 1 > 0$ or $K > 1$, $K > 0$

Hence for stability $\infty > K > 1$, i.e. system is not stable for all values of K.

Now auxiliary equation is, $s^2 + K = 0$

$$s^2 = -K = -1 \ [\text{Since for sustained oscillation}$$
$$K - 1 = 0, \text{ or } K = 1]$$

Which gives $s = \pm j1$

Hence, ω (frequency for oscillation) $= 1$ rad/sec

Given t_{ss} (setling time) $= \dfrac{4}{r\,\omega_n} = 4$ sec

$\Rightarrow$ $r\,\omega_n = 1$

From the root locus $\omega_d = \sqrt{2}$ rad/sec.

and $\omega_d = \sqrt{1-r^2}\ \omega_n$

or $\sqrt{2} = \sqrt{1-r^2}\ 1/r$

$$r = \frac{1}{\sqrt{3}}, \ \omega_n = \sqrt{3} \ \text{rad/sec.}$$

Now, roots are, $s_{1,2} = -r\,\omega_n \pm \sqrt{1-r^2}\ \omega_n = -1 \pm \sqrt{2}\,j$

Problem 6.30.   The signal flow graph of a control system is shown in Fig. Comment upon the stability of this system when the switch s is open.

With the switch s closed, draw the root locus plot of the system with a as a varying parameter. Show that the complex-root branches are part of a circle. From the root locus plot, determine the value of a such that the resulting system has a damping ratio of 0.5. For this value of a, find the overall transfer function in factored form.

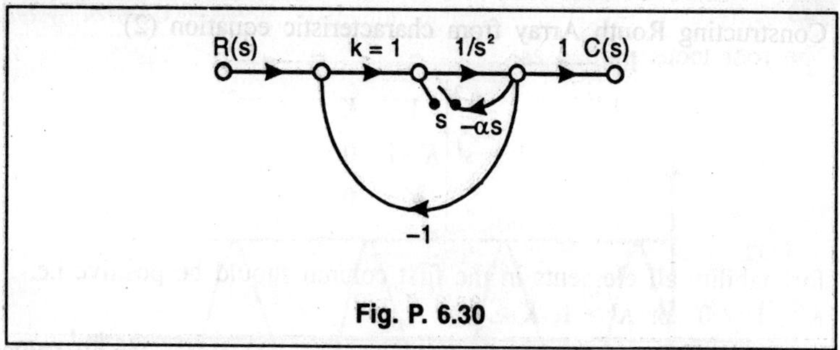

Fig. P. 6.30

Solution:

Case a when *s* is open,

$$T(s) = \frac{C(s)}{R(s)} = \frac{1/s^2}{1+\dfrac{1}{s^2}} = \frac{1}{1+s^2}$$

Pole are at $s = \pm j$

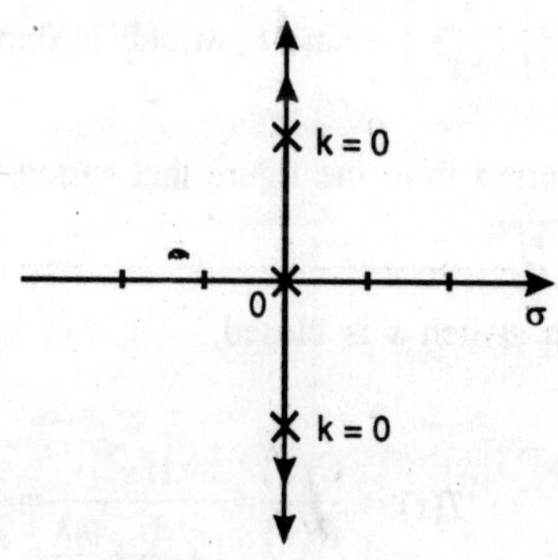

Fig. P. 6.30

$$\phi(\text{angle of Asymptotes}) = \frac{180°(2q+1)}{n-m} = \frac{180°(2q+1)}{2} = 90°, 270°$$

$$\text{Centroid } \sigma_A = 0$$

The root locus plotted as

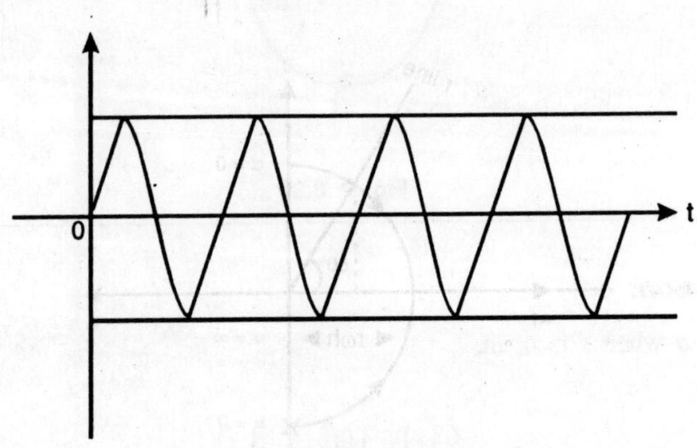

Fig. P. 6.30 (b)

Impulse response of the given transfer function = $\mathcal{L}^{-1} T(s)$

$$\mathcal{L}\left[\frac{1}{1+s^2}\right] = \sin\omega t, \text{ which is drawn above.}$$

It can be inferred from the figure that system is limitedly stable when switch s is open.

Case b when switch s is closed.

$$T(s) = \frac{C(s)}{R(s)} = \frac{1/s^2}{1+\dfrac{1}{s^2}+\dfrac{\alpha s}{s^2}} = \frac{1}{s^2+\alpha s+1}$$

Poles are given by the equation,

$$s^2+\alpha s+1 = 0$$

Rearranging, $$1+\frac{\alpha s}{1+s^2} = 0$$

The root locus of the equation $1 + \dfrac{\alpha s}{1+s^2} = 0$ with α varying parameter is drawn below.

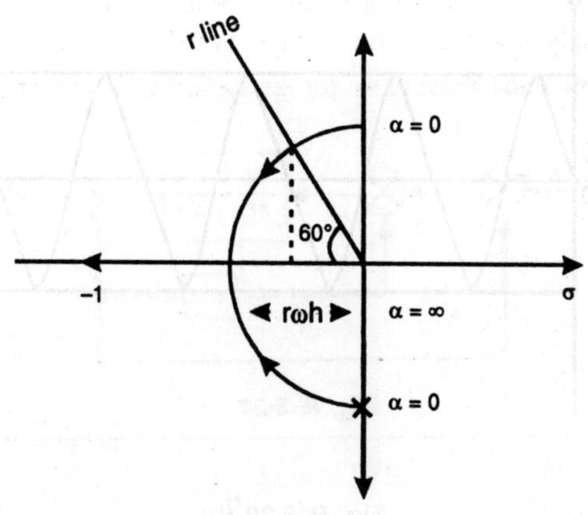

Fig. P. 6.30 (c)

Now from the root locus

$$\cos 60° = \frac{r\omega_n}{1}$$

$$0.5 = r\omega_n$$

As, $\qquad \alpha = 2\, r\omega_n = 2 \times 0.5 = 1$

Now. $\qquad s^2 + \alpha s + 1 = 0$

When $\qquad \alpha = 1$, then $s^2 + s + 1 = 0$

or $\qquad (s+0.5-j0.866)(s+0.5+j0.866) = 0$

Hence, the overall transfer functionin factored form is,

$$\frac{1}{(s+0.5-j0.866)(s+0.5+j0.866)} = 0.$$

Problem 6.31. The block diagram of a control system is shown in Fig. Draw the root locus plot of the system with α as varying parameter.

(a) Determine the steady-state error to the unit-ramp input, damping ratio and setling time for the system without derivative feedback, i.e., $\alpha = 0$

(b) Discuss the effect of derivative feedback on transient as well as steady-state behaviour of the system assuming $\alpha = 0.2$.

(c) Determine the value of α for the system to be critically damped.

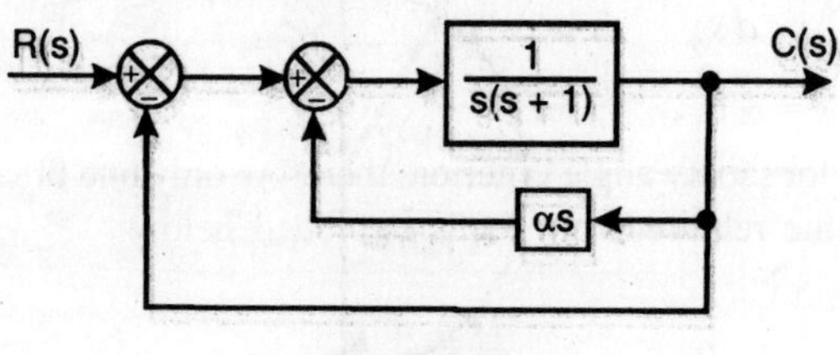

Fig. P. 6.31

Solution:

The transfer function for the given block diagram,

$$T(s) = \frac{1}{1+\dfrac{1}{s^2+(\alpha+1)s}} = \frac{1}{s^2+(\alpha+1)s+1} \qquad (1)$$

The poles can be determined by the characteristic equation

$$s^2+s(\alpha+1)+1 = 0$$

In reduced form, $\qquad 1+\dfrac{s\alpha}{s^2+s+1} = 0$

Now the poles of equation (1) are determined by root locus of $\dfrac{s\alpha}{s^2+s+1}$

$$\Rightarrow \qquad G'(s)H'(s) = \frac{s\alpha}{s^2+s+1}$$

Zeros, $s = 0$

Poles, $s = \dfrac{-1\pm j\sqrt{3}}{2} = -0.5\pm j0.866$

Angle of asymptote, $\phi = \dfrac{180°(2q+1)}{1} = 180°$ (Since $q=0$)

$$-\sigma_A \ (\text{Centroid}) = \frac{-\dfrac{1}{2}-\dfrac{1}{2}}{2} = -\frac{1}{2}$$

For breakaway point, equating

$$\frac{d\alpha}{ds}=0 = \frac{d}{ds}\frac{\left(s^2+s+1\right)}{s} = \frac{(2s+1)s-\left(s^2+s+1\right)}{s^2}=0$$

$$\Rightarrow \qquad s = \pm 1$$

$s = +1$ does not satisfy angle criterion, therefore only one breaway point i.e. $s = -1$ the required root locus is plotted below

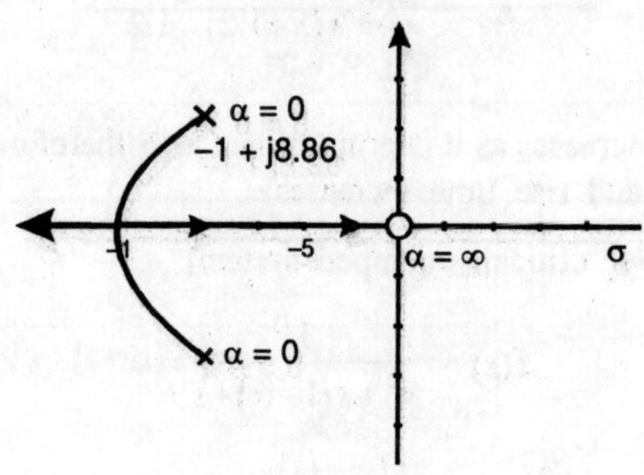

Fig. P. 6.31 (a)

(a) Given $\alpha=0$

and
$$T(s) = \frac{1}{s^2+s+1} = \frac{1(s+1)}{1+\dfrac{1}{s(s+1)}}$$

which gives, $\quad G(s) = \dfrac{1}{s(s+1)}$

$$H(s) = 1, \ = \ \omega_n^2 \ = 1$$

$$\Rightarrow \qquad 2r\omega_n = 1$$
$$r = 0.5$$

Type-1 system

$$K_v = \underset{s\to 0}{\text{Lim}}\, s\, G(s)=1$$

$$e_{ss} \text{ (Steady-state error)} = \frac{1}{K_v} = 1$$

(b) Given $\alpha = 0.2$

$$T(s) = \frac{1}{s^2 + 1.2s + 1}$$

$$\omega_n^2 = 1$$

$$\Rightarrow \qquad 2r\omega_n = 1.2, \quad r = 0.6$$

Now, $\qquad G(s) = \dfrac{1}{s(s+1.2)}$

$$K_v = \lim_{s \to 0} \frac{s \times 1}{s(s+1.2)} = \frac{1}{1.2}$$

$$e_{ss} = 1.2$$

Here, r increases as it is proportional to α therefore, settling time reduces and rise time increases.

(c) $r = 1$ [For critically damped system]

Now, $\qquad T(s) = \dfrac{1}{s^2 + s(1+\alpha) + 1}$

Then $\qquad \omega_n^2 = 1$

$$2r\omega_n = (1+\alpha)$$

$$\Rightarrow \qquad \alpha = 1$$

Problem 6.32. A unity feedback system has an open-loop transfer function

$$G(s) = K/s^2(s+2)$$

(a) By sketching a root locus plot, show that the system is unstable for all values of K.

(b) Add a zero at $s = -a\,(0 \le a < 2)$ and show that the addition of zero stabilizes the system.

(c) If $a = 1$, sketch the root locus plot and determine approximately the value of K which gives the greatest damping ratio for the oscillatory mode. Find also the value of this damping ratio and the corresponding undamped natural frequency.

Solution:

Given,
$$G(s) = \frac{K}{s^2(s+2)}$$

Two poles are $s = 0$ and one at $s = -2$

$$-\sigma_A \,(\text{Centroid}) = -\frac{2}{3}$$

$$\phi_A \,(\text{angle of asymptotes}) = \frac{(2q+1)180°}{3} = 60°, 120°, 180°$$

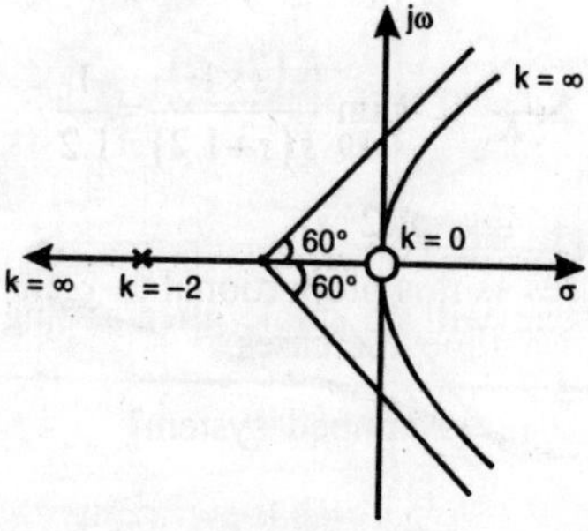

Fig. P. 6.32

Characteristic equation is,

$$1 + G(s) = 0$$
$$\Rightarrow \qquad s^3 + 2s^2 + K = 0$$

$$K = -\left(s^3 + 2s^2\right)$$

$$\frac{dK}{ds} = 0 = -\left(3s^2 + 4s\right)$$

Which gives,
$$s = \frac{-4}{3}, 0$$

Since no values of s statisfies the angle angle criterion so there is no breakway point

(b) now $G_1(s) = G(s)\,(s + a) = \dfrac{K(s+a)}{s^2(s+2)}$

 Characteristic equation is, $1 + G_1(s) = 0$

or
$$s^3 + 2s^2 + Ks + Ka = 0$$

Constructing Routh Array,

$$
\begin{array}{c|cc}
s^3 & 1 & K \\
s^2 & 2 & Ka \\
s^1 & \dfrac{2K - 2Ka}{2} & 0 \\
s^0 & Ka & 0
\end{array}
$$

For stability, $Ka > 0$ and $\dfrac{K(2-a)}{2} > 0$

$\Rightarrow a > 0$ and $a < 2$

Thus, the system will be stable after adding a zero at $s = -a$, where $0 \le a \le 2$

(c)   If $a = 1$

$$G(s) = \frac{K(s+1)}{s^2(s+2)}$$

Two poles at $s = 0$ and one at

$$s = -2 \text{ one zero at } s = -1.$$

$$-\sigma_A(\text{centroid}) = \frac{-2-(-1)}{2} = -0.5$$

Angle of asymptotes, $\phi_A = \dfrac{(2q+1)180°}{3} = 90°, 270°$

For breakaway point, $\dfrac{dK}{ds} = 0$

which gives, $s\left(2s^2 + 5s + 4\right) = 0$

or, $s = 0,\ s = -2.5 \pm j\,0.67$

Now $s = -2.5 \pm j\,0.67$ does not lie on root locus, so it cannot be breakaway point. So breakaway point at $s = 0$

The root locus is drawn in the figure.

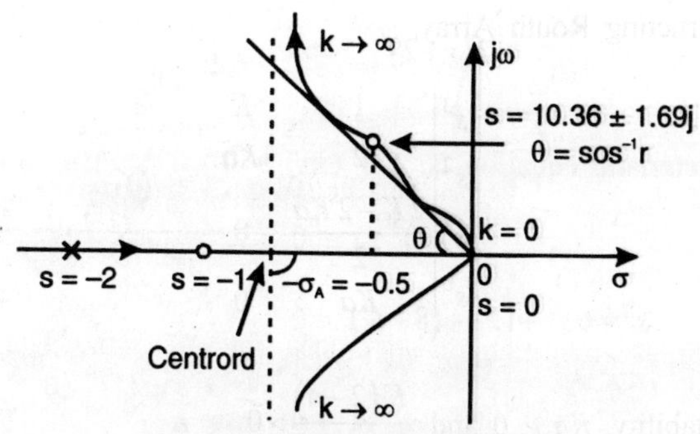

Fig. P. 6.32 (a)

Problem 6.33. Open-loop transfer function of a unity feedback system is

$$G(s) = K/(s+2)^3$$

Sketch the root locus plot and determine the following:

(a) Static loop sensitivity for which the root locus crosses the $j\omega$-axis and the corresponding frequency of sustained oscillations.

(b) The position error constant corresponding to a damping ratio of 0.5. Also determine the peak overshoot, time to peak overshoot and settling time considering the effect of dominant poles only.

(Note: The static loop sensitivity is defined to be the gain in pole zero form.)

Solution:

Given

$$G(s) = \frac{K}{(s+2)^3}$$

There poles are at $s = -2$

Angle of asymptotes,

$$\phi = \frac{(2q+1)180°}{3} = 60°, 180°, 300°$$

$$-\sigma_A (\text{Centroid}) = \frac{-6}{3} = -2$$

For brekaway points, equating $\dfrac{dK}{ds} = 0$

$$\Rightarrow \quad \dfrac{dK}{ds} = 3(s+2)^2 = \Rightarrow$$

$$s = -2$$

The characteristic equation is, $1 + G(s) = 0$

$$\Rightarrow \quad (s+2)^3 + 1 = 0$$

$$\Rightarrow \quad s^3 + 6s^2 + 12s + (8+K) = 0$$

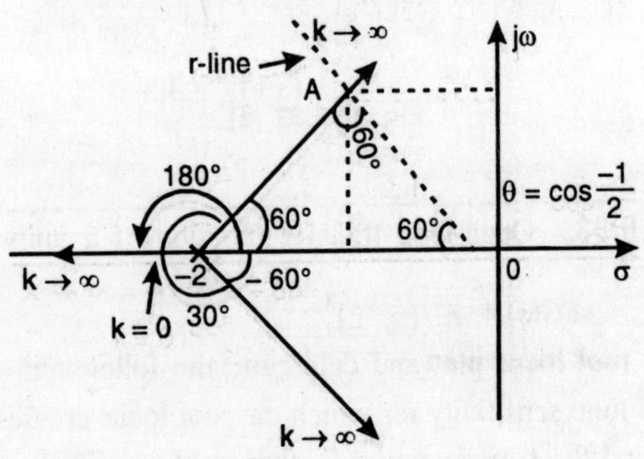

Fig. P. 6.33

Constructing Routh array,

$$
\begin{array}{c|cc}
s^3 & 1 & 12 \\
s^2 & 6 & 8+K \\
s^1 & 64-K & 0 \\
s^0 & 8+K &
\end{array}
$$

The root locus crosses the $j\omega$-axis at $64 - K = 0$

or $\qquad\qquad K = 64$

at a frequency of $\quad 6s^2 + 72 = 0$

$$s = \pm\sqrt{72/6} = \pm j3.46$$

$$\Rightarrow \qquad \omega = 3.46 \text{ rad/sec.}$$

(b) $r = 0.5 \quad \theta = \cos^{-1} r = 60°$

Now drawing r-line on root locus. From, the root locus it can be inferred that $\triangle OAP$ is equilateral. Hence, the point A is $s = -1 + \sqrt{3}\, j$

Dominant equation, $\qquad s^2 + 2r\omega_n s + \omega_n^2 = 0$

$$r = 0.5$$

Dominant poles are, $\qquad\qquad s = -1 \pm \sqrt{3}\, j$

Yields, $\left[s - \left(-1 + j\sqrt{3}\right)\right]\left[s - \left(-1 - j\sqrt{3}\right)\right] = 0$

$$\Rightarrow \qquad (s+1)^2 + 3 = 0$$

$$\Rightarrow \qquad s^2 + 2s + 4 = 0$$

Which gives $2r\omega_n = 2$

or, $\qquad \omega_n = 2$

Now peak overshoot

$$M_p = 100\, e^{-\sqrt{r}/\sqrt{1-r^2}} = 16.2\%$$

$$t_p = \frac{\pi}{\omega_n \sqrt{1-r^2}} = 1.81\,\text{sec}$$

Setlling time, $\qquad t_p = \frac{4}{r\omega_n} = 4\,\text{sec}$

CHAPTER 7

State Variable Analysis

Problem 7.1.   Write the state variable formulation of the parallel RLC network shown in fig. 7.1 below.

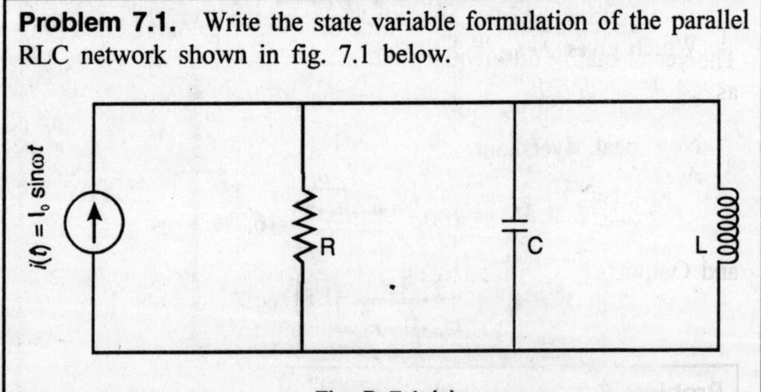

Fig. P. 7.1 (a)

Solution:

Redrawing the network given,

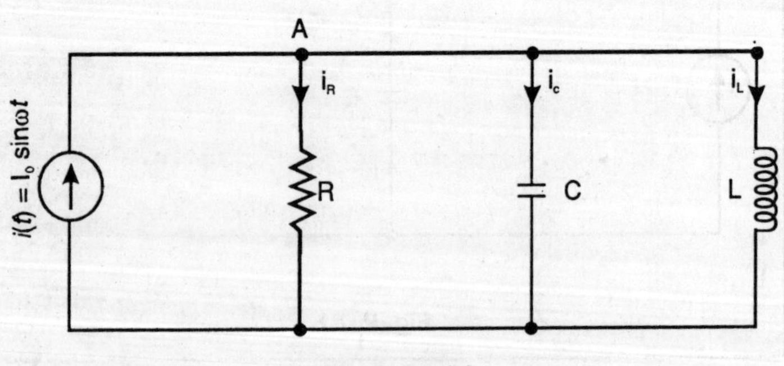

Fig. P. 7.1 (b)

Applying KCL at node A,

$$i_R + i_C + i_L = i$$

or

$$C\frac{dv}{dt} + \frac{1}{R}v + \frac{1}{L}\int v(C)dt = I_0 \sin\omega t$$

Differentiating both sides and dividing by C,

$$\frac{d^2v}{dt^2} + \frac{1}{RC}\frac{dV}{dt} + \frac{1}{LC}V = \frac{\omega}{C}I_o \cos\omega t$$

Now, Let us choose,

$$v(t) = x_1(t)$$

and

$$\dot{x}_1 = x_2$$

then,

$$\dot{x}_2 = -\frac{1}{LC}x_1 - \frac{1}{RC}x_2 + I_o\left(\frac{\omega}{C}\right)\cos\omega t$$

The vector matrix differential form of the state equation can be written as,

$$\frac{d}{dt}\begin{bmatrix} x_1 \\ x_2 \end{bmatrix} = \begin{bmatrix} 0 & 1 \\ -1/LC & -1/RC \end{bmatrix}\begin{bmatrix} x_1 \\ x_2 \end{bmatrix} + \begin{bmatrix} 0 \\ I_0(\omega/c)\cos\omega t \end{bmatrix}$$

and Output

$$V = \begin{bmatrix} 1 & 0 \end{bmatrix}\begin{bmatrix} x_1 \\ x_2 \end{bmatrix}$$

Problem 7.2. Consider the network shown in fig. below. Obtain the state equation of the system.

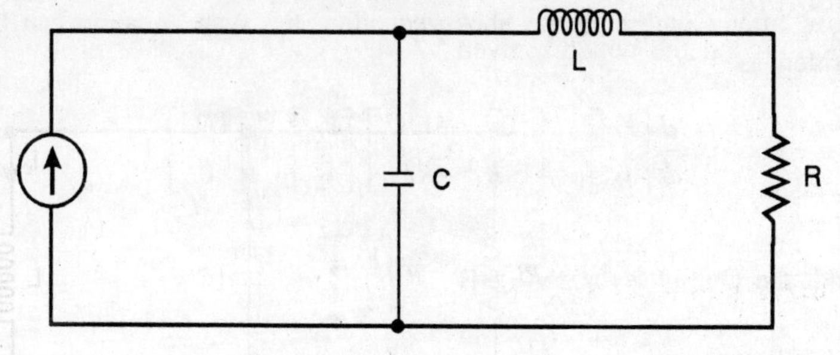

Fig. P. 7.2

Solution:

Redrawing the network given

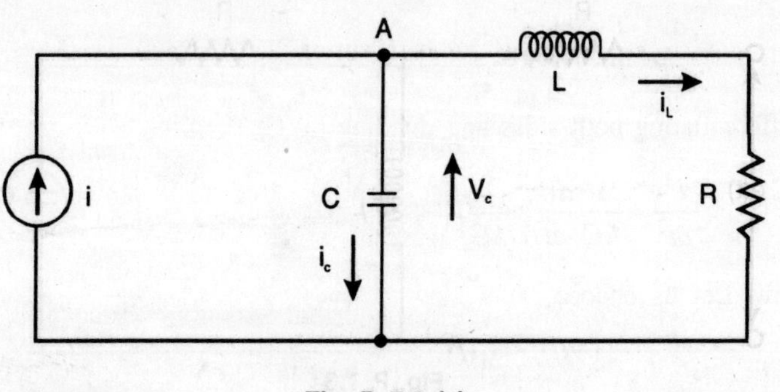

Fig. P. 7.2 (a)

Applying Kirchhoff's current law at node A,

$$i = i_C + i_L$$

$$= C\frac{dV_c}{dt} + i_L$$

$$\frac{dV_c}{dt} = (0)V_c + \left(-\frac{1}{C}\right)i_L + \left(\frac{1}{C}\right)i$$

Now, we will use mesh equation (KVL),

$$V_c = L\frac{di_L}{dt} + Ri_L$$

or

$$\frac{di_L}{dt} = \left(\frac{1}{L}\right)V_C + \left(-\frac{R}{L}\right)i_L + (0)i$$

Now, after Combining the above equation, the state equation can be written as

$$\frac{d}{dt}\begin{bmatrix} V_C \\ i_L \end{bmatrix} = \begin{bmatrix} 0 & -1/C \\ 1/L & -R/L \end{bmatrix}\begin{bmatrix} V_C \\ i_L \end{bmatrix} + \begin{bmatrix} 1/C \\ 0 \end{bmatrix}i$$

And, the Output is $V_R = Ri_L = \begin{bmatrix} 0 & R \end{bmatrix}\begin{bmatrix} V_C \\ i_L \end{bmatrix}$

Note that:

usually in the circuit problem, the current through the inductor and the voltage across the Capacitor are chosen as the state variables.

Problem 7.3. Derive the state-space representation of the network shown below:

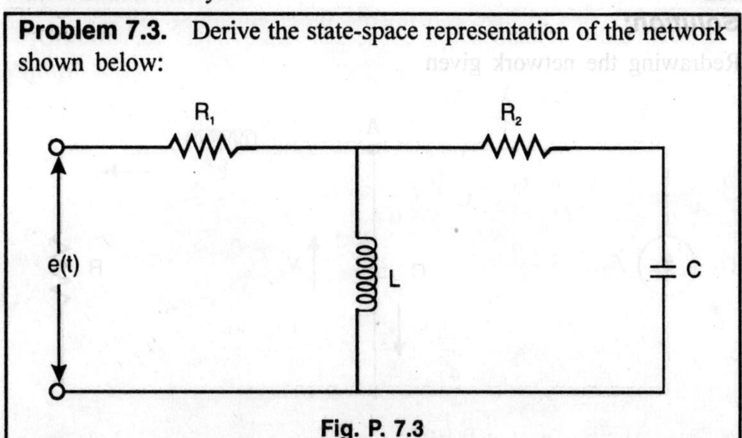

Fig. P. 7.3

Solution:

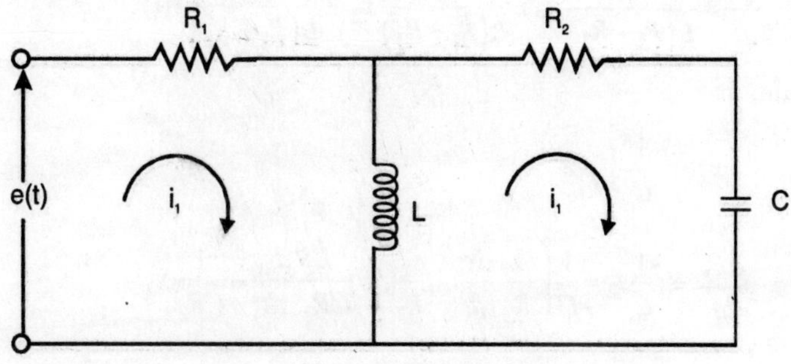

Fig. P. 7.3 (a)

Let the current through the inductor L be,

$$x_1 = i_1 - i_2$$

The voltage across the capacitor be x_2 and *e(t)* is the input to the system. Usually, the current through the inductance and the voltage across the capacitance are the choice of the states. Then, x_1 and x_2 are the states of the system.

Now, we can write the mesh equations as,

$$L\frac{d}{dt}(i_1 - i_2) + R_1 i_1 = e(t)$$

and

$$\frac{1}{C}\int i_2 dt + R_2 i_2 + L\frac{d}{dt}(i_2 - i_1) = 0$$

In terms o x_1 and x_2, we can rewrite these equations as,

$$L\frac{dx_1}{dt} + R_1\left(x_1 + i_2\right) = e(t)$$

as, $\quad x_1 = i_1 - i_2, x_2 = \dfrac{1}{C}\displaystyle\int i_2 dt$

Then, $x_2 + R_2 i_2 - L\dfrac{dx_1}{dt} = 0$

or $\qquad\qquad i_2 = \dfrac{L dx_1}{R_2} - \dfrac{x_2}{R_2}$

Therefore, $\dfrac{dx_1}{dt} + R_1 x_1 + \left(\dfrac{R_1 L}{R_2}\right)\dfrac{dx_1}{dt} - \left(\dfrac{R_1}{R_2}\right)x_2 = e(t)$

or $\dfrac{dx_1}{dt} = \dfrac{-R_1 R_2}{L\left(R_1 + R_2\right)}x_1 + \dfrac{R_1}{L\left(R_1 + R_2\right)}x_2 + \dfrac{R_2}{L\left(R_1 + R_2\right)}e(t)$ (1)

Again, as

$$x_2 = \frac{1}{C}\int i_2\, dt$$

$$\frac{dx_2}{dt} = \frac{1}{C}i_2 = \frac{1}{C}\left(\frac{L}{R_2}\frac{dx_1}{dt} - \frac{x_2}{R_2}\right) = \frac{L}{CR_2}\frac{dx_1}{dt} - \frac{1}{CR_2}x_2$$

$$= \frac{L}{CR_2}\left[\left(-\frac{R_1 R_2}{L\left(R_1 + R_2\right)}\right)x_1 + \frac{R_1}{L\left(R_1 + R_2\right)}x_2\right.$$

$$\left. + \frac{R_2}{L\left(R_1 + R_2\right)}e(t)\right] - \frac{1}{CR_2}x_2$$

$$\frac{dx_2}{dt} = -\frac{R_1}{C\left(R_1 + R_2\right)}x_1 - \frac{1}{C\left(R_1 + R_2\right)}x_2 + \frac{1}{C\left(R_1 + R_2\right)}e(t)$$ (2)

Now, from equations (1) and (2), we can write the state space representation in vector matrix differential equation as,

$$\frac{d}{dt}\begin{bmatrix} x_1 \\ x_2 \end{bmatrix} = \begin{bmatrix} \dfrac{-R_1 R_2}{L\left(R_1 + R_2\right)} & \dfrac{R_1}{L\left(R_1 + R_2\right)} \\[2ex] \dfrac{-R_1}{C\left(R_1 + R_2\right)} & \dfrac{-1}{C\left(R_1 + R_2\right)} \end{bmatrix}\begin{bmatrix} x_1 \\ x_2 \end{bmatrix} + \begin{bmatrix} \dfrac{R_2}{L\left(R_1 + R_2\right)} \\[2ex] \dfrac{1}{C\left(R_1 + R_2\right)} \end{bmatrix}e(t)$$

and Output
$$Y = \begin{bmatrix} 0 & 1 \end{bmatrix} \begin{bmatrix} x_1 \\ x_2 \end{bmatrix}$$

Problem 7.4. Write the state variable formulation of the network shown in the fig. below

$$R_1 = R_2 = 1\Omega,$$
$$L = 1H,$$
$$C_1 = C_2 = 1F$$

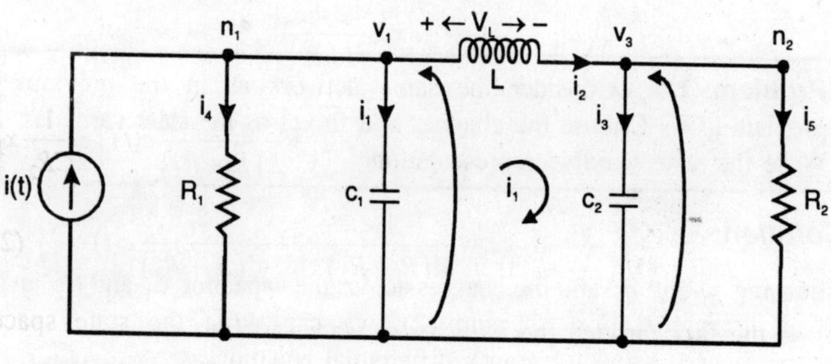

Fig. P 7.4

Solution:

Fig. P 7.4 (a)

Choosing V_1, V_3 and i_2 as the state variables, where V_1 and V_3 are the voltages across capacitors C_1 and C_2. i_2 is the current through the inductance.

Now writing the independent node and loop equations;

At node n_1; $i = i_4 + i_1 + i_2 = V_1 + \dfrac{dV_1}{dt} + i_2$

At node n_2; $i_2 = i_3 + i_5 = \dfrac{dV_3}{dt} + V_3$

Loop L_1, $V_1 = V_2 + V_3 = \dfrac{di_2}{dt} + V_3$

Rearranging these equations;

$$\frac{dV_1}{dt} = -V_1 - i_2 + i$$

$$\frac{dV_3}{dt} = -V_3 + i_2$$

$$\frac{di_2}{dt} = \left(V_1 - V_2\right)$$

Now, the state-variable representation in vector matrix differential equation,

$$\frac{d}{dt}\begin{bmatrix} V_1 \\ i_2 \\ V_3 \end{bmatrix} = \begin{bmatrix} -1 & -1 & 0 \\ 1 & 0 & -1 \\ 0 & 1 & -1 \end{bmatrix}\begin{bmatrix} V_1 \\ i_2 \\ V_3 \end{bmatrix} + \begin{bmatrix} 1 \\ 0 \\ 0 \end{bmatrix} i(t)$$

Problem 7.5. Consider the same network as in the previous problem (7.4). Choose the changes and fluxes as the state variables. Write the state variable representation.

Solution:

Choosing q_1 and q_3 and the charges across the capacitor C_1 and C_2 and ϕ_2 is the flux through the inductor as state variable;

Now writing the independent node and loop equations:

At node n_1: $i = i_4 + i_1 + i_2$

At node n_2: $i_2 = i_3 + i_5$

and loop L_1; $V_1 = V_L + V_3$

After putting the parameter values, we have

$$
\begin{bmatrix}
V_1 = \dfrac{q_1}{C_1} = q_1 & ; & i_1 = \left(\dfrac{dq_1}{dt}\right) \\[2ex]
i_2 = \dfrac{\phi_2}{L} = \phi_2 & ; & i_3 = \left(\dfrac{dq_3}{dt}\right) \\[2ex]
V_3 = \dfrac{q_3}{C_2} = q_3 & ; & V_3 = \dfrac{d\phi_2}{dt} \\[2ex]
i_4 = \dfrac{V_1}{R_1} = V_1 & ; & i_5 = \dfrac{V_3}{R_2} = V_3
\end{bmatrix}
$$

After putting in respective equations and then rearranging, we find the state space model as

$$
\frac{d}{dt}\begin{bmatrix} q_1 \\ \phi_2 \\ q_3 \end{bmatrix} =
\begin{bmatrix} -1 & -1 & 0 \\ 1 & 0 & -1 \\ 0 & 1 & -1 \end{bmatrix}
\begin{bmatrix} q_1 \\ \phi_2 \\ q_3 \end{bmatrix} +
\begin{bmatrix} 1 \\ 0 \\ 0 \end{bmatrix} i
$$

Problem 7.6. Consider the RLC-network shown in fig. below. Write the state variable representation.

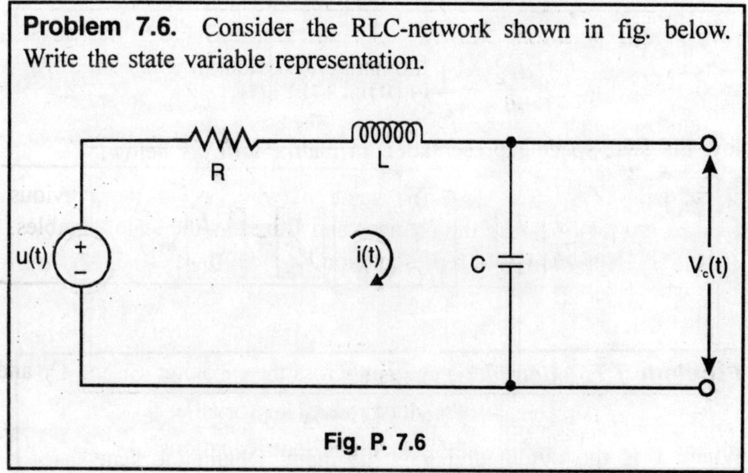

Fig. P. 7.6

Solution:

The state of the network for $t \geq t_0$ is completely determined by $i(t)$ and $V_c(t)$, together with the input excitation $u\,(t)$.

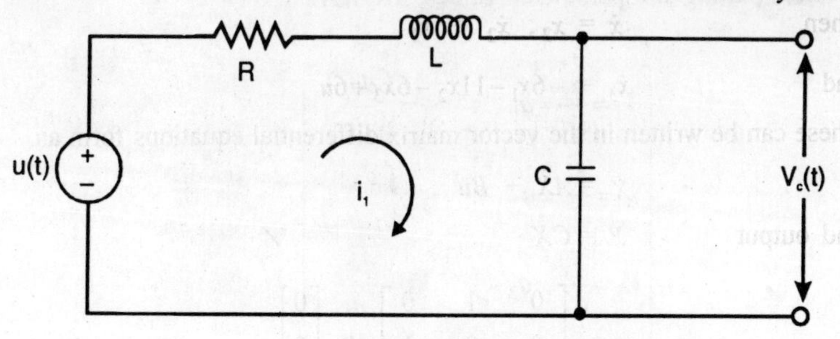

Fig. P. 7.6 (a)

Let us take the current $i(t)$ through the inductors and the voltage $V_C(t)$ across the capacitor as the state variable.

The equations describing the dynamics of the system are,

$$L\frac{di}{dt} + Ri + V_C(t) = u(t)$$

$$C\frac{dV_C}{dt} = i$$

$\therefore$ Now Rearranging the equations

$$\frac{di}{dt} = -\frac{R}{L}i - \frac{1}{L}V_C + \frac{1}{L}u(t)$$

$$\frac{dV_c}{dt} = \frac{1}{C}i + (0)V_C + (0).u(t)$$

Now the state-space representation in matrix form as below:

$$\frac{d}{dt}\begin{bmatrix} i \\ V_C \end{bmatrix} = \begin{bmatrix} -\dfrac{R}{L} & -\dfrac{1}{L} \\ 1/C & 0 \end{bmatrix}\begin{bmatrix} i \\ V_C \end{bmatrix} + \begin{bmatrix} 1/L \\ 0 \end{bmatrix}u(t)$$

Problem 7.7. Consider the system,

$$\dddot{y} + 6\ddot{y} + 11\dot{y} + 6y = 6u$$

Where y is the output and u is the input. Obtain the state-space representation of the system.

Solution:

Let us choose the variables as

$$x_1 = y, \quad x_2 = \dot{y}, \quad x_3 = \ddot{y}$$

Then
$$\dot{x} = x_2, \quad \dot{x}_2 = x_3$$

and
$$\dot{x}_3 = -6x_1 - 11x_2 - 6x_3 + 6u$$

These can be written in the vector matrix differential equations form as,
$$\dot{X} = AX + Bu$$

and output
$$y = CX$$

where
$$A = \begin{bmatrix} 0 & 1 & 0 \\ 0 & 0 & 1 \\ -6 & -11 & -6 \end{bmatrix}; \quad B = \begin{bmatrix} 0 \\ 0 \\ 6 \end{bmatrix}$$

and
$$C = \begin{bmatrix} 1 & 0 & 0 \end{bmatrix}$$

Problem 7.8. The system is represented by the differential equation
$$\ddot{y} + 5\dot{y} + 6y$$

Find the transfer function from state variable representation. Determine the roots of the system.

Solution:

The system can be represented by the Vector matrix differential equation as,
$$\dot{X} = AX + Bu$$

and output
$$y = CX$$

where,
$$A = \begin{bmatrix} 0 & 1 \\ -6 & -5 \end{bmatrix},$$

$$B = \begin{bmatrix} 0 \\ 1 \end{bmatrix},$$

$$C = \begin{bmatrix} 1 & 0 \end{bmatrix}$$

Transfer function $= C[sI - A]^{-1} B = \dfrac{c \, \text{adj}[sI - A] B}{|sI - A|}$

$$= \frac{[1 \; 0]\begin{bmatrix} s+5 & 1 \\ -6 & s \end{bmatrix}\begin{bmatrix} 0 \\ 1 \end{bmatrix}}{\begin{vmatrix} s & -1 \\ 6 & s+5 \end{vmatrix}} = \left(\frac{1}{s^2 + 5s + 6} \right)$$

The characteristic equation

$$|sI - A| = \begin{vmatrix} s & -1 \\ 6 & s+5 \end{vmatrix}$$

$$= s^2 + 5s + 6 = 0$$

Then the eigen values are -2 and -3.

Problem 7.9.   Find the state space representation and State transition Matrix of the network as below:

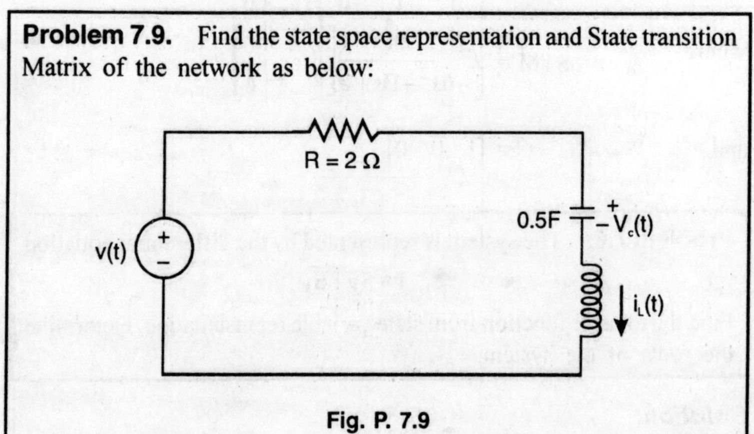

Fig. P. 7.9

Solution:

Now considering the V_C, voltage across the capacitor and i_L, the current through the inductor, as the state variables:

$$V(t) = i_L(t)R + V_C(t) + L\frac{di_L}{dt}$$

$$\therefore \qquad \frac{di_L}{dt} = -\left(\frac{R}{L}\right)i_L - \frac{1}{L}V_C(t) + \frac{1}{L}V(t) \tag{1}$$

and

$$i_L(t) = C\frac{dV_C}{dt}$$

or

$$\frac{dV_C}{dt} = \frac{1}{C}i_L + 0(V_c) + 0(V) \tag{2}$$

Equations (1) and (2) gives the state-space representation

$$\frac{d}{dt}\begin{bmatrix} i_L(t) \\ V_C(t) \end{bmatrix} = \begin{bmatrix} -R/L & -\dfrac{1}{L} \\ 1/C & 0 \end{bmatrix}\begin{bmatrix} i_L(t) \\ V_c(t) \end{bmatrix} + \begin{bmatrix} 1/L \\ 0 \end{bmatrix}V(t)$$

Putting the values, we find the system matrix as,

$$A = \begin{bmatrix} -R/L & -1/L \\ 1/C & 0 \end{bmatrix} \begin{bmatrix} -3 & -1 \\ 2 & 0 \end{bmatrix}$$

Then the State Transition Matrix (S.T.M.)

$$e^{At} = \mathcal{L}^{-1}[sI - A]^{-1} = \mathcal{L}^{-1}\begin{bmatrix} s+3 & 1 \\ -2 & s \end{bmatrix}^{-1}$$

$$\text{STM} = \mathcal{L}^{-1}\frac{1}{s^2+3s+2}\begin{bmatrix} s & -1 \\ 2 & s+3 \end{bmatrix}$$

$$= \begin{bmatrix} \mathcal{L}^{-1}\left(\dfrac{s}{s^2+3s+2}\right) & \mathcal{L}^{-1}\left(\dfrac{-1}{s^2+3s+2}\right) \\ \mathcal{L}^{-1}\left(\dfrac{2}{s^2+3s+2}\right) & \mathcal{L}^{-1}\left(\dfrac{s+3}{s^2+3s+2}\right) \end{bmatrix}$$

$$= \begin{bmatrix} -e^{-t}+2e^{-2t} & -e^{-t}+e^{-2t} \\ 2e^{-t}-2e^{-2t} & 2e^{-t}-e^{-2t} \end{bmatrix}$$

Problem 7.10. Consider the vector matrix differential equation given as

$$\frac{d}{dt}\begin{bmatrix} X_1 \\ X_2 \end{bmatrix} = \begin{bmatrix} 0 & 1 \\ -6 & -5 \end{bmatrix}\begin{bmatrix} X_1 \\ X_2 \end{bmatrix}$$

Also given, $X(0) = \begin{bmatrix} 1 \\ 0 \end{bmatrix}$

Find the state transition matrix. Determine *X(t)*

Solution:

Here $A = \begin{bmatrix} 0 & 1 \\ -6 & -5 \end{bmatrix}$

and then, $(sI - A) = \begin{bmatrix} s & -1 \\ 6 & s+5 \end{bmatrix}$

and
$$[sI-A]^{-1} = \frac{1}{s^2+5s+6}\begin{bmatrix} s+5 & 1 \\ -6 & s \end{bmatrix}$$

Now, the STM,

$$\phi(t) = e^{At} = \mathcal{L}^{-1}[sI-A]^{-1}$$

$$= \mathcal{L}^{-1}\begin{bmatrix} \dfrac{s+5}{s^2+5s+6} & \dfrac{1}{s^2+5s+6} \\ \dfrac{-6}{s^2+5s+6} & \dfrac{5}{s^2+5s+6} \end{bmatrix} = \begin{bmatrix} \phi_{11}(t) & \phi_{12}(t) \\ \phi_{21}(t) & \phi_{22}(t) \end{bmatrix}$$

where
$$\phi_{11}(t) = \mathcal{L}^{-1}\frac{s+5}{(s+2)(s+3)} = \mathcal{L}^{-1}\left(\frac{3}{s+2}\right) - \mathcal{L}^{-1}\left(\frac{2}{s+3}\right)$$

$$= \left(3e^{-2t} - 2e^{-3t}\right)$$

$$\phi_{12}(t) = \mathcal{L}^{-1}\frac{1}{(s+2)(s+3)} = \mathcal{L}^{-1}\frac{1}{(s+2)} - \mathcal{L}^{-1}\left(\frac{1}{s+3}\right)$$

$$= \left(e^{-2t} - e^{-3t}\right)$$

$$\phi_{21}(t) = \mathcal{L}^{-1}\frac{-6}{(s+2)(s+3)} = -6\left[\mathcal{L}^{-1}\left(\frac{1}{s+2}\right) - \mathcal{L}^{-1}\left(\frac{1}{s+3}\right)\right]$$

$$= -6\left(e^{-2t} - e^{-3t}\right)$$

$$\phi_{22}(t) = \mathcal{L}^{-1}\frac{s}{(s+3)(s+2)} = \mathcal{L}^{-1}\left(\frac{3}{s+3}\right) - \mathcal{L}^{-1}\left(\frac{2}{s+2}\right)$$

$$= \left(3e^{-3t} - 2e^{-2t}\right)$$

Hence, the state transition Matrix,

$$\phi(t) = \begin{bmatrix} \left(3e^{-2t} - 2e^{-3t}\right) & \left(e^{-2t} - e^{-3t}\right) \\ -6\left(e^{-2t} - e^{-3t}\right) & \left(3e^{-3t} - 2e^{-2t}\right) \end{bmatrix}$$

Now,
$$X(t) = \phi(t).X(0) = \phi(t).\begin{bmatrix} 1 \\ 0 \end{bmatrix}$$

$$X(t) = \begin{bmatrix} X_1(t) \\ X_2(t) \end{bmatrix} = \begin{bmatrix} 3e^{-2t} - 2e^{-3t} \\ -6\left(e^{-2t} - e^{-3t}\right) \end{bmatrix}$$

Problem 7.11. Find the time response of the system,

$$\dot{X} = \begin{bmatrix} 0 & 1 \\ -6 & -5 \end{bmatrix} X + \begin{bmatrix} 0 \\ 1 \end{bmatrix} u$$

and $\qquad y = \begin{bmatrix} 1 & 0 \end{bmatrix} X$

$u(t)$ is the unit-step input and the initial conditions

$$X_1(0)=0, \ X_2(0)=0$$

Solution:

Here $\qquad A = \begin{bmatrix} 0 & 1 \\ -6 & -5 \end{bmatrix}$

Then the state transition matrix, [from Problem 7.10]

$$\phi(t) = \begin{bmatrix} \left(3e^{-2t} - 2e^{-3t}\right) & \left(e^{-2t} - e^{-3t}\right) \\ -6\left(e^{-2t} - e^{-3t}\right) & \left(3e^{-3t} - 2e^{-2t}\right) \end{bmatrix}$$

Then $\qquad X(t) = \phi(t).X(0) + \int_0^t \phi(t-\tau) B.u(\tau)d\tau$

$$= \begin{bmatrix} \phi_{11}(t) & \phi_{12}(t) \\ \phi_{21}(t) & \phi_{22}(t) \end{bmatrix} + \int_0^t \begin{bmatrix} \phi_{11}(t-\tau) & \phi_{12}(t-\tau) \\ \phi_{21}(t-\tau) & \phi_{22}(t-\tau) \end{bmatrix} \times \begin{bmatrix} 0 \\ 1 \end{bmatrix}.1.d\tau$$

Since $\qquad \begin{bmatrix} X_1(0) \\ X_2(0) \end{bmatrix} = \begin{bmatrix} 0 \\ 0 \end{bmatrix}$

Then $\qquad X(t) = \begin{bmatrix} \int_0^t \phi_{12}(t-\tau)d\tau \\ \int_0^t \phi_{22}(t-\tau)d\tau \end{bmatrix}$

$$= \begin{bmatrix} \int\limits_0^t \exp\{-2(t-\tau)\}d\tau & -\int\limits_0^t \exp\{-3(t-\tau)\}d\tau \\ 3\int\limits_0^t \exp\{-3(t-\tau)\}d\tau & -2\int\limits_0^t \exp\{-2(t-\tau)\}d\tau \end{bmatrix}$$

$$\therefore \quad \begin{bmatrix} X_1(t) \\ X_2(t) \end{bmatrix} = \begin{bmatrix} \dfrac{1}{6} - \dfrac{1}{2}e^{-2t} + 1/3\,e^{-3t} \\ e^{-2t} - e^{-3t} \end{bmatrix}$$

Hence the output,

$$y(t) = \begin{bmatrix} 1 & 0 \end{bmatrix} X(t) = \begin{bmatrix} 1 & 0 \end{bmatrix} \begin{bmatrix} 1/6 - 1/2\,e^{-2t} + 1/3\,e^{-3t} \\ e^{-2t} - e^{-3t} \end{bmatrix}$$

$$= \left(\frac{1}{6} - \frac{1}{2}e^{-2t} + \frac{1}{3}e^{-3t} \right) \quad \textbf{Ans.}$$

Problem 7.12. Write the state variable formulation of the system shown in figure below.

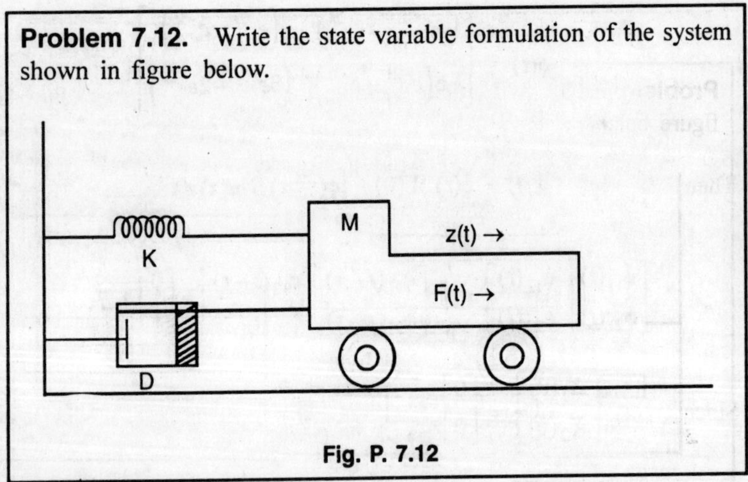

Fig. P. 7.12

Solution:

The force balance equation can be written as,

$$M\frac{d^2Z}{dt^2} = F - KZ - D\frac{dZ}{dt}$$

or
$$\frac{d^2Z}{dt} = \frac{1}{M}F - \frac{K}{M}Z - \frac{D}{M}\frac{dZ}{dt}$$

Let the state variables be x_1 and x_2 choose,

$$x_1 = Z$$
$$u = F$$
$$y = Z$$

Let
$$\frac{dx_1}{dt} = x_2$$

$$\frac{dx_2}{dt} = -\frac{k}{M}x_1 - \frac{D}{M}x_2 + \frac{1}{M}F$$

Hence, the vector matrix differential equation is

$$\frac{d}{dt}\begin{bmatrix} x_1 \\ x_2 \end{bmatrix} = \begin{bmatrix} 0 & 1 \\ -k/M & -D/M \end{bmatrix}\begin{bmatrix} x_1 \\ x_2 \end{bmatrix} + \begin{bmatrix} 0 \\ 1/M \end{bmatrix}F$$

and the output, $y = \begin{bmatrix} 1 & 0 \end{bmatrix}\begin{bmatrix} x_1 \\ x_2 \end{bmatrix}$

Problem 7.13. Write state variable formulation of the system of figure below.

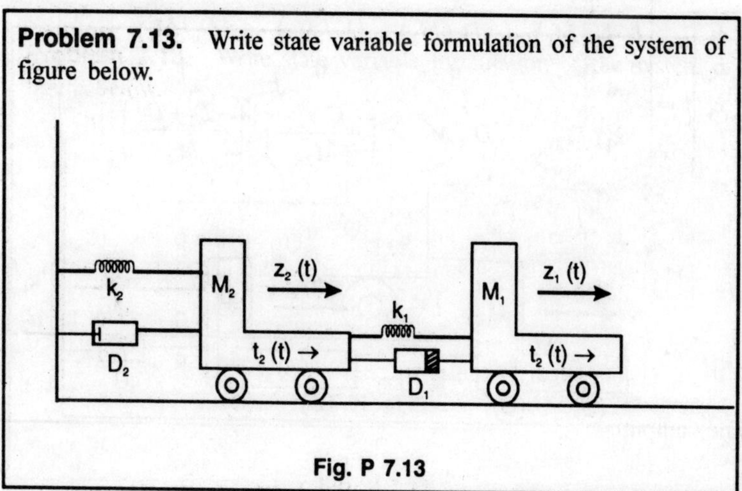

Fig. P 7.13

Solution:

The force balance equation of two masses M_1 and M_2 can be written as,

$$M_1 \frac{d^2 Z_2}{dt^2} = J_2 - k_1 (Z_2 - Z_1) - k_2 Z_2 - D_1 \left(\frac{dZ_2}{dt} - \frac{dZ_1}{dt} \right) - D_2 \frac{dZ_2}{dt}$$

The system is having two inputs and two outputs.

Let us choose the state variables as,

$$x_1 = Z_1$$

$$x_2 = \frac{dZ_1}{dt} \qquad x_3 = Z_2$$

$$x_4 = \frac{dZ_2}{dt}$$

and $\qquad u_1 = J_1$

$$u_2 = J_2$$

$$y_1 = z_1$$

$$y_2 = z_2$$

The vector-matrix differential equation of the multivariable system is

$$\begin{bmatrix} \dot{x}_1 \\ \dot{x}_2 \\ \dot{x}_3 \\ \dot{x}_4 \end{bmatrix} = \begin{bmatrix} 0 & 1 & 0 & 0 \\ (-k_1 / M_1) & (-D_1 / M_1) & (k_1 / M_1) & (D_1 / M_1) \\ 0 & 0 & 0 & 1 \\ k_1 / M_2 & D_1 / M_2 & \left(\dfrac{-k_1 + k_2}{M_2} \right) & \left(\dfrac{-D_1 + D_2}{M_2} \right) \end{bmatrix} \begin{bmatrix} x_1 \\ x_2 \\ x_3 \\ x_4 \end{bmatrix}$$

$$+ \begin{bmatrix} 0 & 0 \\ 1/M_1 & 0 \\ 0 & 0 \\ 0 & 1/M_2 \end{bmatrix} \begin{bmatrix} u_1 \\ u_2 \end{bmatrix}$$

and, output

$$\begin{bmatrix} y_1 \\ y_2 \end{bmatrix} = \begin{bmatrix} 1 & 0 & 0 & 0 \\ 0 & 0 & 1 & 0 \end{bmatrix} \begin{bmatrix} x_1 \\ x_2 \\ x_3 \\ x_4 \end{bmatrix}$$

Problem 7.14. Write the state equation for a series R, L, C, circuit.

Solution:

Corresponding to two storage elements L and C, we may choose i_L and V_C as state variables and write dynamical equations by employing KVL, KCl.

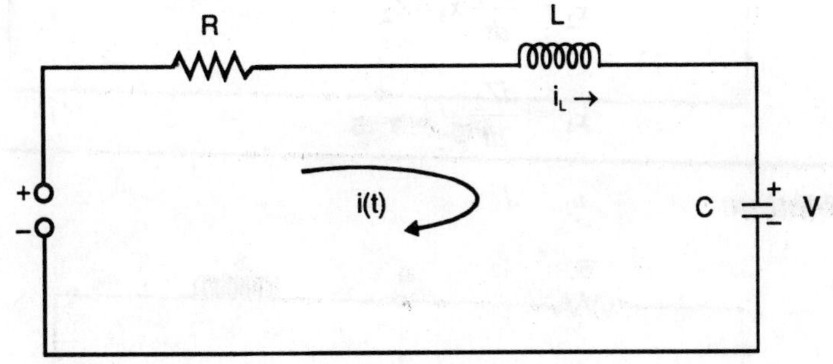

Fig. P. 7.14

$$L\frac{di_L}{dt} + R_i + V_C = V \text{ and,}$$

$$C\frac{dV_C}{dt} = i.$$

Rearranging, by putting chosen state variables on LHS and network variables on RHS of equation

$$\frac{di_L}{dt} = -\frac{R_i}{L} - \frac{1}{L}V_C + \frac{V}{L};$$

$$\frac{dV_C}{dt} = \frac{i}{C}$$

we can rewrite the two equations in matrix form as,

$$\begin{bmatrix} \dfrac{di_L}{dt} \\ \dfrac{dV_C}{dt} \end{bmatrix} = \begin{bmatrix} -R/L & -1/L \\ 1/C & 0 \end{bmatrix} \begin{bmatrix} i \\ V_C \end{bmatrix} + \begin{bmatrix} 1/L \\ 0 \end{bmatrix} V$$

Problem 7.15. Write the state equation in matrix form for the circuit.

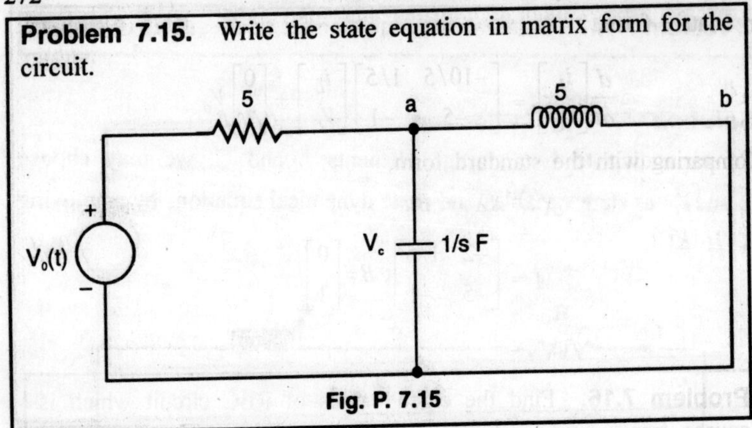

Fig. P. 7.15

Solution:

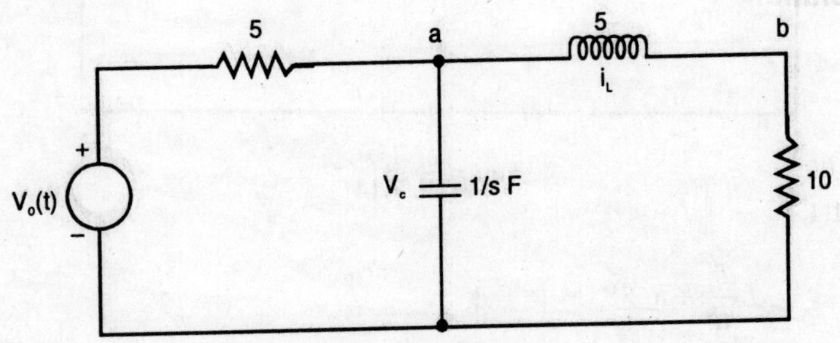

Fig. P. 7.15 (a)

$$\frac{L di_L}{dt} + i_L R = V_C$$

$$\frac{di_L}{dt} = \frac{-10}{5} i_L + \frac{V_C}{5}$$

$$\frac{C dV_C}{dt} = \frac{V_o - V_C}{5} - i_L$$

$$\frac{dV_C}{dt} = \frac{5(V_o - V_C)}{5} - 5i_L$$

$$\frac{dV_C}{dt} = (V_o - V_C) - 5i_L = -5i_L - V_C + V_o$$

In Matrix form,

$$\frac{d}{dt}\begin{bmatrix} i_L \\ V_C \end{bmatrix} = \begin{bmatrix} -10/5 & 1/5 \\ -5 & -1 \end{bmatrix}\begin{bmatrix} i_L \\ V_C \end{bmatrix} + \begin{bmatrix} 0 \\ 1 \end{bmatrix} V_o$$

Comparing with the standard form,

$$\dot{X} = AX + Bu$$

$$A = \begin{bmatrix} -2 & 1/5 \\ -5 & -1 \end{bmatrix}; \ B = \begin{bmatrix} 0 \\ 1 \end{bmatrix}$$

Problem 7.16. Find the eigen values of RLC circuit which is source free.

Solution:

$$s\overline{I} = s\begin{vmatrix} 1 & 0 \\ 0 & 1 \end{vmatrix} = \begin{vmatrix} s & 0 \\ 0 & s \end{vmatrix};$$

det $$\|s\overline{I} - \overline{A}I\| = \begin{vmatrix} s+R/L & 1/L \\ -1/C & s \end{vmatrix} = 0$$

or $$s^2 + s\frac{R}{L} + \frac{1}{LC} = 0$$

$$s_1, s_2 = \frac{-R}{2L} \pm \sqrt{(R/2L)^2 - 1/LC}$$

Problem 7.17. Write the state equations of the Circuit and find the eigen values.

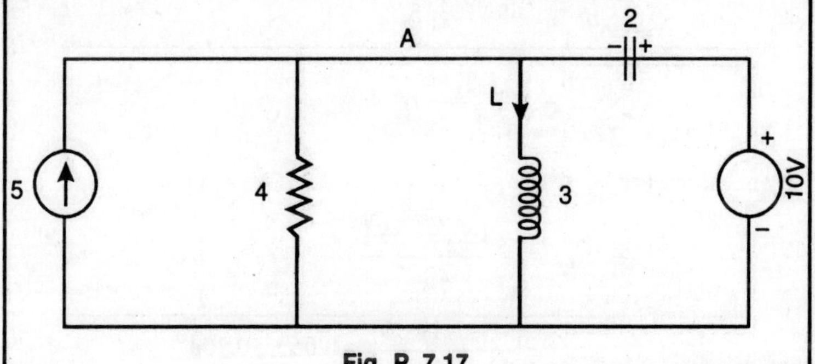

Fig. P. 7.17

Solution:

$$3\frac{di_L}{dt} - 10 + V_C = 0$$

At the node A,

$$2\frac{dv_C}{dt} = i_L + \frac{-V_C + 10}{4} - 5$$

$$\frac{di_L}{dt} = \frac{-V_C}{3} + \frac{10}{3}$$

$$\frac{dV_C}{dt} = \frac{i_L}{2} - \frac{1}{8}V_C - \frac{5}{2} + \frac{10}{8}$$

$$\frac{d}{dt}\begin{bmatrix} i_L \\ V_C \end{bmatrix} = \begin{bmatrix} 0 & -1/3 \\ 1/2 & -1/8 \end{bmatrix}\begin{bmatrix} i_L \\ V_C \end{bmatrix} +$$

$$\begin{bmatrix} 1/3 & 0 \\ 1/8 & -1/2 \end{bmatrix}\begin{bmatrix} 10 \\ 5 \end{bmatrix}$$

This is of the form.

$$\frac{d}{dt}x(t) = Ax(t) + Bx(t)$$

Eigen Values $\quad\quad A' = \begin{bmatrix} 0 & -1/3 \\ 1/2 & -1/8 \end{bmatrix}$

$$\text{Det}[sI - A] = s\begin{vmatrix} 1 & 0 \\ 0 & 1 \end{vmatrix} - \begin{vmatrix} 0 & -1/3 \\ 1/2 & -1/8 \end{vmatrix}$$

$$= \begin{vmatrix} s & 0 \\ 0 & s \end{vmatrix} - \begin{vmatrix} 0 & -1/3 \\ 1/2 & -1/8 \end{vmatrix} = \begin{vmatrix} s & 1/3 \\ -1/2 & s+1/8 \end{vmatrix}$$

$$s^2 + \frac{1}{8}s + \frac{1}{6} = 0$$

$$24s^2 + 3s + 4 = 0$$

Eigen values are

$$s_1, \; s_2 = \frac{-3 \pm \sqrt{9 - 384}}{48}$$

$$= \frac{-3 \pm j19.36}{48} = -0.06 \pm j0.403$$

Problem 7.18. Obtain the A, B, C, and D matrices of the Circuit.

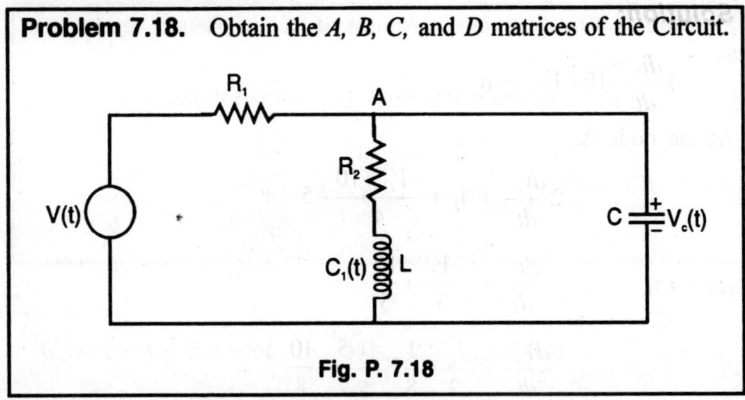

Fig. P. 7.18

Solution:

$$\frac{V - V_C}{R_1} = i_1 + C\frac{dV_C}{dt}$$

$$\frac{dV_C}{dt} = \frac{-V_C}{CR_1} - \frac{i_1}{C} + \frac{V}{CR_1}$$

$$L\frac{di_1}{dt} + i_1 R_2 = V_C$$

$$\frac{di_1}{dt} = \frac{V_C}{L} - \frac{i_1 R_2}{L}$$

$$\frac{d}{dt}\begin{bmatrix} V_C \\ i_1 \end{bmatrix} = \begin{bmatrix} -1/CR_1 & -1/C \\ 1/L & -R_2/L \end{bmatrix}\begin{bmatrix} V_C \\ i_1 \end{bmatrix} + \begin{bmatrix} 1/CR_1 \\ 0 \end{bmatrix} V$$

$$A = \begin{bmatrix} -1/CR_1 & -1/C \\ 1/L & -R_2/L \end{bmatrix}; B = \begin{bmatrix} 1/CR_1 \\ 0 \end{bmatrix}$$

As the output variable is also the state variable V_c; the output equation is,

$$C(t) = \begin{bmatrix} 1 & 0 \end{bmatrix}\begin{bmatrix} V_C \\ i_1 \end{bmatrix} + \begin{bmatrix} 0 \end{bmatrix} V$$

$$C = \begin{bmatrix} 1 & 0 \end{bmatrix}; D = \begin{bmatrix} 0 \end{bmatrix}$$

Problem 7.19. The state equations of LTIV system is represented by $\dot{X}(t) = Ax(t) + Bu(t)$.

Find the state transition matrix $\phi(t)$ of A of the following:

$$A = \begin{bmatrix} -2 & 0 & 0 \\ 0 & -1 & 1 \\ 0 & 0 & -1 \end{bmatrix} \quad B = \begin{bmatrix} 0 \\ 1 \\ 0 \end{bmatrix}$$

Solution:

$$\phi(t) = L^{-1}\left[(sI - A)\right]^{-1}. \text{ It does not depend on } B.$$

$$sI - A = s\begin{bmatrix} 1 & 0 & 0 \\ 0 & 1 & 0 \\ 0 & 0 & 1 \end{bmatrix} - \begin{bmatrix} -2 & 0 & 0 \\ 0 & -1 & 1 \\ 0 & 0 & -1 \end{bmatrix}$$

$$= \begin{bmatrix} s+2 & 0 & 0 \\ 0 & s+1 & -1 \\ 0 & 0 & s+1 \end{bmatrix}$$

det of $\quad (sI - A) = (s+2)(s+1)^2$

$$(sI - A)^{-1} = \frac{\text{adjoint}\,(sI - A)}{\det\,(sI - A)}$$

$$\text{adj}(sI - A) = \begin{bmatrix} (s+1)^2 & 0 & 0 \\ 0 & (s+2)(s+1) & (s+2) \\ 0 & 0 & (s+2)(s+1) \end{bmatrix}$$

$$\phi(s) = (sI - A)^{-1} = \begin{bmatrix} \dfrac{(s+1)^2}{(s+2)(s+1)^2} & 0 & 0 \\ 0 & \dfrac{(s+2)(s+1)}{(s+2)(s+1)^2} & \dfrac{(s+2)}{(s+2)(s+1)^2} \\ 0 & 0 & \dfrac{(s+2)(s+1)}{(s+2)(s+1)^2} \end{bmatrix}$$

$$\phi(s) = \begin{bmatrix} 1/s+2 & 0 & 0 \\ 0 & 1/s+1 & 1/(s+1)^2 \\ 0 & 0 & 1/s+1 \end{bmatrix}$$

$$\phi(t) = L^{-1}\,\phi(s) \begin{bmatrix} e^{-2t} & 0 & 0 \\ 0 & e^{-t} & te^{-t} \\ 0 & 0 & e^{-t} \end{bmatrix}$$

Problem 7.20. The transfer function block diagram of a feedback control system is shown in the figure below. For each of the following inputs determine the steady state value of $e(t)$. (a) $r(t) = u(t)$ – unit step, (b) $r(t) = \sin 2t$.

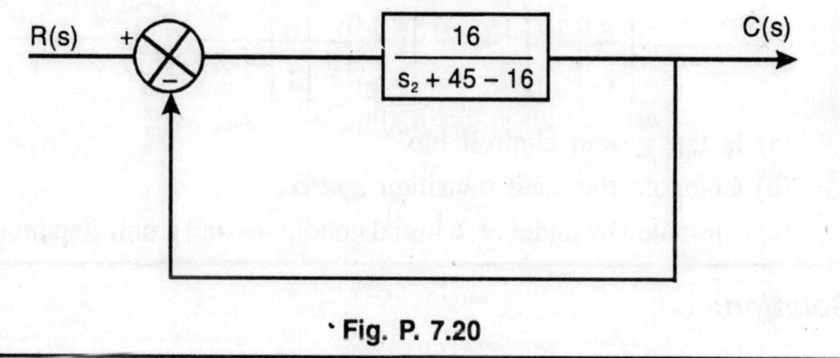

Fig. P. 7.20

Solution:

$$G(s) = \frac{C(s)}{R(s)} = \frac{\dfrac{16}{s^2+4s-16}}{1+\dfrac{16}{s^2+4s-16}}$$

$$= \frac{16}{s^2+4s-16+16} = \frac{16}{(s+4)s}$$

For $r(t) = u(t)$, $R(s) = 1/s$

$$G(s) = \frac{16\,R(s)}{s(s+4)} = \frac{16}{s^2(s+4)}$$

Δs value as $s \to 0$ $\lim\limits_{S \to 0} sG(s) = \dfrac{16}{4} = 4$

For $\qquad\qquad r(t) = \sin 2t,$

$$R(s) = \frac{2}{s^2 + 4}$$

$$G(s) = \frac{16}{s(s+4)} \times \frac{2}{(s^2+4)} = \frac{32}{s(s+4)(s^2+4)}$$

Steady state value as $s \to 0$ $\lim\limits_{S \to 0} s\, G(s) = \dfrac{32}{16} = 2$

Problem 7.21.   The state equations of a LTIV system are as given below:

$$\begin{bmatrix} \dot{x}_1 \\ \dot{x}_2 \end{bmatrix} = \begin{bmatrix} -2 & 0 \\ 1 & -1 \end{bmatrix} \begin{bmatrix} x_1 \\ x_2 \end{bmatrix} + \begin{bmatrix} 0 \\ 1 \end{bmatrix} u(t); \, t > 0$$

(a) Is the system controllable?

(b) Compute the state transition matrix

(c) Compute $x(t)$ under zero initial conditions and a unit step input.

Solution:

(a) $\qquad\qquad A = \begin{bmatrix} -2 & 0 \\ 1 & -1 \end{bmatrix}; \; B = \begin{bmatrix} 0 \\ 1 \end{bmatrix}$

For controllability $s = \begin{bmatrix} B & AB \end{bmatrix}$ must be non singular, that means, determinant value should not be equal to zero.

$$\det s = \det \begin{bmatrix} 0 & 0 \\ 1 & -1 \end{bmatrix} = 0$$

Hence, the system is not state Controllable as it is singular.

(b) $\qquad (sI - A) = s\begin{bmatrix} 1 & 0 \\ 0 & 1 \end{bmatrix} \begin{bmatrix} -2 & 0 \\ 1 & -1 \end{bmatrix}$

$$= \begin{bmatrix} s+2 & 0 \\ -1 & s+1 \end{bmatrix}$$

$$\det (sI - A) = (s+2)(s+1)$$

$$\text{adj } (sI - A) = \begin{vmatrix} s+1 & 0 \\ 1 & s+2 \end{vmatrix}$$

$$(sI - A)^{-1} = \begin{bmatrix} 1/s+2 & 0 \\ 1/(s+1)(s+2) & 1/s+1 \end{bmatrix}$$

$$\phi(t) = L^{-1}(sI - A)^{-1} = \begin{bmatrix} e^{-2t} & 0 \\ e^{-t} - e^{-2t} & e^{-t} \end{bmatrix}$$

is the transition matrix in t domain.

(c) For a unit step function applied at $t = 0$, the complete solution
of the state equation at $t > 0$ is,

$$n(t) = e^{At} n(0^+) + \int_0^t e^{A(t-\tau)} Br(\tau)d\tau$$

$$= \begin{vmatrix} e^{-2t} & 0 \\ e^{-t} - e^{-2t} & e^{-t} \end{vmatrix} \begin{bmatrix} x_1(0) \\ x_2(0) \end{bmatrix} +$$

$$\int_0^t \begin{vmatrix} e^{-2(t-\tau)} & 0 \\ e^{-(t-\tau)} - e^{-2(t-\tau)} & e^{-(t-\tau)} \end{vmatrix} \begin{bmatrix} 0 \\ 1 \end{bmatrix} [1] dr$$

$$= 0 + \int_0^t e^{-(t-\tau)} d\tau = 1 - e^{-t}$$

Problem 7.22. A state variable formulation of a system is given
by the expression

$$\begin{bmatrix} \dot{x}_1 \\ \dot{x}_2 \end{bmatrix} = \begin{bmatrix} 1 & 0 \\ 0 & -1 \end{bmatrix} \begin{bmatrix} x_1 \\ x_2 \end{bmatrix} + \begin{bmatrix} 0 \\ 1 \end{bmatrix} u(t)$$

$$y = \begin{bmatrix} 1 & 1 \end{bmatrix} \begin{bmatrix} x_1 \\ x_2 \end{bmatrix}$$

what is the transfer fucntion of the system?

Solution:

The transfer function is given by

$$y(s) = \frac{0U(s)}{s-1} + \frac{1U(s)}{s+1}$$

$$TF = \frac{y(s)}{U(s)} = \frac{1}{s+1}$$

Problem 7.23. A system is characterised by the following state space equations,

$$\begin{bmatrix} \dot{x}_1 \\ \dot{x}_2 \end{bmatrix} = \begin{bmatrix} -3 & 1 \\ -2 & 0 \end{bmatrix}\begin{bmatrix} x_1 \\ x_2 \end{bmatrix} + \begin{bmatrix} 0 \\ 1 \end{bmatrix}\dot{u}; \, t>0 \tag{1}$$

$$y = \begin{bmatrix} 1 & 0 \end{bmatrix}\begin{bmatrix} x_1 \\ x_2 \end{bmatrix} \tag{2}$$

(a) Find the transfer function of the system
(b) Solve the state equation for a unit step input under zero initial condition.

Solution:

(a) $\qquad G(s) = \dfrac{C(s)}{R(s)} = D(sI - A)^{-1} B + E$

A, B, D, E as per equation (1) and (2)

$$G(s) = \begin{bmatrix} 1 & 0 \end{bmatrix}\begin{bmatrix} s+3 & -1 \\ 2 & s \end{bmatrix}^{-1}\begin{bmatrix} 0 \\ 1 \end{bmatrix}$$

$$= \begin{bmatrix} 1 & 0 \end{bmatrix}\frac{1}{s(s+3)+2}\begin{bmatrix} s & 1 \\ -2 & s+3 \end{bmatrix}\begin{bmatrix} 0 \\ 1 \end{bmatrix}$$

$$= \begin{bmatrix} 1 & 0 \end{bmatrix}\begin{bmatrix} \dfrac{s}{s^2+3s+2} & \dfrac{1}{s^2+3s+2} \\ \dfrac{-2}{s^2+3s+2} & \dfrac{s+3}{s^2+3s-2} \end{bmatrix}\begin{bmatrix} 0 \\ 1 \end{bmatrix}$$

$$= \begin{bmatrix} 1 & 0 \end{bmatrix} \begin{bmatrix} \dfrac{s}{(s+2)(s+1)} & \dfrac{1}{(s+2)(s+1)} \\[2ex] \dfrac{-2}{(s+2)(s+1)} & \dfrac{s+3}{(s+2)(s+1)} \end{bmatrix} \begin{bmatrix} 0 \\ 1 \end{bmatrix}$$

$$= \frac{1}{(s+2)(s+1)} \left|(s+0)0 + (1+0)1\right|$$

$$= \frac{1}{(s+2)(s+1)}$$

(b) State equation for a unit step input under zero initial condition:

$$X(t) = e^{At} X(0^+) + \int_0^t e^{A(t-\tau)} . B.u(\tau)d\tau$$

$$= 0 + \int_0^t \begin{bmatrix} -e^{-(t-\tau)} + 2e^{-2(t-\tau)} & e^{-(t-\tau)} - e^{-2(t-\tau)} \\ -2e^{-(t-\tau)} + 2e^{-2(t-\tau)} & 2e^{-(t-\tau)} - e^{-2(t-\tau)} \end{bmatrix} \begin{bmatrix} 0 \\ 1 \end{bmatrix} .u(1).d\tau$$

$$= \int_0^t \begin{bmatrix} e^{-(t-\tau)} - e^{-2(t-\tau)} \\ 2e^{-(t-\tau)} - e^{-2(t-\tau)} \end{bmatrix} d\tau = \begin{bmatrix} 1 - e^{-t} - \dfrac{1}{2} + \dfrac{1}{2}e^{-2t} \\[2ex] 2 - 2e^{-t} - \dfrac{1}{2} + \dfrac{1}{2}e^{-2t} \end{bmatrix}$$

$$= \begin{bmatrix} \dfrac{1}{2} - e^{-1} + \dfrac{1}{2}e^{-2t} \\[2ex] \dfrac{3}{2} - 2e^{-t} + \dfrac{1}{2}e^{-2t} \end{bmatrix} \quad \textbf{Ans.}$$

Problem 7.24. A second order linear system is described by,

$$\dot{X}_1 = -3/2\,x_1 + 1/2\,x_2 + u$$

$$\dot{X}_2 = -\frac{1}{2}(x_1 + x_2) + u$$

$$y = x_1 + x_2$$

obtain the T.F. $\dfrac{Y}{U(s)}$ and also calculate the zero input response $y(t)$

of $x_1(0)=1$ and $x_2(0)=-1$

Solution:

The required state space representation as below;

$$\begin{bmatrix} \dot{X}_1 \\ \dot{X}_2 \end{bmatrix} = \begin{bmatrix} -3/2 & 1/2 \\ -1/2 & -1/2 \end{bmatrix}\begin{bmatrix} x_1 \\ x_2 \end{bmatrix} + \begin{bmatrix} 1 \\ 1 \end{bmatrix} u$$

and

$$y = \begin{bmatrix} 1 & 1 \end{bmatrix}\begin{bmatrix} x_1 \\ x_2 \end{bmatrix}$$

Now

$$(sI - A) = \begin{bmatrix} s+3/2 & -1/2 \\ 1/2 & s+1/2 \end{bmatrix}$$

del

$$(sI - A) = (s+3/2)(s+1/2)+1/4$$

$$= \frac{4s^2+8s+4}{4} = \left(s^2+2s+1\right) = (s+1)^2$$

Then,

$$(sI - A)^{-1} = \frac{1}{(s+1)^2} \times \begin{bmatrix} s+1/2 & 1/2 \\ -1/2 & s+3/2 \end{bmatrix}$$

Now, the transfer function is given by,

$$\frac{Y(s)}{U(s)} = \begin{bmatrix} 1 & 1 \end{bmatrix}\frac{1}{(s+1)^2}\begin{bmatrix} s+1/2 & \dfrac{1}{2} \\ -1/2 & s+3/2 \end{bmatrix}\begin{bmatrix} 1 \\ 1 \end{bmatrix}$$

After Simplifying,

$$\text{T.F.} = \frac{(s+1/2-1/2).1+(s+3/2+1/2).1}{(s+1)^2}$$

$$= \frac{2s+2}{(s+1)^2} = \left(\frac{2}{s+1}\right)$$

Now

$$\phi(s) = \begin{bmatrix} \dfrac{2s+1}{2(s+1)^2} & \dfrac{1}{2(s+1)^2} \\ \dfrac{-1}{2(s+1)^2} & \dfrac{2s+3}{2(s+1)^2} \end{bmatrix}$$

$$L^{-1}\phi(s) = \phi(t) = \begin{bmatrix} L^{-1}\dfrac{2s+1}{2(s+1)^2} & L^{-1}\dfrac{1}{2(s+1)^2} \\[3mm] L^{-1}\dfrac{-1}{2(s+1)^2} & L^{-1}\dfrac{2s+3}{2(s+1)^2} \end{bmatrix}$$

Now

$$\phi(t) = \begin{bmatrix} e^{-t}-\dfrac{1}{2}te^{-t} & \dfrac{1}{2}te^{-t} \\[3mm] -\dfrac{1}{2}te^{-t} & e^{-t}+\dfrac{1}{2}te^{-t} \end{bmatrix}$$

Now

$$\begin{bmatrix} X_1(t) \\ X_2(t) \end{bmatrix} = \begin{bmatrix} e^{-t}-1/2te^{-t} & 1/2te^{-t} \\ -1/2te^{-t} & e^{-t}+1/2te^{-t} \end{bmatrix} \begin{bmatrix} 1 \\ -1 \end{bmatrix}$$

$$= \begin{bmatrix} e^{-t}-1/2te^{-t}-1/2te^{-t} \\ -1/2te^{-t}-e^{-t}-1/2te^{-t} \end{bmatrix} = \begin{bmatrix} e^{-t}-te^{-t} \\ -e^{-t}-te^{-t} \end{bmatrix}$$

Problem 7.25. Obtain the response of the following system

$$\begin{bmatrix} \dot{X}_1 \\ \dot{X}_2 \end{bmatrix} = \begin{bmatrix} 0 & 1 \\ -2 & -3 \end{bmatrix} \begin{bmatrix} X_1 \\ X_2 \end{bmatrix} + \begin{bmatrix} 2 \\ 0 \end{bmatrix}(u)$$

where $u = 0$; $t < 0 = e^{-t}$; $t \geq 0$

with $X_1(0)=X_2(0)=0$; using Laplace transform method.

(I.E.S.-1992)

Solution:

Using the Laplace transform method,

$$X(s) = (sI-A)^{-1}.X(0)+(sI-A)^{-1}.B.u(s)$$

As the

$$X(0) = \begin{bmatrix} X_1(0) \\ X_2(0) \end{bmatrix} = \begin{bmatrix} 0 \\ 0 \end{bmatrix};$$

$\therefore$

$$X(s) = (sI-A)^{-1}B.u(s)$$

As,

$$u(t) = e^{-t}; t\geq 0$$
$$= 0; t < 0$$

Therefore,
$$X(s) = \left(sI - A\right)^{-1} B \cdot \left(\frac{1}{s+1}\right)$$

$$= \begin{bmatrix} s & -1 \\ 2 & s+3 \end{bmatrix}^{-1} \cdot \begin{bmatrix} 2 \\ 0 \end{bmatrix} \cdot \left(\frac{1}{s+1}\right)$$

$$= \begin{bmatrix} s+3 & 1 \\ -2 & s \end{bmatrix} \cdot \begin{bmatrix} 2 \\ 0 \end{bmatrix} \cdot \frac{1}{(s+1) \cdot \left(s^2 + 3s + 2\right)}$$

$$= \begin{bmatrix} 2(s+3) \\ -4 \end{bmatrix} \cdot \frac{1}{(s+1)^2 (s+2)}$$

Now expanding in partial fraction,

$$\frac{2(s+3)}{(s+1)^2 (s+2)} = \frac{A}{s+2} + \frac{B_1}{s+1} + \frac{B_2}{(s+1)^2}$$

$$A = \left. \frac{2(s+3)}{(s+1)^2} \right|_{s=-2} = 2$$

$$B_2 = \left. \frac{2(s+3)}{(s+2)} \right|_{s=-1} = 4$$

$$B_1 = \left. \frac{d}{ds}\left[\frac{2(s+3)}{(s+2)}\right]\right|_{s=-1} = \left. \frac{(s+2) - (2s+6)}{(s+2)^2} \right|_{s=-1} = -2$$

Hence

$$\frac{2(s+3)}{(s+1)^2 (s+2)} = \frac{2}{s+2} - \frac{2}{s+1} + \frac{4}{(s+1)^2}$$

Similarly,

$$\frac{-4}{(s+1)^2 (s+2)} = \frac{A_1}{s+2} - \frac{B_3}{s+1} + \frac{B_4}{(s+1)^2}$$

$$\therefore \qquad A_1 = \left. \frac{-4}{(s+1)^2} \right|_{s=-2} = -4$$

$$B_4 = \frac{-4}{s+2}\bigg|_{s=-1} = -4$$

$$B_3 = \frac{d}{ds}\left(\frac{-4}{s+2}\right)\bigg|_{s=-1} = \frac{-4}{(s+2)^2}\bigg|_{s=-1} = 4$$

Hence
$$\frac{-4}{(s+2)(s+1)^2} = \frac{-4}{s+2} + \frac{4}{s+1} - \frac{4}{(s+1)^2}$$

Therefore
$$X(s) = \begin{bmatrix} \dfrac{2}{s+2} - \dfrac{2}{s+1} + \dfrac{4}{(s+1)^2} \\[3mm] \dfrac{-4}{s+2} + \dfrac{4}{s+1} - \dfrac{4}{(s+1)^2} \end{bmatrix}$$

and
$$X(t) = \begin{bmatrix} \left(2e^{-2t} - 2e^{-t} + 4te^{-t}\right)u(t) \\[2mm] \left(-4e^{-2t} + 4e^{-t} - 4te^{-t}\right)u(t) \end{bmatrix} \quad \textbf{Ans.}$$

Problem 7.26. For the system descibed by

$$\begin{bmatrix} \dot{X}_1 \\ \dot{X}_2 \end{bmatrix} = \begin{bmatrix} 0 & 1 \\ 0 & -2 \end{bmatrix}\begin{bmatrix} X_1 \\ X_2 \end{bmatrix}.$$

Determine the state transition matrix (STM). [I.E.S.-1991]

Solution:

As
$$\dot{X} = AX$$

Here
$$A = \begin{bmatrix} 0 & 1 \\ 0 & -2 \end{bmatrix}$$

$$\phi(t, t_o) = I + A(t - t_0) + \frac{A^2}{2!}(t - t_0)^2 + \frac{A^3}{3!}(t - t_0)^3 + \ldots$$

$$= \begin{pmatrix} 1 & 0 \\ 0 & 1 \end{pmatrix} + (t - t_0)\begin{pmatrix} 0 & 1 \\ 0 & -2 \end{pmatrix} + \frac{(t - t_0)^2}{2!}\begin{bmatrix} 0 & 1 \\ 0 & -2 \end{bmatrix} \times$$

$$\begin{bmatrix} 0 & 1 \\ 0 & -2 \end{bmatrix} + \frac{(t - t_0)^2}{3!}\begin{bmatrix} 0 & 1 \\ 0 & -2 \end{bmatrix}\begin{bmatrix} 0 & 1 \\ 0 & -2 \end{bmatrix}\begin{bmatrix} 0 & 1 \\ 0 & -2 \end{bmatrix} + \ldots$$

$$= \begin{bmatrix} 1 & 0 \\ 0 & 1 \end{bmatrix} + (t-t_0)\begin{bmatrix} 0 & 1 \\ 0 & -2 \end{bmatrix} + \frac{(t-t_0)^2}{2!}\begin{bmatrix} 0 & -2 \\ 0 & 4 \end{bmatrix} +$$

$$\frac{(t-t_0)^3}{3!}\begin{bmatrix} 0 & 8 \\ 0 & -4 \end{bmatrix} + \ldots\ldots$$

$$= \begin{bmatrix} a_{11} & a_{12} \\ a_{21} & a_{22} \end{bmatrix} \quad \text{where} \quad a_{11} = 1$$

$$1 - 2a_{12} = 1 - 2(t-t_0) + \frac{2^2(t-t_0)^2}{2!} = e^{-2(t-t_0)}$$

or

$$a_{12} = \left(\frac{1 - e^{-2(t-t_0)}}{2}\right)$$

$$a_{21} = 0$$

$$a_{22} = 1 - 2(t-t_0) + \frac{2^2}{2!}(t-t_0)^2 + \ldots = e^{-2(t-t_0)}$$

Hence, S.T.M.

$$\phi(t) = \begin{bmatrix} 1 & \left(\dfrac{1 - e^{-2(t-t_0)}}{2}\right) \\ 0 & e^{-2(t-t_0)} \end{bmatrix}$$

Problem 7.27. For the system shown in fig. P. 7.27 obtain a state variable representation in the form

$$\dot{X} = AX + bu$$

$$Y = CX + du$$

The elements X_1 and X_2 of the state vector are shown in figure. Making use of the state variable representation obtain the transfer function $\dfrac{Y(s)}{U(s)}$ for the system.

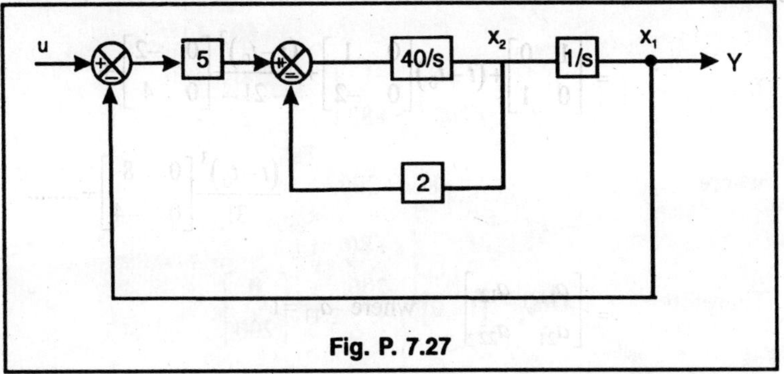

Fig. P. 7.27

Solution:

From the fig. we get,

$$X_1(s) = \frac{X_2(s)}{s} \tag{1}$$

and

$$X_2(s) = \left[s\{U(s) - X_1(s)\} - 2X_2(s) \right]\frac{40}{s} \tag{2}$$

and

$$Y(s) = X_1(s)$$

Rearranging equation (1) & (2)

$$X_2 = sX_1(s)$$

$$sX_2(s) = 200U(s) - 200X_1(s) - 80X_2(s)$$

Inverse transformation and writing in differential form,

$$\begin{bmatrix} \dot{X_1} \\ \dot{X_2} \end{bmatrix} = \begin{bmatrix} 0 & 1 \\ -200 & -80 \end{bmatrix}\begin{bmatrix} X_1 \\ X_2 \end{bmatrix} + \begin{bmatrix} 0 \\ 200 \end{bmatrix}u(t)$$

$$\dot{y}(t) = \begin{bmatrix} 1 & 0 \end{bmatrix}\begin{bmatrix} X_1 \\ X_2 \end{bmatrix} + \begin{bmatrix} 0 \end{bmatrix}u(t)$$

Hence

$$[A] = \begin{bmatrix} 0 & 1 \\ -200 & -80 \end{bmatrix}$$

$$[B] = \begin{bmatrix} 0 \\ 200 \end{bmatrix} \text{ and } C = \begin{bmatrix} 1 & 0 \end{bmatrix}$$

The transfer function,

$$G(s) = C(sI - A)^{-1}B$$

Now
$$(sI - A)^{-1} = \begin{bmatrix} s & -1 \\ 200 & s+80 \end{bmatrix} = \frac{\begin{bmatrix} s+80 & 1 \\ -200 & s \end{bmatrix}}{\Delta}$$

where
$$\Delta = s^2 + 80s + 200$$

Therefore,
$$G(s) = \begin{bmatrix} 1 & 0 \end{bmatrix} \frac{\begin{bmatrix} s+80 & 1 \\ -200 & s \end{bmatrix}}{\Delta} \begin{bmatrix} 0 \\ 200 \end{bmatrix}$$

$$= \begin{bmatrix} 1 & 0 \end{bmatrix} \begin{bmatrix} 200 \\ 200s \end{bmatrix} \cdot \frac{1}{\Delta} = \frac{200}{\Delta} = \left(\frac{200}{s^2 + 80s + 200} \right)$$

Problem 7.28.   A feedback system is characterised by the closed-loop

$$T(s) = \frac{s^2 + 3s + 3}{s^3 + 2s^2 + 3s + 1}$$

Draw a suitable signal flow graph and there from construct a state model of the system.

Solution:

Given T.F. is
$$T(s) = \frac{s^2 + 3s + 3}{s^3 + 2s^2 + 3s + 1} \qquad (1)$$

Dividing into two parts on shown below,

$$T(s) = \frac{Y(s)}{U(s)} = \frac{X_1(s)}{U(s)} \cdot \frac{Y(s)}{X_1(s)}$$

where
$$\frac{X_1(s)}{U(s)} = \frac{1}{s^3 + 2s^2 + 3s + 1}$$

and
$$\frac{Y(s)}{X_1(s)} = s^2 + 3s + 3$$

$$U(s) \longrightarrow \boxed{\frac{1}{s^3 + 2s^2 + 3s + 1}} \xrightarrow{\ X(s)\ } \boxed{s^2 + 3s + 3} \longrightarrow Y$$

Now
$$\frac{X_1(s)}{U(s)} = \frac{1}{s^3 + 2s^2 + 3s + 1}$$

$$\Rightarrow \quad \dddot{x}_1 + 2\ddot{x}_1 + 3\dot{x}_1 + x_1 = u \tag{2}$$

Let
$$\dot{x}_1 = x_2 \tag{3}$$

$$\ddot{x}_1 = \dot{x}_2 = x_3 \tag{4}$$

Then equation (2) becomes,

$$\dot{x}_3 = U - 2x_2 - 3x_2 + x_1 \tag{5}$$

From equation (3), (4) and (5), we get its state equations

$$\begin{bmatrix} \dot{x}_1 \\ \dot{x}_2 \\ \dot{x}_3 \end{bmatrix} = \begin{bmatrix} 0 & 1 & 0 \\ 0 & 0 & 1 \\ -1 & -3 & -2 \end{bmatrix} \begin{bmatrix} x_1 \\ x_2 \\ x_3 \end{bmatrix} + \begin{bmatrix} 0 \\ 0 \\ 1 \end{bmatrix} u \tag{6}$$

Now
$$\frac{Y(s)}{X_1(s)} = s^2 + 3s + 3$$

$$\Rightarrow \quad y = \ddot{x}_1 + 3\dot{x}_1 + 3x_1 = x_3 + 3x_2 + 3x_1 \tag{7}$$

Then output equation is,

$$y = \begin{bmatrix} 3 & 3 & 1 \end{bmatrix} \begin{bmatrix} x_1 \\ x_2 \\ x_3 \end{bmatrix} \tag{8}$$

The signal flow graph corresponding to equation (6) and (8) is drawn below

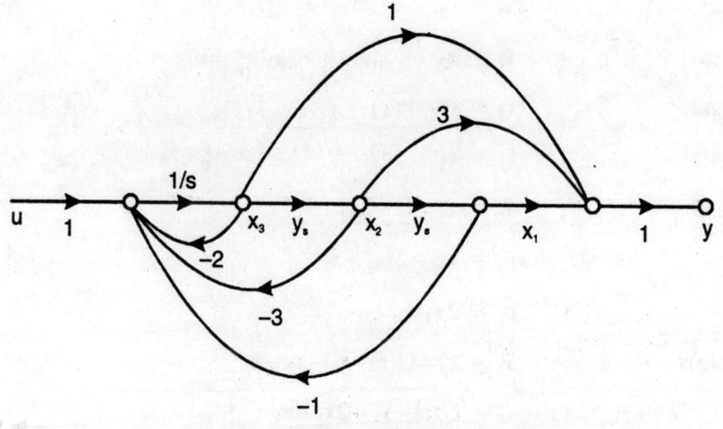

Fig. P. 7.28

Problem 7.29. A feedback system is represented by a signal flow graph shown in the figure below.

(a) Construct a state model of the system

(b) Diagnotize the coefficient Matrix A of the state model of part (a).

(c) Determine the stability of the system.

(d) Determine the transfer function C(s)/U(s) from the state model of part (a)

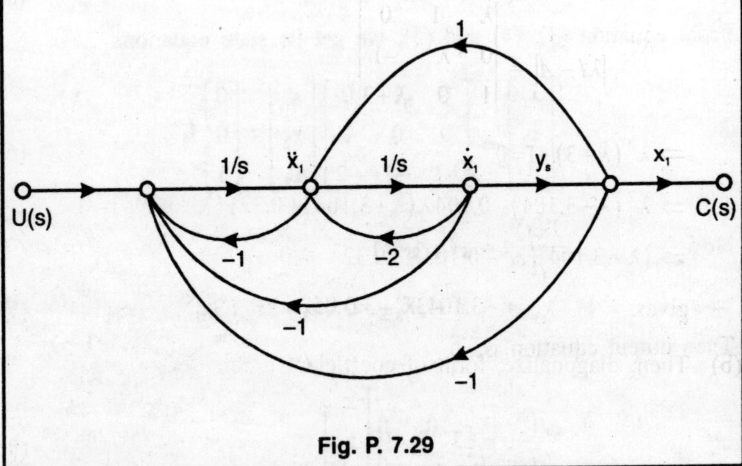

Fig. P. 7.29

Solution:

From the signal flow graph

$$\Rightarrow \qquad sx_1 = x_2$$

$$\Rightarrow \qquad \dot{x}_1 = x_2$$

$$\text{and} \qquad x_3 = sx_2$$

$$\therefore \qquad \dot{x}_2 = x_3$$

$$\text{and} \qquad \ddot{x}_1 = x_3 = x_1 - 2x_2 + \dot{x}_3/s$$

$$\therefore \qquad sx_3 = sx_1 - 2sx_2 + \dot{x}_3$$

$$x_2 = 2x_3$$

$$\text{Now} \qquad \dddot{x}_1 = \dot{x}_3 = U(t) - x_3 - x_2 - x_1$$

$$= U(t) - x_3 - 2x_3 - x_1$$

$$= U(t) - 3x_3 - x_1$$

Then, the state equation are,

$$\begin{bmatrix} \dot{x}_1 \\ \dot{x}_2 \\ \dot{x}_3 \end{bmatrix} = \begin{bmatrix} 0 & 1 & 0 \\ 0 & 0 & 1 \\ -1 & 0 & -3 \end{bmatrix} + \begin{bmatrix} 0 \\ 0 \\ 1 \end{bmatrix} u(t)$$

and $\qquad y = x_1$

The characteristic equation is,

$$|\lambda I - A| = \begin{vmatrix} \lambda & -1 & 0 \\ 0 & \lambda & -1 \\ 1 & 0 & \lambda+3 \end{vmatrix}$$

$$\Rightarrow \lambda^2(\lambda+3)+1=0$$

$$\Rightarrow \lambda^2(\lambda+3.104)-0.104\lambda(\lambda+3.104)+0.32)(\lambda+3.104)=0$$

$$\Rightarrow (\lambda+3.104)(\lambda^2-0.104\lambda+0.32) = 0$$

gives, $\qquad \lambda_3 = -3.104, \lambda_{1,2}=0.052 \pm j0.565$

(b) Then, diagonalize, form of coefficient matrix are,

$$A = \begin{vmatrix} \lambda_1 & 0 & 0 \\ 0 & \lambda_2 & 0 \\ 0 & 0 & \lambda_3 \end{vmatrix}$$

(c) $\qquad T(s) = \dfrac{C(s)}{U(s)} = C(sI-A)BU(s)$

$$= [1 \quad 0 \quad 0]\left\{\dfrac{\text{adj}(sI-A)}{\det(sI-A)}\right\}\begin{bmatrix} 0 \\ 0 \\ 1 \end{bmatrix}\cdot\dfrac{1}{s}$$

$$= [1 \quad 0 \quad 0]\dfrac{\begin{bmatrix} s(s+3) & -1 & s \\ -s(s+3) & -s(s+3) & 1 \\ 1 & s & s^2 \end{bmatrix}\begin{bmatrix} 0 \\ 0 \\ 1 \end{bmatrix}[1/s]}{s^3+3s^2+1}$$

$$= \dfrac{1}{s^3+3s^2+1}$$

(d)   From transfer function, characteristic equation is, $s^3 + 3s^2 + 1 = 0$

Constructing Routh array

$$
\begin{array}{c|cc}
s^3 & 1 & 0 \\
s^2 & 3 & 1 \\
s^1 & -1/30 & \\
s^0 & 1 &
\end{array}
$$

Since, one element in the first column is –ve, therefore, according to Routh Hurwtz criterion, system is unstable

---

Problem 7.30.   The state equation of a linear time invarient system is given by

$$
\begin{bmatrix} x_1 & (t) \\ x_2 & (t) \end{bmatrix} = \begin{bmatrix} 0 & 1 \\ -1 & -2 \end{bmatrix} \begin{bmatrix} x_1 & (t) \\ x_2 & (t) \end{bmatrix} + \begin{bmatrix} 0 \\ 1 \end{bmatrix} r(t)
$$

(a)   Find state transistion Matrix $\phi(t)$ and the characteristic equation of the system.

(b)   Determine the state vector $x_2(t)$ for $t \geq 0$ when $r(t) = u(t)$. Assume values of state initially to be zero.   [GATE'95]

Solution:

(a)   The state transistion matrix $\phi(t,0)$ is

$$
\phi = \left[(sI - A)^{-1} \right]
$$

where

$$
sI = \begin{bmatrix} s & 0 \\ 0 & s \end{bmatrix} - \begin{bmatrix} 0 & 1 \\ -1 & -2 \end{bmatrix} = \begin{bmatrix} s & -1 \\ 1 & s+2 \end{bmatrix}
$$

$$
(sI - A)^{-1} = \frac{\begin{bmatrix} s+2 & -1 \\ 1 & s \end{bmatrix}^T}{s(s+2)+1} = \frac{\begin{bmatrix} s+2 & 1 \\ -1 & s \end{bmatrix}}{s^2 + 2s + 1}
$$

$$
\phi(t,0) = \frac{s+2}{s^2 + 2s + 1} = \frac{s+2}{(s+1)^2}
$$

expanding in fartical fraction.

$$\phi(t,0) = \frac{K_1}{(s+1)^2}+\frac{K_2}{s+1}$$

$$\therefore \qquad K_1 = s|_{s=2}=1 \text{ and } K_2=\frac{d}{ds}(s)=1$$

$$\frac{s+2}{(s+1)^2} = \frac{1}{(s+1)^2}+\frac{1}{(s+1)}$$

$$\frac{s}{(s+1)^2} = \frac{K}{(s+1)^2}+\frac{K_2}{s+1}$$

$$K_1 = s|_{s=-1}=-1, \quad K_2=\frac{d}{ds}(s)=1$$

Hence $\quad \dfrac{-s}{(s+1)^2} = \dfrac{-1}{(s+1)^2}+\dfrac{1}{s+1}$

$$\phi(s)=\begin{bmatrix} \dfrac{1}{(s+1)^2}+\dfrac{1}{(s+1)} & \dfrac{1}{(s+1)^2} \\ \dfrac{-1}{(s+1)^2} & \dfrac{-1}{(s+1)^2}+\dfrac{1}{(s+2)} \end{bmatrix}$$

$$\phi(t,0) = \begin{bmatrix} te^{-t}+e^{-t} & te^{-t} \\ -te^{-t} & -te^{-t}+e^{-t} \end{bmatrix} u(t)$$

Characteristic equation is, $\det |sI - A| = s^2+2s+1=0$

(b) $\qquad X(s) = (sI-A)^{-1} X(0)+(sI-A)^{-1} BU(s)$

Since $\qquad X(0) = \begin{bmatrix} 0 \\ 0 \end{bmatrix}$, therefore

$$X(s) = (sI-A)^{-1} BU(s)$$

$$= \frac{\begin{bmatrix} s+2 & 1 \\ -1 & 3 \end{bmatrix}\begin{bmatrix} 0 \\ 1 \end{bmatrix}1/8}{s^2+2s+1}$$

$$= \frac{1}{(s+1)^2}[1/3].\frac{1}{8} = \begin{bmatrix} \dfrac{1}{s(s+1)^2} \\[3mm] \dfrac{1}{(s+1)^2} \end{bmatrix}$$

$$= \frac{1}{s(s+1)^2} = \frac{K_1}{s} + \frac{K_2}{(s+1)^2} + \frac{K_3}{s+1}$$

$$K_1 = \left.\frac{1}{(s+1)^2}\right|_{s=-1} = 1, \quad K_2 = \left.\frac{1}{s}\right|_{s=-1} = 1$$

$$K_3 = \left.\frac{d}{ds}\left[\frac{1}{s}\right]\right|_{s=-1} = -1, \quad K_4 = \left.\frac{1}{s^2}\right|_{s=-1} = -1$$

Hence $\quad \dfrac{1}{s(s+1)^2} = \dfrac{1}{s} - \dfrac{1}{(s+1)^2} - \dfrac{1}{s+1}$

$$\therefore \qquad x(s) = \begin{bmatrix} \dfrac{1}{s} - \dfrac{1}{(s+1)^2} - \dfrac{1}{s+1} \\[4mm] \dfrac{1}{(s+1)^2} \end{bmatrix}$$

The State vector is

$$x(t) = \begin{bmatrix} 1 - te^{-t} & -e^{-t} \\[2mm] te^{-t} \end{bmatrix} u(t)$$

$$X_2(t) = te^{-t}U(t)$$

Problem 7.31.   Obtain a state space representation in diagram form for the following system:

$$\frac{d^3 y}{dt^3} + 6\frac{d^2 y}{dt^2} + 11\frac{dy}{dt} + 6y = 6u(t)$$

GATE' 95]

Solution:

$$\frac{d^3y}{dt^3}+6\frac{d^2y}{dt^2}+11\frac{dy}{dt}+6y=6u(t)$$

Let $\qquad x_1(t)=y(t)$

$$x_2(t)=\frac{dx_1(t)}{dt}=x_1(t)=\frac{dy}{dt}$$

$$x_3(t)=\frac{dx_2(t)}{dt}=x_2(t)=\frac{d^2y}{dt^2}$$

$$x_4(t)=\frac{dx_3(t)}{dt}=x_3(t)=\frac{d^3y}{dt^3}$$

$$\dot{x}_3(t)=\frac{d^3y}{dt^3}=-6\frac{d^2y}{dt^2}-11\frac{dy}{dt}-6y+6u(t)$$

$$\begin{bmatrix}\dot{x}_1(t)\\\dot{x}_2(t)\\\dot{x}_3(t)\end{bmatrix}=\begin{bmatrix}0 & 1 & 0\\0 & 0 & 1\\-6 & -11 & -6\end{bmatrix}\begin{bmatrix}x_1(t)\\x_2(t)\\x_3(t)\end{bmatrix}-\begin{bmatrix}0\\0\\6\end{bmatrix}u(t)$$

$$A=\begin{bmatrix}0 & 1 & 0\\0 & 0 & 1\\-6 & -11 & -6\end{bmatrix};\; B=\begin{bmatrix}0\\0\\6\end{bmatrix}$$

Output $\qquad y=X_1(t)=\begin{bmatrix}1 & 0 & 0\end{bmatrix}\begin{bmatrix}x_1\\x_2\\x_3\end{bmatrix}+6u(t)$

$$C=\begin{bmatrix}1 & 0 & 0\end{bmatrix}$$

The characteristic equation is

$$|aI-A|=s^3+6s^2+11s+6=0$$

The eigenvalues are $s_1=1,\ s_2=-2,\ s_3=-3$

Transformation Matrix:

$$P=\begin{bmatrix}1 & 1 & 1\\s_1 & s_2 & s_3\\s_1^2 & s_2^2 & s_3^2\end{bmatrix}=\begin{bmatrix}1 & 1 & 1\\-1 & -2 & -3\\1 & 4 & 9\end{bmatrix}$$

As $|P| \neq 0, |P|^{-1}$ exists

$$|P|^{-1} = \frac{-1}{2}\begin{bmatrix} -6 & -5 & -1 \\ 6 & 8 & 2 \\ -2 & -3 & -1 \end{bmatrix}$$

$\therefore$ Diagonal Matrix, $A = P^{-1}AP$

$$= \frac{-1}{2}\begin{bmatrix} -6 & -5 & -1 \\ 6 & 8 & 2 \\ -2 & -3 & -1 \end{bmatrix}\begin{bmatrix} 0 & 1 & 0 \\ 0 & 0 & 1 \\ -6 & -11 & -6 \end{bmatrix}\begin{bmatrix} 1 & 1 & 1 \\ -1 & -2 & -3 \\ 1 & 4 & 9 \end{bmatrix}$$

$$= \begin{bmatrix} -1 & 0 & 0 \\ 0 & -2 & 0 \\ 0 & 0 & -3 \end{bmatrix}$$

$$\hat{B} = P^{-1}B = -\frac{1}{2}\begin{bmatrix} -6 & -5 & -1 \\ 6 & 8 & 2 \\ -2 & -3 & -1 \end{bmatrix}\begin{bmatrix} 0 \\ 0 \\ 6 \end{bmatrix} = \begin{bmatrix} 3 \\ -6 \\ 3 \end{bmatrix}$$

$$\hat{0} = OP = \begin{bmatrix} 1 & 0 & 0 \end{bmatrix}\begin{bmatrix} 1 & 1 & 1 \\ -1 & -2 & -3 \\ 1 & 4 & 9 \end{bmatrix} = \begin{bmatrix} 1 & 1 & 1 \end{bmatrix}$$

Hence, the vector matrix differential equation of the transformed system can be written as

$$\frac{d}{dt}\begin{bmatrix} Z_1 \\ Z_2 \\ Z_3 \end{bmatrix} = \begin{bmatrix} -1 & 0 & 0 \\ 0 & -2 & 0 \\ 0 & 0 & -3 \end{bmatrix}\begin{bmatrix} Z_1 \\ Z_2 \\ Z_3 \end{bmatrix} + \begin{bmatrix} 3 \\ -6 \\ 3 \end{bmatrix}u$$

and output $y = \begin{bmatrix} 1 & 1 & 1 \end{bmatrix}\begin{bmatrix} Z_1 \\ Z_2 \\ Z_3 \end{bmatrix}$

Problem 7.32. From the signal flow graph shown in the figure below, obtain the state space model with x_1, x_2, x_3 and x_4 as state variable and write the transfer function directly from the state space model.

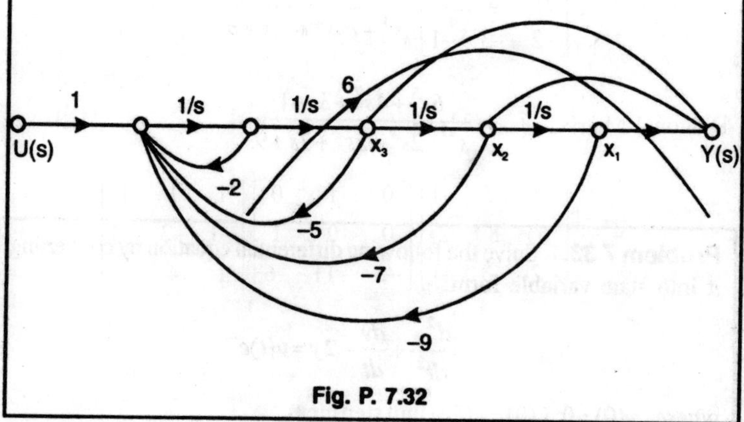

Fig. P. 7.32

Solution:

From the signal flow graph, we have

$$\dot{x}_1 = x_2$$

$$\dot{x}_2 = x_3$$

$$\dot{x}_3 = x_4$$

$$\dot{x}_4 = u - 2x_4 - 5x_3 - 7x_2 - 9x_1$$

and

$$y = 3x_2 + 4x_3 + 6x_4 + x_1$$

Writing in matrix form:

$$\begin{bmatrix} \dot{x}_1 \\ \dot{x}_2 \\ \dot{x}_3 \\ \dot{x}_4 \end{bmatrix} = \begin{bmatrix} 0 & 1 & 0 & 0 \\ 0 & 0 & 1 & 0 \\ 0 & 0 & 0 & 1 \\ -9 & -7 & -5 & -5 \end{bmatrix} \begin{bmatrix} x_1 \\ x_2 \\ x_3 \\ x_4 \end{bmatrix} + \begin{bmatrix} 0 \\ 0 \\ 0 \\ 1 \end{bmatrix} u$$

and

$$y = \begin{bmatrix} 1 & 3 & 4 & 6 \end{bmatrix} \begin{bmatrix} x_1 \\ x_2 \\ x_3 \\ x_4 \end{bmatrix}$$

Transfer function $H(s) = \dfrac{Y(s)}{U(s)}$

$$= \frac{b_2 s^3 + b_2 s^2 + b_1 s + b_0}{s^4 + a_3 s^3 + a_2 s^2 + a_1 s + a_0}$$

$$= \frac{6s^3 + 4s^2 + 3s + 1}{s^4 + 2s^3 + 5s^2 + 7s + 9}$$

Problem 7.33. Solve the following differential equation by converting it into state variable form:

$$\frac{d^2 y}{dt^2} + \frac{dy}{dt} - 2y = u(t)e^{-t}$$

where $y(0)=0; \dot{y}(0)=u(t)=$ unit step input

[GATE' 95]

Solution:

$$\frac{d^2 y}{dt^2} + \frac{dy}{dt} - 2y = u(t)e^{-t}$$

Defining the state variables

$$x_1 = y$$
$$x_2 = \dot{y} = \dot{x}_1$$

$$\therefore \qquad \dot{x}_1 = x_2 \text{ and } \dot{x}_2 = -x_2 + 2x_1 + r(t)$$

Putting in Matrix form:

$$\begin{bmatrix} \dot{x}_1 \\ \dot{x}_2 \end{bmatrix} = \begin{bmatrix} 0 & 1 \\ 2 & -1 \end{bmatrix} \begin{bmatrix} x_1 \\ x_2 \end{bmatrix} + \begin{bmatrix} 0 \\ 1 \end{bmatrix} r(t)$$

and
$$y = \begin{bmatrix} 1 & 0 \end{bmatrix} \begin{bmatrix} x_1 \\ x_2 \end{bmatrix}$$

$$X(s) = (sI - A)^{-1} x(0) + (sI - A)^{-1} BR(s)$$

$$X(0) = \begin{bmatrix} x_1 & (0) \\ x_2 & (0) \end{bmatrix} = \begin{bmatrix} y & (0) \\ \dot{y} & (0) \end{bmatrix} = \begin{bmatrix} 0 \\ 0 \end{bmatrix}$$

Hence
$$X(s) = (sI - A)^{-1} BR(s)$$

$$(sI - A) = \begin{bmatrix} s & -1 \\ -2 & s+1 \end{bmatrix}$$

$$(sI-A)^{-1} = \frac{\begin{bmatrix} s+1 & 2 \\ 1 & 8 \end{bmatrix}^T}{s^2+s-2} = \frac{\begin{bmatrix} s+1 & 1 \\ 2 & 8 \end{bmatrix}}{(s-1)(s+2)}$$

$$X(s) = \begin{bmatrix} s+1 & 1 \\ 2 & 8 \end{bmatrix} \frac{1}{(s-1)(s+2)} \begin{bmatrix} 0 \\ 1 \end{bmatrix}$$

$$= \frac{1}{(s-1)(s+2)(s+1)} \begin{bmatrix} 1 \\ s \end{bmatrix}$$

Expanding in Partial fraction,

$$\frac{1}{(s-1)(s+2)(s+1)} = \frac{K_1}{(s-1)} - \frac{K_2}{(s+2)} + \frac{K_3}{s+1}$$

$$K_1 = \frac{1}{(s+2)(s+1)} \bigg|_{s=1} = \frac{1}{6}$$

$$K_2 = \frac{1}{(s-1)(s+1)} \bigg|_{s=-2} = \frac{1}{3}$$

$$K_3 = \frac{1}{(s-1)(s+2)} \bigg|_{s=-1} = -\frac{1}{2}$$

Hence

$$\frac{1}{(s-1)(s+2)(s+1)} = \frac{K_1}{(s-1)} + \frac{K_2}{(s+2)} + \frac{K_3}{s+1}$$

$$= \frac{1/6}{(s-1)} + \frac{1/3}{(s+2)} - \frac{1/2}{s+1}$$

similarly
$$\frac{s}{(s-1)(s+2)(s+1)} = \frac{K_1}{(s-1)} + \frac{K_2}{(s+2)} + \frac{K_3}{s+1}$$

$$K_1 = \frac{s}{(s+2)(s+1)}\Bigg|_{s=1} = \frac{1}{6}$$

$$K_2 = \frac{s}{(s-1)(s+1)}\Bigg|_{s=-2} = -\frac{2}{3}$$

$$K_3 = \frac{s}{(s-1)(s+2)}\Bigg|_{s=-1} = \frac{1}{2}$$

Hence,
$$\frac{s}{(s-1)(s+2)(s+1)} = \frac{\frac{1}{6}}{(s-1)} - \frac{\frac{2}{3}}{(s+2)} + \frac{\frac{1}{2}}{(s+1)}$$

and
$$X(s) = \begin{bmatrix} \dfrac{\frac{1}{6}}{s-1} + \dfrac{\frac{1}{3}}{s+2} - \dfrac{\frac{1}{2}}{s+1} \\[4mm] \dfrac{\frac{1}{6}}{s-1} - \dfrac{\frac{2}{3}}{s+2} + \dfrac{\frac{1}{2}}{s+1} \end{bmatrix}$$

$$X(t) = \begin{bmatrix} \dfrac{1}{6}e^t + \dfrac{1}{3}e^{-2t} - \dfrac{1}{2}e^{-t} \\[4mm] \dfrac{1}{6}e^t + \dfrac{2}{3}e^{-2t} + \dfrac{1}{2}e^{-t} \end{bmatrix} u(t)$$

$$y(t) = x_1 = \left[\frac{1}{6}e^t + \frac{1}{3}e^{-2t} - \frac{1}{2}e^{-t} \right] u(t)$$

Problem 7.34. Following in the state-space dynamic model of a linear system whose eignvalues are $-3, -2, -1$.

$$X = \begin{bmatrix} 0 & 1 & 0 \\ 0 & 0 & 1 \\ -6 & -11 & -6 \end{bmatrix} x + \begin{bmatrix} 0 \\ 0 \\ 2 \end{bmatrix} u$$

Given that $u=0$, $x(0)=(0\ 0\ 1)$; Find $x(1)$ **[GATE' 94]**

Solution:

The state vector $x(t)$ is given by

$$x(t) = e^{At} = C(0)$$

State transition matrix, $e^{At} = Te^{\Delta t}T^{-1}$

Let the eigen vector be

$$e_1 = \begin{bmatrix} a_1 \\ a_2 \\ a_3 \end{bmatrix}, \quad e_2 = \begin{bmatrix} b_1 \\ b_2 \\ b_3 \end{bmatrix}, \quad e_3 = \begin{bmatrix} c_1 \\ c_2 \\ c_3 \end{bmatrix}$$

Then

$$Ae_1 = \lambda_1 C_1$$

or

$$\begin{bmatrix} 0 & 1 & 0 \\ 0 & 0 & 1 \\ -6 & -11 & -6 \end{bmatrix} \begin{bmatrix} a_1 \\ a_2 \\ a_3 \end{bmatrix} = -1 \begin{bmatrix} a_1 \\ a_2 \\ a_3 \end{bmatrix}$$

which gives the equations

$$a_2 = -a_1, \; a_3 = -a_2$$

$$\bullet \; -6a_1 - 11a_2 - 6a_3 = -a_3$$

Putting $a_1 = 1$, we have $a_2 = -1, a_3 = 1$

or

$$e_1 = \begin{bmatrix} 1 \\ -1 \\ 1 \end{bmatrix}$$

Similarly

$$Ae_2 = \lambda_2 e_1$$

or

$$\begin{bmatrix} 0 & 1 & 0 \\ 0 & 0 & 1 \\ -6 & -11 & -6 \end{bmatrix} \begin{bmatrix} b_1 \\ b_2 \\ b_3 \end{bmatrix} = -2 \begin{bmatrix} b_1 \\ b_2 \\ b_3 \end{bmatrix}$$

or

$$b_2 = -2b_1; \; b_3 = -2b_2$$

$$-6b_1 - 11b_2 - 6b_3 = -2b_3$$

Putting $b_1 = -2$, we have $b_2 = 2, b_3 = 4$

or
$$e_2 = \begin{bmatrix} 1 \\ -2 \\ 4 \end{bmatrix}$$

Similarly, $\qquad Ae_3 = \lambda_3 e_3$

$$\begin{bmatrix} 0 & 1 & 0 \\ 0 & 0 & 1 \\ -6 & -11 & -6 \end{bmatrix} \begin{bmatrix} C_1 \\ C_2 \\ C_3 \end{bmatrix} = -3 \begin{bmatrix} C_1 \\ C_2 \\ C_3 \end{bmatrix}$$

or
$$C_2 = -3C_1,\ C_3 = -3C_2$$

$$-6C_1 - 11C_2 - 6C_3 = -3C_3$$

Putting $\qquad C_1 = 1$, we have $C_2 = -3,\ C_3 = 9$

or
$$C_3 = \begin{bmatrix} 1 \\ -3 \\ 9 \end{bmatrix}$$

Hence
$$T = \begin{bmatrix} e_1 & e_2 & e_3 \end{bmatrix} = \begin{bmatrix} 1 & 1 & 1 \\ -1 & -2 & -3 \\ 1 & 4 & 9 \end{bmatrix}$$

and
$$T^{-1} = \frac{\begin{bmatrix} -6 & 6 & -2 \\ -5 & 8 & -2 \\ -1 & 2 & -1 \end{bmatrix}^T}{\Delta}$$

where $\Delta = 1(-18+12) - 1(-9+3) + 1(-4+2) = -6 + 6 - 2 = -2$

Hence
$$T^{-1} = \begin{bmatrix} 3 & 2.5 & 0.5 \\ -3 & -4 & -1 \\ 1 & 1.5 & 0.5 \end{bmatrix}$$

and $e^{At} = T^{\lambda t} T^{-1}$; $x(t) = e^{At} X(0)$

$$= \begin{bmatrix} 1 & 1 & 1 \\ -1 & -2 & -3 \\ 1 & 4 & 9 \end{bmatrix} \begin{bmatrix} e^{-t} & 0 & 0 \\ 0 & e^{-2t} & 0 \\ 0 & 0 & e^{-3t} \end{bmatrix} \begin{bmatrix} 3 & 2.5 & 0.5 \\ -3 & -4 & -1 \\ 1 & 1.5 & 0.5 \end{bmatrix} \begin{bmatrix} 0 \\ 0 \\ 1 \end{bmatrix}$$

$$= \begin{bmatrix} 1 & 1 & 1 \\ -1 & -2 & -3 \\ 1 & 4 & 9 \end{bmatrix} \begin{bmatrix} e^{-t} & 0 & 0 \\ 0 & e^{-2t} & 0 \\ 0 & 0 & e^{-3t} \end{bmatrix} \begin{bmatrix} 0.5 \\ -1 \\ 0.5 \end{bmatrix}$$

$$= \begin{bmatrix} 1 & 1 & 1 \\ -1 & -2 & -3 \\ 1 & 4 & 9 \end{bmatrix} \begin{bmatrix} 0.5e^{-t} \\ -e^{-2t} \\ 0.5e^{-3t} \end{bmatrix}$$

$$= \begin{bmatrix} 0.5e^{-t} & -e^{-2t} & 0.5e^{-3t} \\ -0.5e^{-t} & 2e^{-2t} & -1.5e^{-3t} \\ 0.5e^{-t} & -4e^{-2t} & 4.5e^{-3t} \end{bmatrix}$$

and
$$X(1) = X(t)\Big|_{t=1} = \begin{bmatrix} 0.0735 \\ 0.9121 \\ -0.1334 \end{bmatrix}$$

Problem 7.35. A control system is described by the differential equation

$$\frac{d^2 y(t)}{dt^3} = u(t)$$

where $y(t)$ is the observed output and $u(t)$ is the input

(a) Describe the system in the state variable form, i.e.,

$$\dot{x} = Ax + Bu, \quad \dot{y} = Cx + Du$$

(b) Calculate the state transition matrix e^{At} of the system.

(c) Is the system controllable?

Solution:

Differential equation is $\dfrac{d^3 y(t)}{dt^3} = u(t)$

(a) Putting $\quad x_1(t) = y(t)$

$$x_2(t) = \frac{dx_1}{dt}$$

$$x_3(t) = \frac{dx_2}{dt}$$

we have, $\dfrac{dx_1}{dt} = x_2(t),\ \dfrac{dx_2}{dt} = x_3(t),\ \dfrac{dx_3}{dt} = u(t)$

and $\qquad y(t) = x_1(t)$

In matrix form, we have

$$\begin{bmatrix} \dot{x}_1 \\ \dot{x}_2 \\ \dot{x}_3 \end{bmatrix} = \overset{\mathbf{A}}{\begin{bmatrix} 0 & 1 & 0 \\ 0 & 0 & 1 \\ 0 & 0 & 0 \end{bmatrix}} \begin{bmatrix} x_1 \\ x_2 \\ x_3 \end{bmatrix} + \overset{\mathbf{B}}{\begin{bmatrix} 0 \\ 0 \\ 1 \end{bmatrix}} u(t)$$

and $\qquad y(t) = \overset{\mathbf{C}}{\frac{1}{3}[1 \quad 0 \quad 0]} + \overset{\mathbf{D}}{[0]} u(t)$

(b) The state transistion matrix is given by

$$\phi(t) = \mathcal{L}^{-1}\left[(sI - A)^{-1}\right]$$

$$sI - A = \begin{bmatrix} s & 0 & 0 \\ 0 & s & 0 \\ 0 & 0 & s \end{bmatrix} - \begin{bmatrix} 0 & 1 & 0 \\ 0 & 0 & 1 \\ 0 & 0 & 0 \end{bmatrix} = \begin{bmatrix} s & -1 & 0 \\ 0 & s & -1 \\ 0 & 0 & s \end{bmatrix}$$

$$(sI - A) \text{ adjoint} = \begin{bmatrix} s^2 & 0 & 0 \\ +s & s^2 & 0 \\ 1 & +s & s^2 \end{bmatrix} = \begin{bmatrix} s^2 & s & 1 \\ 0 & s^2 & s \\ 0 & 0 & s^2 \end{bmatrix}$$

$$\text{Determinant, } \Delta = \begin{vmatrix} s & -1 & 0 \\ 0 & s & -1 \\ 0 & 0 & s \end{vmatrix} = s^3$$

$$\therefore \quad (sI-A)^{-1} = \frac{\mathrm{adj}(sI-A)}{\Delta} \begin{bmatrix} \dfrac{1}{s} & \dfrac{1}{s^2} & \dfrac{1}{s^3} \\ 0 & \dfrac{1}{s} & \dfrac{1}{s^2} \\ 0 & 0 & \dfrac{1}{s} \end{bmatrix}$$

and
$$\phi(t) = \mathcal{L}^{-1}\left[(sI-A)^{-1}\right] = \begin{bmatrix} u(t) & tu(t) & \dfrac{t^2}{2}u(t) \\ 0 & u(t) & tu(t) \\ 0 & 0 & u(t) \end{bmatrix}$$

(c) System controllability

$$B = \begin{bmatrix} 0 \\ 0 \\ 1 \end{bmatrix}$$

$$AB = \begin{bmatrix} 0 & 1 & 0 \\ 0 & 0 & 1 \\ 0 & 0 & 0 \end{bmatrix}\begin{bmatrix} 0 \\ 0 \\ 1 \end{bmatrix} = \begin{bmatrix} 0 \\ 1 \\ 0 \end{bmatrix}$$

$$A^2B = \begin{bmatrix} 0 & 1 & 0 \\ 0 & 0 & 1 \\ 0 & 0 & 0 \end{bmatrix}\begin{bmatrix} 0 & 1 & 0 \\ 0 & 0 & 1 \\ 0 & 0 & 0 \end{bmatrix}\begin{bmatrix} 0 \\ 0 \\ 1 \end{bmatrix}$$

$$= \begin{bmatrix} 0 & 0 & 1 \\ 0 & 0 & 0 \\ 0 & 0 & 0 \end{bmatrix}\begin{bmatrix} 0 \\ 0 \\ 1 \end{bmatrix} = \begin{bmatrix} 1 \\ 0 \\ 0 \end{bmatrix}$$

$$\begin{bmatrix} B & AB & A^2B \end{bmatrix} = \begin{bmatrix} 0 & 0 & 1 \\ 0 & 1 & 0 \\ 1 & 0 & 0 \end{bmatrix}$$

Determinant $\Delta = 1(-1) = -1 \neq 0$

Hence the system is totally controllable

Problem 36.   A linear system is described by the following transfer function

$$G(s) = \frac{s+1}{(s+2)^2\,(s+5)}$$

Draw state transition signal flow graphs using canonical forms of decomposition.

Solution:

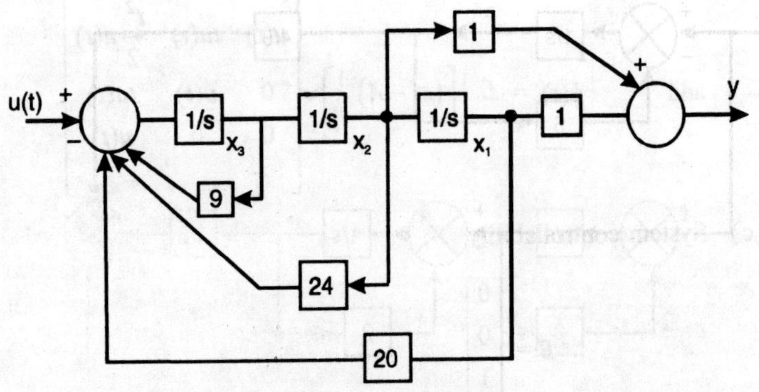

Fig. P. 7.36

I.   Control Canonical form of decomposition

Note: 1/s block denotes an integrator $\int$.

II.   Observer canonical form

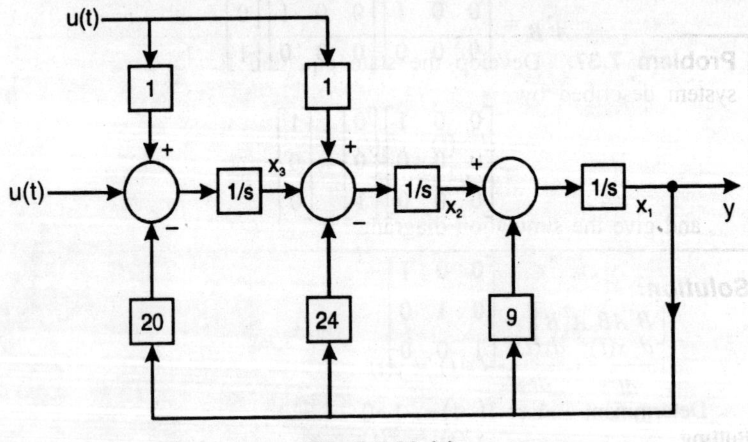

Fig. P 7.36 (a)

Note that, $1/s$ block denotes an integrator $\int$.

III. By Partial fraction Equation

$$G(s) = \frac{4}{9}\frac{1}{s+2} - \frac{1}{3}\frac{1}{(s+2)^2} - \frac{4}{9}\frac{1}{s+5}$$

Jordan form,

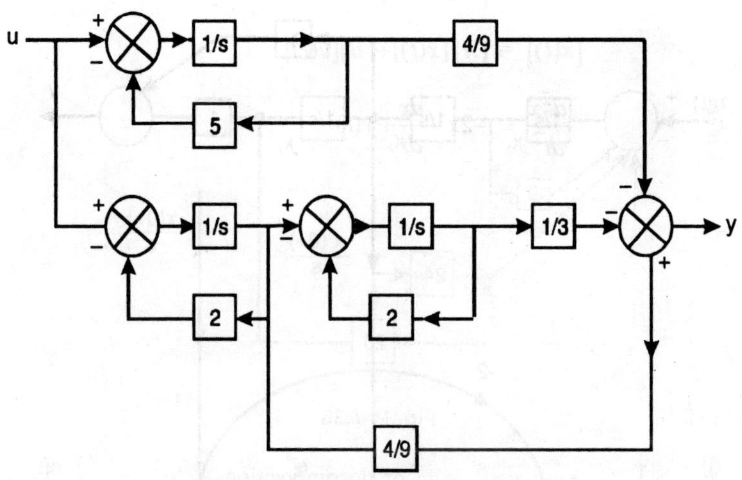

Fig. P. 7.36 (b)

Problem 7.37. Develop the state equation formulation for the system described by

$$\frac{d^2x(t)}{dt^2} + \frac{dx(t)}{dt} + 2x(t) = y(t)$$

and give the simulation diagram.

Solution:

$$\frac{d^2x(t)}{dt^2} + \frac{dx(t)}{dt} + 2x(t) = y(t)$$

Putting
$$x_1(t) = x(t)$$

$$x_2(t) = \frac{dx(t)}{dt}$$

Thus,
$$\frac{dx_2(t)}{dt} = -x_2(t) - 2x_1(t) + y(t)$$

$$= -2x_1(t) - x_2(t) + y(t)$$

$$A = \begin{bmatrix} 0 & 1 \\ -2 & -1 \end{bmatrix}, \quad B = \begin{bmatrix} 0 \\ 1 \end{bmatrix}$$

$$[\dot{x}(t)] = [A][x(t)] + [B][u(t)]$$

$$\frac{d^2x}{dt^2} = -2x - \frac{dx}{dt} + y(t)$$

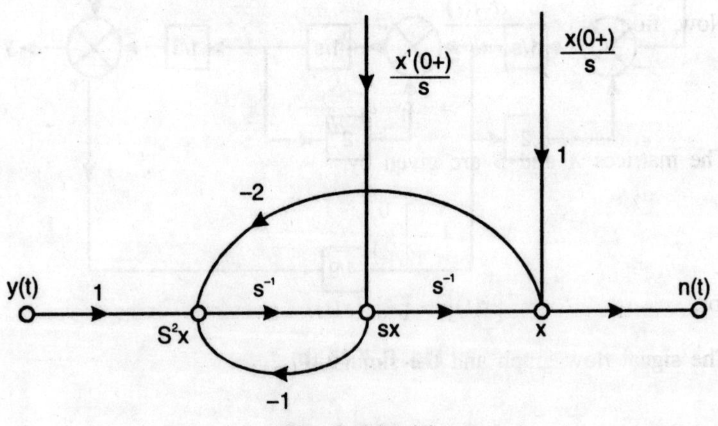

Fig. P. 7.37 (a)

The analog Simulation diagram is

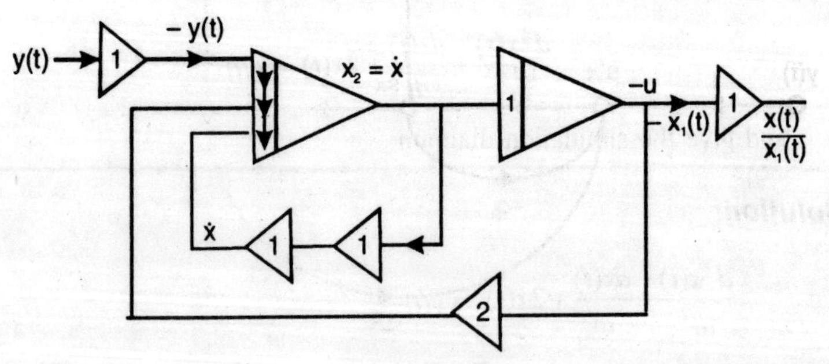

Fig. P. 7.37 (b)

Problem 7.38. Develop the state equation formulation for a system described by

$$\ddot{x} + a_1 x + x = y$$ and give an analog simulation diagram.

Solution:

The system equation is,

$$\ddot{x} + a_1 \dot{x} + x = y \tag{1}$$

i.e.,

$$\frac{d^2 x}{dt^2} + a_1 \frac{dx}{dt} + x = y$$

Putting

$$x_1(t) = x(t)$$

$$x_2(t) = \frac{dx(t)}{dt}$$

Now, from (1)

$$\frac{dx_2(t)}{dt} = -a_1 \frac{dx_1}{dt} - x_1 + y$$

$$= -x_1 - a_1 \frac{dx_1}{dt} + y$$

The matrices A and B are given by,

$$A = \begin{bmatrix} 0 & 1 \\ -1 & -a_1 \end{bmatrix}, \quad B = \begin{bmatrix} 0 \\ -1 \end{bmatrix}$$

for,

$$[\dot{x}(t)] = [A][x(t)] + [B][u(t)]$$

The signal flow graph and the simulation diagrams are given below,

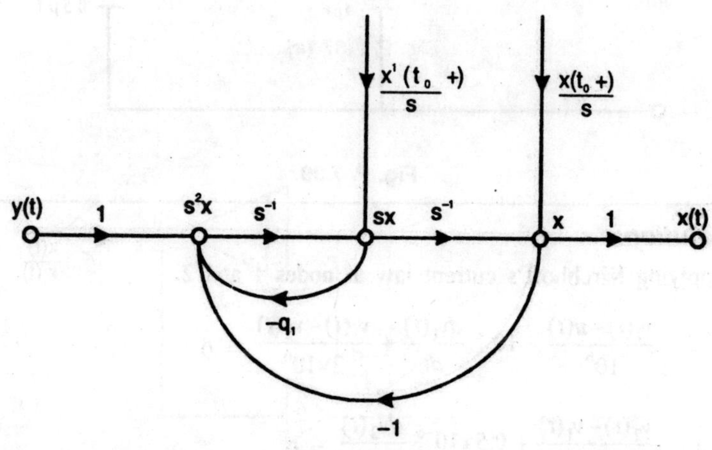

Fig. 7.38 (a)

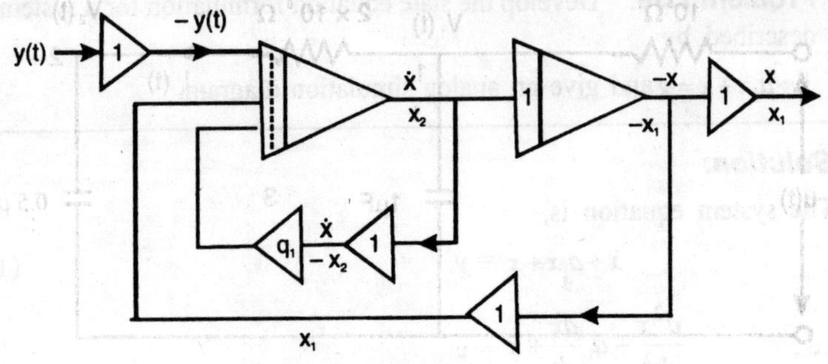

Fig. 7.38 (b)

Problem 7.39.   Write the differential equations characterising the network given in the figure, and hence obtain the state equation $\dot{x} = An + Bu$ and the output equation $y = Cn + Du$.

Take the voltage across the Capacitor, $v_1(t)$ and $v_2(t)$ as the two state variables and the current through the 2×10^6 ohm resistor as the output variable.

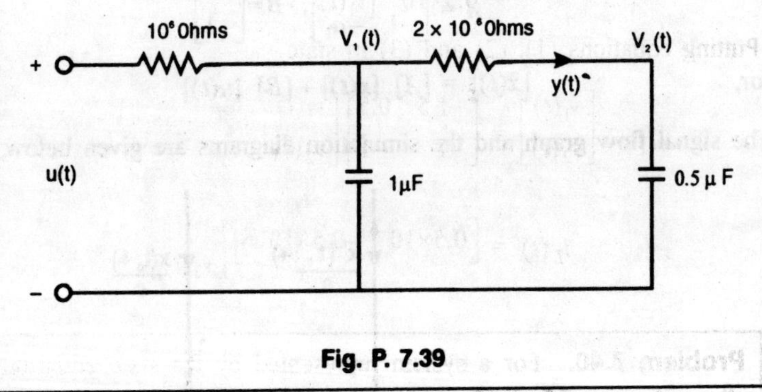

Fig. P. 7.39

Solution:

Applying Kirchhoff's current law at nodes 1 and 2.

$$\frac{v_1(t) - u(t)}{10^6} + 10^{-6}\frac{dv_1(t)}{dt} + \frac{v_1(t) - v_2(t)}{2 \times 10^6} = 0$$

and $$\frac{v_2(t) - v_1(t)}{2 \times 10^6} + 0.5 \times 10^{-6}\frac{dv_2(t)}{dt} = 0$$

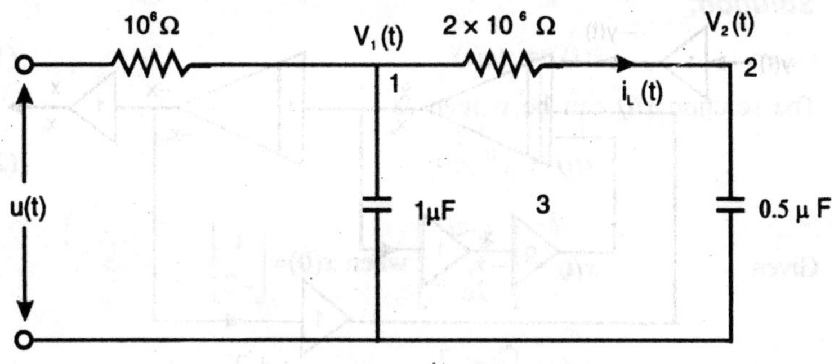

Fig. P. 7.39 (a)

Now, rearranging the above two equations we have,

$$\frac{dv_1(t)}{dt} = -1.5v_1(t) + 0.5v_2(t) + u(t) \tag{1}$$

and

$$\frac{dv_2(t)}{dt} = v_1(t) - v_2(t) \tag{2}$$

and the output $i_L(t)$ is,

$$i_L(t) = 0.5 \times 10^{-6} \frac{dv_2(t)}{dt} \tag{3}$$

$$= 0.5 \times 10^{-6} \left[v_1(t) - v_2(t) \right]$$

Putting equations (1), (2) and (3) in state variable form,

$$\begin{bmatrix} \dot{v}_1(t) \\ \dot{v}_2(t) \end{bmatrix} = \begin{bmatrix} -1.5 & 0.5 \\ 1 & -1 \end{bmatrix} \begin{bmatrix} v_1(t) \\ v_2(t) \end{bmatrix} + \begin{bmatrix} 1 \\ 0 \end{bmatrix} u(t)$$

$$i_L(t) = \begin{bmatrix} 0.5 \times 10^{-6} & -0.5 \times 10^{-6} \end{bmatrix} \begin{bmatrix} v_1(t) \\ v_2(t) \end{bmatrix}$$

Problem 7.40. For a system represented by the state equation

$$\dot{x}(t) = Ax(t)$$

the response of $x(t) = \begin{bmatrix} e^{-2t} \\ -2e^{-2t} \end{bmatrix}$ when $x(0) = \begin{bmatrix} 1 \\ -2 \end{bmatrix}$

and $x(t) = \begin{bmatrix} e^{-t} \\ -e^{-t} \end{bmatrix}$ when $x(0) = \begin{bmatrix} 1 \\ -1 \end{bmatrix}$

Determine the system matrix A and the state transition matrix.

Solution:

Given
$$\dot{x}(t) = Ax(t) \tag{1}$$

The solution $x(t)$ can be written as
$$x(t) = e^{At}x(0) \tag{2}$$

Given
$$x(t) = \begin{bmatrix} e^{-2t} \\ -2e^{-2t} \end{bmatrix} \text{ when } x(0) = \begin{bmatrix} 1 \\ -2 \end{bmatrix}$$

and
$$x(t) = \begin{bmatrix} e^{-t} \\ -e^{-t} \end{bmatrix} \text{ when } x(0) = \begin{bmatrix} 1 \\ -1 \end{bmatrix}$$

From equation (2) and assuming $e^{At} = \begin{bmatrix} a_0 & a_1 \\ a_2 & a_3 \end{bmatrix}$

we get
$$\begin{bmatrix} e^{-2t} \\ -2e^{-2t} \end{bmatrix} = \begin{bmatrix} a_0 & a_1 \\ a_2 & a_3 \end{bmatrix} \begin{bmatrix} 1 \\ -2 \end{bmatrix}$$

resulting,
$$a_0 - 2a_1 = e^{-2t} \tag{3}$$
$$a_2 - 2a_3 = -2e^{-2t} \tag{4}$$

and
$$\begin{bmatrix} e^{-t} \\ -e^{-t} \end{bmatrix} = \begin{bmatrix} a_0 & a_1 \\ a_2 & a_3 \end{bmatrix} \begin{bmatrix} 1 \\ -1 \end{bmatrix}$$

yields
$$a_0 - a_1 = e^{-t} \tag{5}$$
$$a_2 - a_3 = -e^{-t} \tag{6}$$

From equation (3) and (5), we get
$$a_0 = 2e^{-t} - e^{-2t}$$
$$a_1 = e^{-t} - e^{-2t}$$

From equation (4) and (6)
$$a_2 = -2e^{-t} - 2e^{-2t}, \quad a_3 = e^{-t} - 2e^{-2t}$$

State transition matrix,
$$e^{At} = \phi(t) = \begin{bmatrix} 2e^{-t} - e^{-2t} & e^{-t} - e^{-2t} \\ e^{-t} + 2e^{-2t} & -e^{-t} + 2e^{-2t} \end{bmatrix}$$

Then
$$\phi(s) = \mathcal{L}[\phi(s)] = \begin{bmatrix} \dfrac{2}{s+1} - \dfrac{1}{s+2} & \dfrac{1}{s+1} - \dfrac{1}{s+2} \\[2mm] \dfrac{-2}{s+1} + \dfrac{2}{s+2} & -\dfrac{1}{s+1} + \dfrac{2}{s+2} \end{bmatrix}$$

$$= \frac{1}{(s+1)(s+2)} \begin{bmatrix} s+3 & 1 \\ -2 & s \end{bmatrix}$$

$$[\phi(s)]^{-1} = \frac{adj[\phi(s)]}{det[\phi(s)]} = \begin{bmatrix} s & -1 \\ 2 & s+3 \end{bmatrix}$$

Since,
$$[sI-A]^{-1} = \phi(s)$$

$$\Rightarrow \qquad (sI - A) = [\phi(s)]^{-1} = \begin{bmatrix} s & -1 \\ 2 & s+3 \end{bmatrix}$$

Then system Matrix

$$A = \begin{bmatrix} s & 0 \\ 0 & s \end{bmatrix} - \begin{bmatrix} s & -1 \\ 2 & s+3 \end{bmatrix} = \begin{bmatrix} 0 & 1 \\ -2 & -3 \end{bmatrix}$$

Problem 7.41. A linear time-invariant system is characterized by the homogenous state equation

$$\begin{bmatrix} \dot{x}_1 \\ \dot{x}_2 \end{bmatrix} = \begin{bmatrix} 1 & 0 \\ 1 & 1 \end{bmatrix} \begin{bmatrix} x_1 \\ x_2 \end{bmatrix}$$

(a) Compute the solution of the homogenous equation, assuming the initial state vector

$$x_0 = \begin{bmatrix} 1 \\ 0 \end{bmatrix}$$

(Employ both the Laplace transform method and canonical transformation method).

(b) Consider now that the system has a forcing function and is represented by the following nonhomogeneous state equation

$$\begin{bmatrix} \dot{x}_1 \\ \dot{x}_2 \end{bmatrix} = \begin{bmatrix} 1 & 0 \\ 1 & 1 \end{bmatrix} \begin{bmatrix} x_1 \\ x_2 \end{bmatrix} + \begin{bmatrix} 1 \\ 0 \end{bmatrix} u$$

Solution:

$$\begin{bmatrix} \dot{x}_1 \\ \dot{x}_2 \end{bmatrix} = \begin{bmatrix} 1 & 0 \\ 1 & 1 \end{bmatrix} \begin{bmatrix} x_1 \\ x_2 \end{bmatrix}$$

Here
$$A = \begin{bmatrix} 1 & 0 \\ 1 & 1 \end{bmatrix}$$

Then, e^{At} (state transistion matrix) $= \mathcal{L}^{-1}\left\{[sI-A]^{-1}\right\}$

$$= \mathcal{L}^{-1}\begin{bmatrix} s-1 & 0 \\ -1 & s-1 \end{bmatrix}^{-1} = \mathcal{L}^{-1}\left[\frac{\begin{bmatrix} s-1 & 0 \\ 1 & s-1 \end{bmatrix}}{(s-1)^2}\right]$$

$$= \mathcal{L}^{-1}\begin{bmatrix} \dfrac{1}{s-1} & 0 \\ \dfrac{-1}{(s-1)^2} & \dfrac{1}{s-1} \end{bmatrix} = \begin{bmatrix} e^t & 0 \\ te^t & e^t \end{bmatrix}$$

Then solution $\quad X(t) = e^{At} X(0)$

$$\bullet \quad \begin{bmatrix} e^t & 0 \\ te^t & e^t \end{bmatrix}\begin{bmatrix} 1 \\ 0 \end{bmatrix} = \begin{bmatrix} e^t \\ te^t \end{bmatrix}$$

(b)
$$X(t) = \phi(t).\left[X(0) \int_0^t \phi(t-\tau) Bu(\tau)d\tau \right]$$

$$= \phi(t)\left[X(0) + \int_0^t \phi(\tau) Bu(\tau)d\tau \right]$$

by using $\phi(t_1+t_2) = \phi(t_1)\cdot\phi(t_2)$

But from (a), $\quad \phi(t) = \begin{bmatrix} e^t & 0 \\ te^t & e^t \end{bmatrix}$

therefore, $X(t) = \phi(t)\left[X(0) + \int_0^t \begin{bmatrix} e^{-\tau} & 0 \\ -\tau e^{-\tau} & e^{-\tau} \end{bmatrix} \right] \begin{bmatrix} 0 \\ 1 \end{bmatrix} u(\tau)\, d\tau$

Since $u(\tau)$ is a unit step input, $u(\tau) = 1$

$$\therefore \quad X(t) = \phi(t)\left[X(0) + \int_0^t \begin{bmatrix} 0 \\ e^{-\tau} \end{bmatrix} d\tau \right]$$

$$= \left\{ \begin{bmatrix} 1 \\ 0 \end{bmatrix} + \begin{bmatrix} 0 \\ 1-e^{-t} \end{bmatrix} \right\} \phi(t)$$

$$= \begin{bmatrix} 1 \\ 1-e^{-t} \end{bmatrix} \begin{bmatrix} e^t & 0 \\ t\,e^t & e^t \end{bmatrix} = \begin{bmatrix} e^t \\ t\,e^t + e^t - 1 \end{bmatrix}$$

$$= \begin{bmatrix} e^t \\ e^t(t+1)-1 \end{bmatrix}$$

Problem. 7.42 Write the state equations of the system in shown Fig.P. 7.42, in which x_1, x_2 and x_3 constitute the state vector. Determine whether the system is completely controllable and observable.

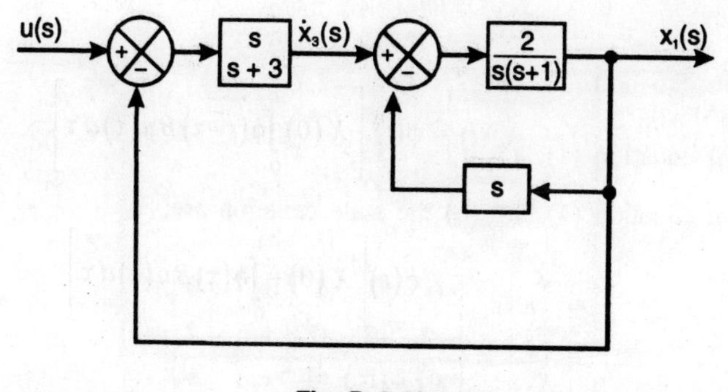

Fig. P. 7.42

Solution:

Constructing signal flow graph for the given block diagram.

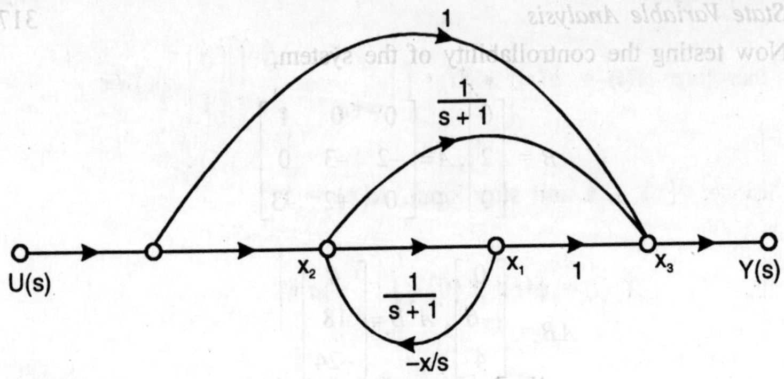

Fig. P. 7.42 (a)

From the graph, $s\,\dot{X}_1(s) = X_3(s)$ (1)

and $\left[U(s) - X_1(s)\right]\dfrac{2}{s+3} = X_2(s)$ (2)

and $X_1(s) = \dfrac{2}{s^2 + 3s}X_2(s)$ (3)

Now from equation (1), $\dot{x}_1(t) = X_3(t)$ (4)

From equation (2),

$\Rightarrow$ $2u(t) - 2x_1 = \dot{x}_2 + 3x_2$

$\Rightarrow$ $\dot{x}_2 = 2u(t) - 2x_1 - 3x_2$ (5)

From equation (3), $s^2 X_1 + 3s\,X_1 = 2X_2$

From equation (1), $\dot{x}_3 + 2x_2 - 3x_3$ (6)

From equation (4) and (6) the state equation are,

$$\begin{bmatrix} \dot{X}_1 \\ \dot{X}_2 \\ \dot{X}_3 \end{bmatrix} = \begin{bmatrix} 0 & 0 & 1 \\ -2 & -3 & 0 \\ 0 & +2 & -3 \end{bmatrix}\begin{bmatrix} x_1 \\ x_2 \\ x_3 \end{bmatrix} + \begin{bmatrix} 0 \\ 2 \\ 0 \end{bmatrix}u(t)$$

$$Y = X_1$$

Now testing the controllability of the system,

$$B = \begin{bmatrix} 0 \\ 2 \\ 0 \end{bmatrix}, A = \begin{bmatrix} 0 & 0 & 1 \\ -2 & -3 & 0 \\ 0 & +2 & -3 \end{bmatrix}$$

$$AB = \begin{bmatrix} 0 \\ -6 \\ 4 \end{bmatrix}, A^2B = \begin{bmatrix} 4 \\ 18 \\ -24 \end{bmatrix}$$

The composite matrix defined by,

$$Q_c = \begin{bmatrix} B & AB & A^2 B \end{bmatrix}$$

$$= \begin{bmatrix} 0 & 0 & 4 \\ 2 & -6 & 18 \\ 0 & 4 & -24 \end{bmatrix}$$

As, $\Delta Q_c \neq 0$ and its rank is $r = n = 3$. So, system is completely controllable. Now, testing the observability

$$C = \begin{bmatrix} 1 & 0 & 0 \end{bmatrix}$$

$$C^T = \begin{bmatrix} 1 \\ 0 \\ 0 \end{bmatrix}, A^T = \begin{bmatrix} 0 & -2 & 0 \\ 0 & -3 & 2 \\ 1 & 0 & -3 \end{bmatrix}$$

$$A^T C^T = \begin{bmatrix} 0 \\ 0 \\ 1 \end{bmatrix}, \left(A^T\right)^2 C^T = \begin{bmatrix} 0 \\ 2 \\ -3 \end{bmatrix}$$

The composite matrix defined by

$$Q_0 = \begin{bmatrix} C^T & A^T c^T & A^{T^2} C^T \end{bmatrix}$$

$$= \begin{bmatrix} 1 & 0 & 0 \\ 0 & 0 & 2 \\ 0 & 1 & -3 \end{bmatrix}$$

$\Delta Q_0 \neq 0$ and rank $r = n = 3$

Hence, the system is also completely observable.

Problem 7.43. Block diagram representation of a linear time-invariant system is given in Fig. 7.43 Check whether the system is completely observable.

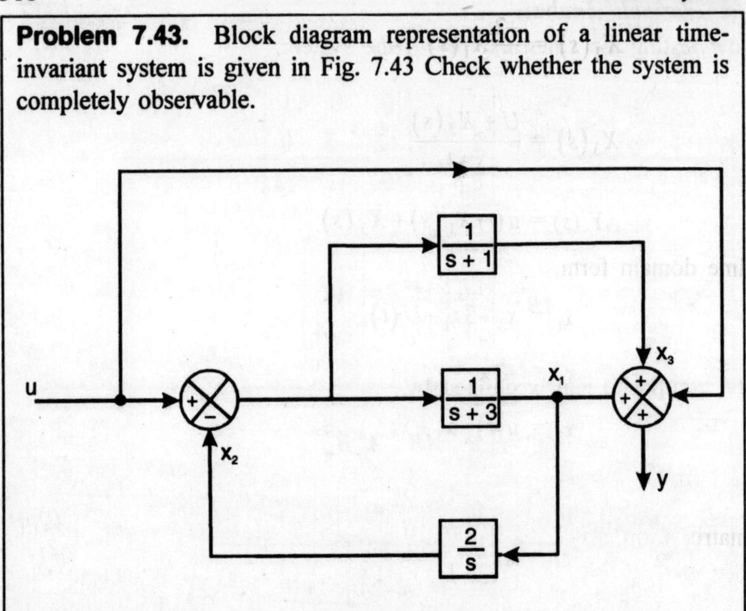

Fig. P. 7.43

Solution:

Constructing the signal flow graph for the above block diagram,
State equations in frequency domain,

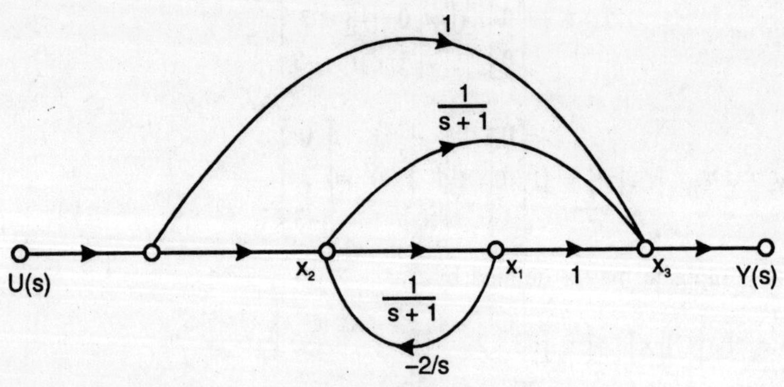

Fig. P 7.43 (a)

$$X_1(s) = \frac{U + X_2(s)}{s+3}$$

$$X_2(s) = \frac{-2}{s} X_1(s)$$

$$X_3(s) = \frac{U + X_2(s)}{s+1}$$

and,
$$Y(s) = u(t)X_1(s) + X_3(s)$$

In time domain form,

$$\dot{x}_1 = x_2 - 3x_1 + U(t)$$

$$\dot{x}_2 = -2x_1$$

$$\dot{x}_3 = u(t) - x_3 + x_3$$

In matrix form,
$$\begin{bmatrix} \dot{x}_1 \\ \dot{x}_2 \\ \dot{x}_3 \end{bmatrix} = \begin{bmatrix} -3 & 1 & 0 \\ -2 & 0 & 0 \\ 0 & 1 & -1 \end{bmatrix} \begin{bmatrix} x_1 \\ x_2 \\ x_3 \end{bmatrix} + \begin{bmatrix} 1 \\ 0 \\ 1 \end{bmatrix} u(t)$$

$$= [A]\begin{bmatrix} x_1 \\ x_2 \\ x_3 \end{bmatrix} + [B]\,u(t)$$

and
$$Y(t) = \begin{bmatrix} 1 & 0 & 1 \end{bmatrix}\begin{bmatrix} x_1 \\ x_2 \\ x_3 \end{bmatrix} + [1]\,u(t) = [C]\begin{bmatrix} x_1 \\ x_2 \\ x_3 \end{bmatrix} + [\Delta]\,u(t)$$

Now,
$$[C][A] = \begin{bmatrix} 1 & 0 & 1 \end{bmatrix}\begin{bmatrix} -3 & 1 & 0 \\ -2 & 0 & 0 \\ 0 & 1 & -1 \end{bmatrix} = \begin{bmatrix} -3 & 2 & 1 \end{bmatrix}$$

$$[C][A][A] = \begin{bmatrix} -3 & 2 & -1 \end{bmatrix}\begin{bmatrix} -3 & 1 & 0 \\ -2 & 0 & 0 \\ 0 & 1 & -1 \end{bmatrix} = \begin{bmatrix} -5 & -4 & 1 \end{bmatrix}$$

then,
$$\Delta = \begin{bmatrix} C \\ CA \\ CA^2 \end{bmatrix} = \begin{bmatrix} 1 & 0 & 1 \\ -3 & 2 & -1 \\ 5 & -4 & 1 \end{bmatrix}$$

rank of this matrix is, $r = 2$ as $\begin{bmatrix} 1 & 0 \\ -3 & 2 \end{bmatrix} \neq 0$ and $\Delta = 0$

So, one of the state variable is unobservable and system is not completely observable.

Problem 7.44. Find the forward path gain k_A and k^T for the system shown in fig. 7.44 such that the closed-loop transfer function is

$$\frac{24}{\left[(s+1)^2 + \left(\sqrt{3}\right)^2\right](s+6)}$$

Draw the root locus plot and comment upon the stability of the compensated system.

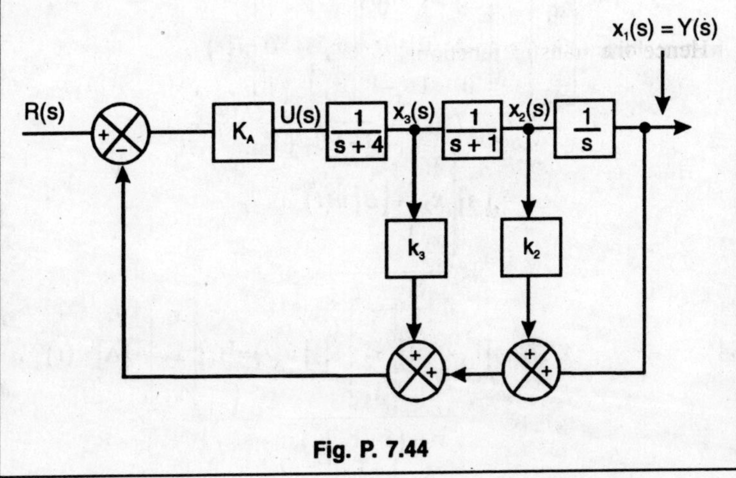

Fig. P. 7.44

Solution:

Drawing signal flow graph for the given block diagram,

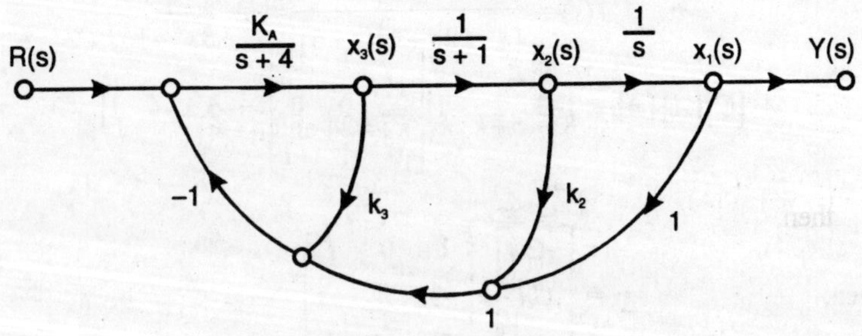

Fig. P. 7.44 (a)

There is single forward path with gain

$$P_1 = \frac{k_A}{s(s+1)(s+4)}$$

and three individual loops bring again,

$$P_{11} = \frac{-k_A k_3}{(s+4)}$$

$$P_{21} = \frac{k_A k_2}{(s+1)(s+4)}$$

$$P_{31} = \frac{-k_A}{s(s+1)(s+4)}$$

Hence the transfer function,

$$T(s) = \frac{P_1 \Delta_1}{1-(P_{11}+P_{21}+P_{31})}$$

$$= \frac{\dfrac{k_A}{s-(s+1)(s+4)}}{1+\dfrac{k_A k_3}{s+4}+\dfrac{k_A k_3}{(s+4)(s+1)}+\dfrac{k_A}{s(s+1)(s+4)}}$$

$$= \frac{k_A}{s^3+(5+k_A k_3)s^2+(4+k_A k_2)s+k_A}$$

Comparing with the given transfer function,

$$T(s) = \frac{24}{(s^2+2s+4)(s+6)} = \frac{24}{s^3+8s^2+16s+24}$$

$$8 = 5+k_A k_3, \quad k_A = 24 \text{ and } 4+k_A(k_2+k_3)=16$$

then,

$$k_3 = \frac{8.5}{k_4} = \frac{3}{8}$$

$$k_2 = \frac{12}{k_A} - k_3 = \frac{1}{8}$$

Hence $$k_A = 24, k_2 = \frac{1}{8}, k_3 = \frac{3}{8}$$

Root locus plot:

$$T(s) = \frac{24}{s^3 + 8s^2 + 16s + 24} = \frac{\dfrac{24}{s^3 + 8s^2 + 16s}}{1 + \dfrac{24}{s^3 + 8s^2 + 16s}}$$

The open loop transfer function,

$$G(s) = \frac{24}{s\left(s^2 + 8s + 16\right)} = \frac{k_A}{s(s+4)^2}$$

There are two poles at $s = 0$ and $s = -4$

$$-\sigma_A \text{ (Centroid)} = \frac{-4 - 4 - 0}{3} = \frac{-8}{3}$$

$$\phi_A \text{ (angle of asymptotes)} = \frac{180(2q+1)}{3} = 60°, 180°, 300°$$

For breakway point, $\dfrac{dk_A}{ds}$

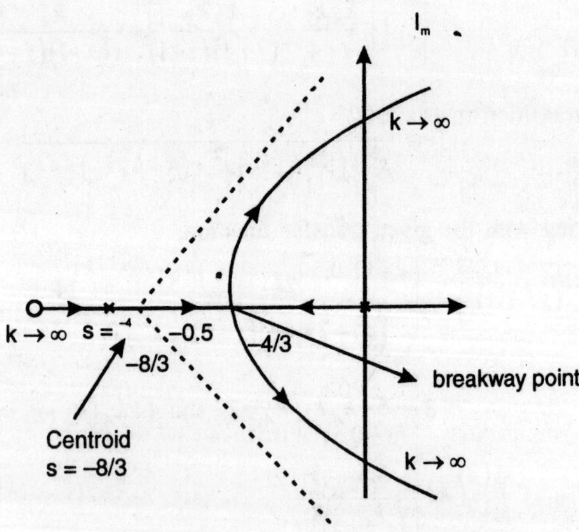

Fig. P. 7.44 (b)

or $\qquad \dfrac{d}{ds}\left(s^3+8s^2+16s\right)=0$

$\therefore \qquad\qquad s=\dfrac{-16\pm\sqrt{64}}{6}=-4,\ \dfrac{-4}{3}$

The breaway point will be at $s=\dfrac{-4}{3}$ as it lies on root locus.

Problem 7.45. Obtain the response $y(t)$ of the following system:-

$$\begin{bmatrix} \dot{x}_1 \\ \dot{x}_2 \end{bmatrix}=\begin{bmatrix} -1 & -0.5 \\ 1 & 0 \end{bmatrix}\begin{bmatrix} x_1 \\ x_2 \end{bmatrix}+\begin{bmatrix} 0.5 \\ 0 \end{bmatrix}u,$$

$$\begin{bmatrix} x_1(0) \\ x_2(0) \end{bmatrix}=\begin{bmatrix} 0 \\ 0 \end{bmatrix},\ y=\begin{bmatrix} 1 & 0 \end{bmatrix}\begin{bmatrix} x_1 \\ x_2 \end{bmatrix}$$

where $u(t)$ is the unit step input occurring at $t=0$ or $u(t)=1(t)$

Solution:

For this system

$$A=\begin{bmatrix} -1 & 0.5 \\ 1 & 0 \end{bmatrix},\ B=\begin{bmatrix} 0.5 \\ 0 \end{bmatrix},\ C\begin{bmatrix} 1 & 0 \end{bmatrix}$$

The state transition matrix $\phi(t)=e^{At}$ can be obtained as follows:

$$\phi(t)=e^{At}=\mathcal{L}^{-1}\left[\left(sI-A\right)^{-1}\right]$$

Since $\qquad \left(sI-A\right)^{-1}=\begin{bmatrix} s+1 & 0.5 \\ -1 & s \end{bmatrix}^{-1}=\dfrac{1}{s^2+s+0.5}\begin{bmatrix} s & -0.5 \\ 1 & s+1 \end{bmatrix}$

$\Rightarrow \qquad \phi(s)=\begin{bmatrix} \dfrac{s+0.5-0.5}{(s+0.5)^2+0.5^2} & \dfrac{-0.5}{(s+0.5)^2+(0.5)^2} \\[4mm] \dfrac{1}{(s+0.5)^2+0.5^2} & \dfrac{s+0.5+0.5}{(s+0.5)^2+0.5^2} \end{bmatrix}$

We have $\qquad \phi(t)=e^{At}=\mathcal{L}^{-1}\left[\left(sI-A\right)^{-1}\right]$

$$\Rightarrow \quad \phi(t) = \begin{bmatrix} e^{-0.5t}(\cos 0.5t - \sin 0.5t) & -e^{-0.5t}\sin 0.5t \\ 2e^{-0.5t}\sin 0.5t & e^{-0.5t}(\cos 0.5t + \sin 0.5t) \end{bmatrix}$$

Since,
$$X_0 = \begin{bmatrix} x_1(0) \\ x_2(0) \end{bmatrix} = \begin{bmatrix} 0 \\ 0 \end{bmatrix}$$

The complete solution of the equations with a forcing vector U is

$$x(t) = \phi(t)x_0 + L^{-1}\big[\phi(s)Bu(s)\big]$$

$$= e^{At}x_0 + A^{-1}\left(e^{At} - 1\right)Bu(t)$$

Thus,
$$x(t) = \begin{bmatrix} 0 & 1 \\ -2 & -2 \end{bmatrix}\begin{bmatrix} 0.5e^{-0.5t}(\cos 0.5t - \sin 0.5t) - 0.5 \\ e^{-0.5t}\sin 0.5t \end{bmatrix}$$

$$= \begin{bmatrix} e^{-0.5t}\sin 0.5t \\ -e^{-0.5t}(\cos 0.5t + \sin 0.5t) + 1 \end{bmatrix}$$

Hence, the output $y(t)$ can be given by

$$y(t) = \begin{bmatrix} 1 & 0 \end{bmatrix}\begin{bmatrix} x_1 \\ x_2 \end{bmatrix} = x_1 = e^{-0.5t}\sin 0.5t$$

Problem 7.46. Obtain the Resolvant matrix and hence state-transition matrix $\phi(t)$ of the following system:

$$\begin{bmatrix} \dot{x}_1 \\ \dot{x}_2 \end{bmatrix} = \begin{bmatrix} 0 & 1 \\ -2 & -3 \end{bmatrix}\begin{bmatrix} x_1 \\ x_2 \end{bmatrix}$$

Also obtain the inverse of the state-transition matrix $\phi^{-1}(t)$

Solution:

For the given system $A = \begin{bmatrix} 0 & 1 \\ -2 & -3 \end{bmatrix}$

The state-transition matrix $\phi(t)$ is given by

$$\phi(t) = e^{At} = \mathcal{L}^{-1}\left[(sI-A)^{-1}\right]$$

Where $(sI-A)^{-1}$ is known as resolvant matrix.

Since

$$(sI-A) = \begin{bmatrix} s & 0 \\ 0 & s \end{bmatrix} - \begin{bmatrix} 0 & 1 \\ -2 & -3 \end{bmatrix} = \begin{bmatrix} s & -1 \\ 2 & s+3 \end{bmatrix}$$

The inverse of $(sI-A)$ is given by

$$(sI-A)^{-1} = \frac{1}{(s+1)(s+2)}\begin{bmatrix} s+3 & 1 \\ -2 & s \end{bmatrix}$$

or, resolvant matrix

$$= (sI-A)^{-1} = \begin{bmatrix} \dfrac{s+3}{(s+1)(s+2)} & \dfrac{1}{(s+1)(s+2)} \\ \dfrac{-2}{(s+1)(s+2)} & \dfrac{s}{(s+1)(s+2)} \end{bmatrix}$$

Hence,

$$\phi(t) = e^{At} = \mathcal{L}^{-1}\left[(sI-A)^{-1}\right]$$

$$= \begin{bmatrix} 2e^{-t}-e^{-2t} & e^{-t}-e^{-2t} \\ -2e^{-t}+2e^{-2t} & -e^{-t}+2e^{-2t} \end{bmatrix}$$

Noting that $\phi^{-1}(t) = \phi(-t)$

Inverse of state-transition matrix

$$\phi^{-1}(t) = e^{-At} = \begin{bmatrix} 2e^{-t}-e^{2t} & e^{t}-e^{2t} \\ -2e^{t}+2e^{2t} & -e^{t}+2e^{2t} \end{bmatrix}$$

Problem 7.47. Consider the system described by

$$\begin{bmatrix} \dot{x}_1 \\ \dot{x}_2 \end{bmatrix} = \begin{bmatrix} 1 & 1 \\ -2 & -1 \end{bmatrix}\begin{bmatrix} x_1 \\ x_2 \end{bmatrix} + \begin{bmatrix} 0 \\ 1 \end{bmatrix}u$$

$$y = \begin{bmatrix} 1 & 0 \end{bmatrix}\begin{bmatrix} x_1 \\ x_2 \end{bmatrix}$$

Is the system controllable and observable?

Solution:

For the given system,

$$A = \begin{bmatrix} 1 & 1 \\ -2 & -1 \end{bmatrix}, \; B = \begin{bmatrix} 0 \\ 1 \end{bmatrix}, \; C = \begin{bmatrix} 1 & 0 \end{bmatrix}$$

$$C^T = \begin{bmatrix} 1 \\ 0 \end{bmatrix}$$

We check the rank of matrix [$B : AB$], if it is equal to order of matrix i.e. 2 in this case the system is state observable.

$$AB = \begin{bmatrix} 1 & 1 \\ -2 & -1 \end{bmatrix}\begin{bmatrix} 0 \\ 1 \end{bmatrix} = \begin{bmatrix} 1 \\ -1 \end{bmatrix}$$

Matrix $\quad [B : AB] = \begin{bmatrix} 0 & 1 \\ 1 & -1 \end{bmatrix}$

Since determinant $\begin{bmatrix} 0 & 1 \\ 1 & -1 \end{bmatrix} \neq 0$

$\therefore$ The rank is 2 = order of matrix,

Thus, the system is completely state controllable.

For output controllability,

We check the rank of the matrix $\begin{bmatrix} CB & : & CAB \end{bmatrix}$

$$\begin{bmatrix} CB & : & CAB \end{bmatrix} = \begin{bmatrix} 0 & 1 \end{bmatrix}$$

the rank of this matrix = 1, Hence the system is completely output controllable.

To test observability,

We check the rank of the matrix

$$\begin{bmatrix} C^T : & A^T C^T & : & A^{T^2} C^T ... \end{bmatrix}$$

Here, $\quad A^T C^T = \begin{bmatrix} 1 & -2 \\ 1 & -1 \end{bmatrix}\begin{bmatrix} 1 \\ 0 \end{bmatrix} = \begin{bmatrix} +1 \\ +1 \end{bmatrix}$

$\therefore \quad \begin{bmatrix} C^T & : & A^T C^T \end{bmatrix} = \begin{bmatrix} 1 & 1 \\ 0 & 1 \end{bmatrix}$

Rank is 2, hence the system is completely observable.

Problem 7.48.   Show that the following system is not completely observable.

$$\dot{x} = A\,x + B\,u$$
$$y = C\,x, \quad \text{where,}$$

$$x = \begin{bmatrix} x_1 \\ x_2 \\ x_3 \end{bmatrix}, A = \begin{bmatrix} 0 & 1 & 0 \\ 0 & 0 & 1 \\ -6 & -11 & -6 \end{bmatrix}, B = \begin{bmatrix} 0 \\ 0 \\ 1 \end{bmatrix}, C = \begin{bmatrix} 4 & 5 & 1 \end{bmatrix}$$

Solution:

Note that the control function u does not affect the complete observability of the system. To examine complete observability, we may simply set $u = 0$.

Now,
$$C^T = \begin{bmatrix} 4 \\ 5 \\ 1 \end{bmatrix}$$

$$A^T C^T = \begin{bmatrix} 0 & 0 & -6 \\ 1 & 0 & -11 \\ 0 & 1 & -6 \end{bmatrix} \begin{bmatrix} 4 \\ 5 \\ 1 \end{bmatrix} = \begin{bmatrix} -6 \\ -7 \\ -1 \end{bmatrix}$$

Similarly,
$$\left(A^T\right)^2 C^T = \begin{bmatrix} 6 \\ 5 \\ -1 \end{bmatrix}$$

Now, to find the rank of matrix $\left[C^T : A^T C^T : \left(A^T\right)^2 C^T \right]$;

$$\left[C^T : A^T C^T : \left(A^T\right)^2 C^T \right] = \begin{bmatrix} 4 & -6 & 6 \\ 5 & -7 & 5 \\ 1 & -1 & -1 \end{bmatrix}$$

Note that,
$$\begin{vmatrix} 4 & -6 & 6 \\ 5 & -7 & 5 \\ 1 & -1 & -1 \end{vmatrix} = 0$$

So, the rank of the matrix is less than 3. Therefore, the system is not completely observable.

Problem 7.49.   Discuss the state controllability of the following system:

$$\begin{bmatrix} \dot{x}_1 \\ \dot{x}_2 \end{bmatrix} = \begin{bmatrix} -3 & 1 \\ -2 & 1.5 \end{bmatrix} \begin{bmatrix} x_1 \\ x_2 \end{bmatrix} + \begin{bmatrix} 1 \\ 4 \end{bmatrix} u$$

Solution:

For this system,

$$A = \begin{bmatrix} -3 & 1 \\ -2 & 1.5 \end{bmatrix}, \; B = \begin{bmatrix} 1 \\ 4 \end{bmatrix}$$

also

$$AB = \begin{bmatrix} -3 & 1 \\ -2 & 1.5 \end{bmatrix} \begin{bmatrix} 1 \\ 4 \end{bmatrix} = \begin{bmatrix} 1 \\ 4 \end{bmatrix}$$

Alternate method of checking state controllability:

We see that vectors B and $A\,B$ are not linearly independent and the rank of the matrix $[\, B : AB]$ is 1.

Therefore, the system is not completely state controllable.

Also
$$\dot{x}_1 = -3x_1 + x_2 + u$$

$$\dot{x}_2 = -2x_1 + 1.5x_2 + 4u$$

elimination of x_2 from the two equations yields

$$\ddot{x}_1 + 1.5\dot{x}_1 - 2.5x_1 = \dot{u} + 2.5u$$

or, in the form of transfer function

$$\frac{X_1(s)}{U(s)} = \frac{(s+2.5)}{(s+2.5)(s-1)}$$

Because of the cancellation of the factor ($s + 2.5$), the system is not completely state controllable.

This is an unstable system. Remember that stability and controllability are quite different things.

Problem 7.50. the transfer function of a control system is given by

$$\frac{Y(s)}{U(s)} = \frac{s+2}{s^3 + 9s^2 + 26s + 24}$$

Check for controllability and observability.

Solution:

The state equations are written below:

$$\dot{x}_1 = x_2$$

$$\dot{x}_2 = x_3$$

$$\dot{x}_3 = -24x_1 - 26x_2 - 9x_3 + u$$

and

$$y = 2x_1 + x_2$$

or,

$$\begin{bmatrix} \dot{x}_1 \\ \dot{x}_2 \\ \dot{x}_3 \end{bmatrix} = \begin{bmatrix} 0 & 1 & 0 \\ 0 & 0 & 1 \\ -24 & -26 & -9 \end{bmatrix} \begin{bmatrix} x_1 \\ x_2 \\ x_3 \end{bmatrix} + \begin{bmatrix} 0 \\ 0 \\ 1 \end{bmatrix} u$$

and

$$y = \begin{bmatrix} 2 & 1 & 0 \end{bmatrix} \begin{bmatrix} x_1 \\ x_2 \\ x_3 \end{bmatrix}$$

The coefficient matrices are:

$$A = \begin{bmatrix} 0 & 1 & 0 \\ 0 & 0 & 1 \\ -24 & -26 & -9 \end{bmatrix}, B = \begin{bmatrix} 0 \\ 0 \\ 1 \end{bmatrix}$$

$$C = \begin{bmatrix} 2 & 1 & 0 \end{bmatrix}$$

(a) To check for controllability:

$$A = \begin{bmatrix} 0 & 1 & 0 \\ 0 & 0 & 1 \\ -24 & -26 & -9 \end{bmatrix}, B = \begin{bmatrix} 0 \\ 0 \\ 1 \end{bmatrix}$$

$$AB = \begin{bmatrix} 0 \\ 1 \\ -9 \end{bmatrix}$$

$$A^2B = \begin{bmatrix} 0 & 1 & 0 \\ 0 & 0 & 1 \\ -24 & -26 & -9 \end{bmatrix}\begin{bmatrix} 0 \\ 1 \\ -9 \end{bmatrix} = \begin{bmatrix} 1 \\ -99 \\ 55 \end{bmatrix}$$

Controllability test matrix

$$Q_c = \begin{bmatrix} B : AB : A^2B \end{bmatrix}$$

$$\Rightarrow \qquad Q_c = \begin{bmatrix} 0 & 0 & 1 \\ 0 & 1 & -9 \\ 1 & -9 & 55 \end{bmatrix}$$

$$|Q_c| = \begin{bmatrix} 0 & 0 & 1 \\ 0 & 1 & -9 \\ 1 & -9 & 55 \end{bmatrix} = -1$$

i.e., $|Q_c| \neq 0$, Thus, the rank of the matrix

$Q_c = 3 = $ order of Q_c.

The system completely state controllable.

(b) To check for observability:

$$A^T = \begin{bmatrix} 0 & 0 & -24 \\ 1 & 0 & -26 \\ 0 & 1 & -9 \end{bmatrix}, \quad C^T = \begin{bmatrix} 2 \\ 1 \\ 0 \end{bmatrix}$$

The observability test matrix

$$Q_0 = \begin{bmatrix} C^T : A^T C^T : \left(A^T\right)^2 C^T \end{bmatrix}$$

$$\therefore \qquad Q_0 = \begin{bmatrix} 2 & 0 & -24 \\ 1 & 2 & -26 \\ 0 & 1 & -7 \end{bmatrix}$$

$|Q_0| = $ determinant of the above matrix $= 0$

Thus, the rank of $Q_0 \neq 3$ and hence the system is not observable.

Problem 7.51. For the system given below:
Obtain (i) Zero input response (ii) Zero state response (iii) Total response

$$\dot{X} = \begin{bmatrix} 0 & 1 \\ -2 & -1 \end{bmatrix}\begin{bmatrix} x_1 \\ x_2 \end{bmatrix} + \begin{bmatrix} 0 \\ 1 \end{bmatrix} u(t)$$

where, $x_1(0) = 1$, $x_2(0)=0$ and $u(t)=1$

Solution:

We know that with a forcing vector, the total response is given by

$$x(t) = \underbrace{\phi(t)x(0)}_{ZIR} + \underbrace{L^{-1}\big[\phi(s)\,Bu(s)\big]}_{ZSR}$$

(i) Zero input response is given by

$$x(t)\Big|_{ZIR} = \phi(t)x(0)$$

where, $\phi(t) = \mathcal{L}^{-1}(sI-A)^{-1}$

$$(sI - A) = \begin{bmatrix} s & -1 \\ 2 & s+1 \end{bmatrix}$$

$$(sI - A)^{-1} = \frac{1}{s^2+s+3}\begin{bmatrix} s+1 & 1 \\ -2 & s \end{bmatrix}$$

$$= \begin{bmatrix} \dfrac{s+1}{s^2+s+3} & \dfrac{1}{s^2+s+3} \\[3mm] \dfrac{-2}{s^2+s+3} & \dfrac{s}{s^2+s+3} \end{bmatrix}$$

$$\phi(t) = \mathcal{L}^{-1}(sI-A)^{-1}$$

and $x(t) = \phi(t)x(0)$

$$\therefore \qquad x(t) = \mathcal{L}^{-1}\begin{bmatrix} \dfrac{s+1}{s^2+s+3} & \dfrac{1}{s^2+s+3} \\[3mm] \dfrac{-2}{s^2+s+3} & \dfrac{s}{s^2+s+3} \end{bmatrix}\begin{bmatrix} 1 \\ 0 \end{bmatrix}$$

$$= \mathcal{L}^{-1}\begin{bmatrix} \dfrac{s+1}{s^2+s+3} \\[3mm] \dfrac{-2}{s^2+s+3} \end{bmatrix}$$

$$= \mathcal{L}^{-1}\begin{bmatrix} \dfrac{(s+0.5)+0.5}{(s+0.5)^2+(1.65)^2} \\[4mm] -\dfrac{2}{1.65}\cdot\dfrac{1.65}{(s+0.5)^2+(1.65)^2} \end{bmatrix}$$

taking Laplace inverse transform, the zero input response is

$$x(t)\Big|_{\text{ZIR}} = \begin{bmatrix} \left(e^{-0.5t}.\cos 1.65t + 0.3\,e^{-0.5t}.\sin 1.65t\right) \\[2mm] -1.21e^{-0.5t}.\sin 1.65t \end{bmatrix}$$

(ii) Zero state Response (ZSR) is given by

$$x(t)\Big|_{\text{ZSR}} = \mathcal{L}^{-1}\left[\phi(s)\,Bu(s)\right]$$

$$= \mathcal{L}^{-1}\begin{bmatrix} \dfrac{s+1}{s^2+s+3} & \dfrac{1}{s^2+s+3} \\[4mm] \dfrac{-2}{s^2+s+3} & \dfrac{s}{s^2+s+3} \end{bmatrix}\begin{bmatrix} 0 \\ 1 \end{bmatrix}\dfrac{1}{s}$$

$$= \mathcal{L}^{-1}\begin{bmatrix} \dfrac{1}{s\left(s^2+s+3\right)} \\[4mm] \dfrac{1}{s^2+s+3} \end{bmatrix}$$

$$= \mathcal{L}^{-1}\begin{bmatrix} \dfrac{1}{3}\cdot\dfrac{1}{s}\cdot\dfrac{3}{\left(s^2+s+3\right)} \\[4mm] \dfrac{1}{1.65}\cdot\dfrac{1.65}{(s+0.5)^2+(1.65)^2} \end{bmatrix}$$

Taking Inverse Laplace transform,

$$x(t)\Big|_{\text{ZSR}} = \begin{bmatrix} 0.33\left\{1-1.04\,e^{-0.5t}.\sin\left(1.65t+\tan^{-1}3.42\right)\right\} \\ 0.6\,e^{-0.5t}.\sin 1.65t \end{bmatrix}$$

The total response

$$x(t) = x(t)\Big|_{\text{ZIR}} + x(t)\Big|_{\text{ZSR}}$$

By adding the matrices of above two responses, we get total response and thus, values of $x_1(t)$ and $x_2(t)$ individually.

CHAPTER 8

Bode Plots Analysis

Problem 8.1. The asymtotic Bode plot of a minimum phase linear system is shown below. Determine its transfer function.

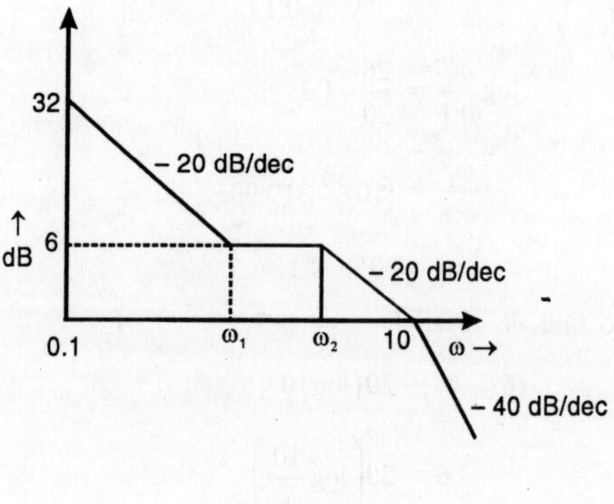

Fig. P. 8.1

Solution:

From the Bode plot, the following transfer function is formed as below; Since the initial slope is −20 dB/dec therefore there will be one pole at the origin.

The corner frequencies are at $\omega = \omega_1, \omega_2, 10$

(i) At $\omega = \omega_1$, slope of the plot is zero i.e. there is a zero at $\omega = \omega_1$

(ii) At $\omega = \omega_2$, the slope of the plot is −20dB/dec hence there is a pole at $\omega = \omega_2$

(iii)   At $\omega = 10$, the slope of the plot is -40 dB/dec hence there is again a pole at $\omega = 10$

Note: For every slope of -20 dB/dec there will be a pole at that particular corner frequency for every slope of $+20$ dB/dec there will be a zero at that particular corner frequency. Therefore, the transfer function formed will be:

$$G(s) = \frac{K\left(1 + \dfrac{s}{\omega_1}\right)}{s\left(1 + \dfrac{s}{\omega_2}\right)\left(1 + \dfrac{s}{10}\right)}$$

Now, from the figure,

$$(32 - 6) = 20\left(\log \omega_1 - \log 0.1\right)$$

$$\Rightarrow \qquad 26 = 20\left(\log \frac{\omega_1}{0.1}\right)$$

$$\Rightarrow \qquad \log \frac{\omega_1}{0.1} = \frac{26}{20} = 1.3$$

$$\Rightarrow \qquad \frac{\omega_1}{0.1} = (10)^{1.3} = 19.995 \square\ 20$$

$$\Rightarrow \qquad \omega_1 = 20 \times 0.1 = 2 \text{ rad/sec.}$$

Again to find ω_2

$$(6 - 0) = 20\left(\log 10 - \log \omega_2\right)$$

$$\Rightarrow \qquad 6 = 20\left(\log \frac{10}{\omega_2}\right)$$

$$\Rightarrow \qquad \log \frac{10}{\omega_2} = \frac{6}{20} = 0.3$$

$$\Rightarrow \qquad \frac{10}{\omega_2} = (10)^{0.3} = 1.995 \square\ 2$$

$$\Rightarrow \qquad \frac{10}{\omega_2} = 2$$

$$\Rightarrow \qquad \omega_2 = \frac{10}{2} = 5 \text{ rad/sec.}$$

To determine K, $20 \log K - 20 \log (0.1) = 32$

$\Rightarrow$ $20 \log K - 20 \times (-1) = 32$

$\Rightarrow$ $20 \log K + 20 = 32$

$\Rightarrow$ $20 \log K = 12$

$\Rightarrow$ $\log K = \dfrac{12}{20} = 0.6$

$\Rightarrow$ $K = (10)^{0.6} = 3.981 \approx 4$

Hence, $G(s) = \dfrac{4\left(1+\dfrac{s}{2}\right)}{s\left(1+\dfrac{s}{5}\right)\left(1+\dfrac{s}{10}\right)}$

$\Rightarrow$ $G(s) = \dfrac{4\left(\dfrac{s+2}{2}\right)}{s\left(\dfrac{5+s}{5}\right)\left(\dfrac{10+s}{10}\right)}$

$\Rightarrow$ $G(s) = \dfrac{4\times5\times10}{2}\dfrac{(s+2)}{s(s+5)(s+10)}$

$\Rightarrow$ $G(s) = \dfrac{100\,(s+2)}{s(s+5)(s+10)}$

Problem 8.2.   The asymtotic approximation of the log-magnitude verses frequency plot of a minimum phase system with real poles and one zero is shown in the figure. Find its transfer function.

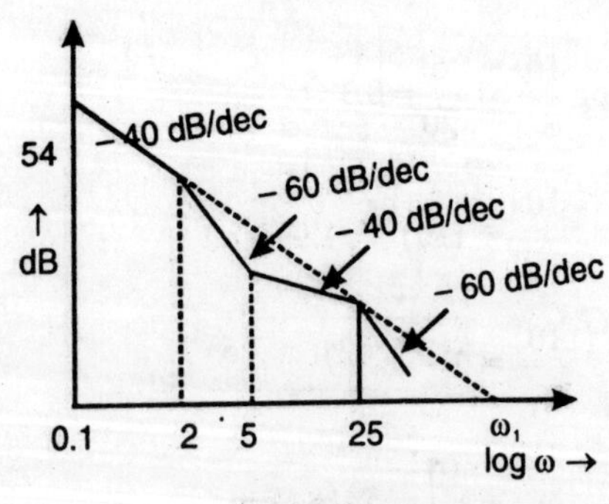

Fig. P. 8.2

Solution:

From the figure we have initial slope of the Bode plot to be -40 dB/dec, hence the transfer function has two poles in the origin. The corner frequencies are at $\omega = 2, 5, 25$ at $\omega = 2$ the slope is -60 dB/dec hence a pole exists. At $\omega = 5$ slop is -40 dB/dec hence there is a zero. Again at $\omega = 25$ the slope is -60 dB/dec hence a pole exists. The transfer function takes the form,

$$G(s) = \frac{K\left(1+\dfrac{s}{5}\right)}{s^2\left(1+\dfrac{s}{2}\right)\left(1+\dfrac{s}{25}\right)}$$

To find K, we have from the Bode plot,

$$(54 - 0) = 40\left(\log\omega_1 - \log 0.1\right)$$

$$\Rightarrow \qquad \log\left(\frac{\omega_1}{0.1}\right) = \frac{54}{40}$$

$$\Rightarrow \qquad \log\frac{\omega_1}{0.1} = 1.35$$

$$\Rightarrow \qquad \omega_1 = 22.387 \times 0.1$$

$$\Rightarrow \qquad \omega_1 = 2.2387.$$

If we extend the initial slope -40 dB/dec line it cuts the frequency $(\log\omega)$ axis at $\omega = \omega_1$ which is equal to $\sqrt{K}$ since two poles exists at the origin. Hence, $\omega_1 = \sqrt{K}$

$$\Rightarrow \qquad K = \omega_1^2 = (2.2387)^2 = 5.0117 \approx 5$$

so,
$$G(s) = \frac{5\left(1+\dfrac{s}{5}\right)}{s^2\left(1+\dfrac{s}{2}\right)\left(1+\dfrac{s}{25}\right)}$$

$$\Rightarrow \qquad G(s) = \frac{5\times 2\times 25\,(s+5)}{5s^2\,(s+2)(s+25)}$$

$$\Rightarrow \qquad G(s) = \frac{50\,(s+5)}{s^2\,(s+2)(s+25)}$$

Problem 8.3.   Obtain the transfer function for the response curves shown below.

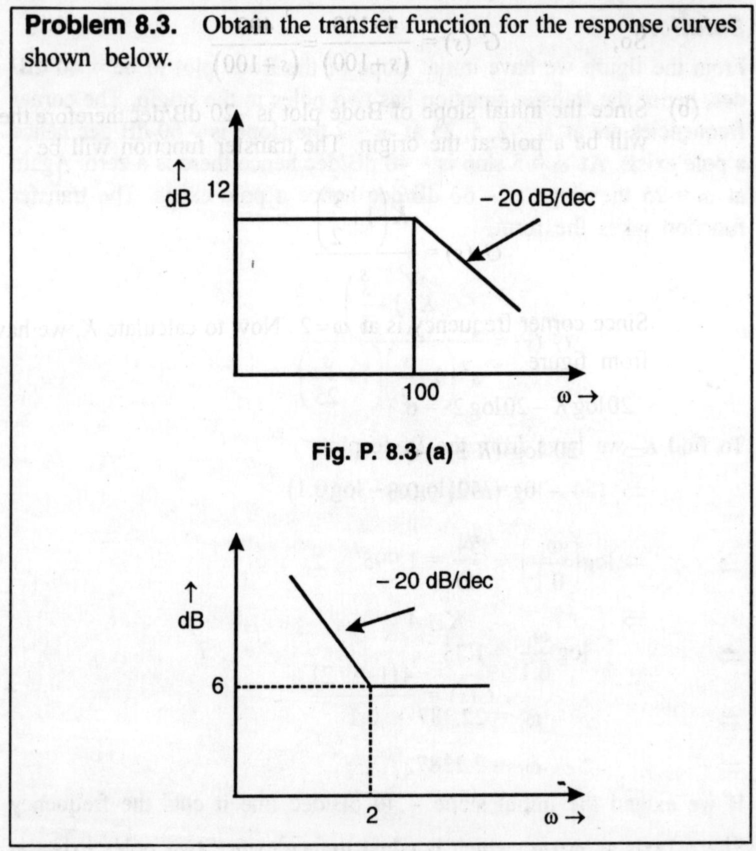

Fig. P. 8.3 (a)

Fig. P. 8.3 (b)

Solution:

(a)   This is the response of first order system. The corner frequency is at $\omega = 100$.

$$G(s) = \frac{K}{\left(1 + \dfrac{s}{100}\right)}$$

To determine K, from figure, $20 \log K = 12$

$$\Rightarrow \qquad K = \frac{12}{20} = 0.6$$

$$\Rightarrow \qquad K = 10^{0.6} = 3.981 \approx 4.$$

So, $$G(s) = \frac{4 \times 100}{(s+100)} = \frac{400}{(s+100)}$$

(b) Since the initial slope of Bode plot is -20 dB/dec therefore there will be a pole at the origin. The transfer function will be

$$G(s) = \frac{K\left(1+\dfrac{s}{2}\right)}{s}$$

Since corner frequency is at $\omega = 2$. Now to calculate K, we have from figure

$$20 \log K - 20 \log 2 = 6$$

$$\Rightarrow \quad 20 \log (K/2) = 0$$

$$\Rightarrow \quad \log (K/2) = 0.3.$$

$$\Rightarrow \quad \frac{K}{2} = 1.995 \approx 2.$$

$$\Rightarrow \quad K \approx 4$$

$$\therefore \quad G(s) = \frac{4(1+s/2)}{s}$$

$$\Rightarrow \quad G(s) = \frac{4(s+2)}{2s}$$

$$\Rightarrow \quad G(s) = \frac{2(s+2)}{s}$$

Problem 8.4. Plot Bode diagrams for the following

(a) 20 (b) $G(s) = \dfrac{10}{s}$ (c) $= \dfrac{100}{10+s}$

Solution:

(a) $$\text{Magnitude} = 20 \log 20$$
$$= 20 \times 1.301$$
$$= 26.02 \approx 26$$

Phase Angle, $\phi = 0°$ (Since phase angle of a constant is zero)

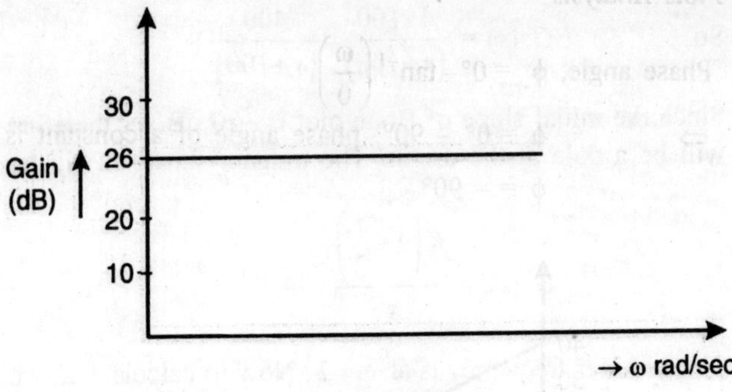

Fig. P. 8.4 (a)

Fig. P. 8.4 (b)

(b) $$G\ (s) = \frac{10}{s}$$

$$\Rightarrow \qquad G(j\omega) = \frac{10}{(j\omega)}$$

$$\text{Magnitude} = 20\log\left|G(j\omega)\right| = 20\log 10 - 20\log(j\omega)$$

$$\Rightarrow \qquad 20\log\left|G(j\omega)\right| = 20 - 20\log\omega$$

Now,

$$\omega = 1, \left|G(j\omega)\right| = 20 \text{ dB}$$

$$\omega = 10, \left|G(j\omega)\right| = 0 \text{ dB}$$

$$\omega = 20, \left|G(j\omega)\right| = -6 \text{ dB}$$

$$\omega = 100, \left|G(j\omega)\right| = -20 \text{ dB}$$

$$\omega = 0.1, \left|G(j\omega)\right| = 40 \text{ dB}$$

Phase angle, $\phi = 0° - \tan^{-1}\left(\dfrac{\omega}{0}\right)$

$\Rightarrow \qquad \phi = 0° - 90°$; phase angle of a constant is zero.

$\qquad\qquad \phi = -90°$

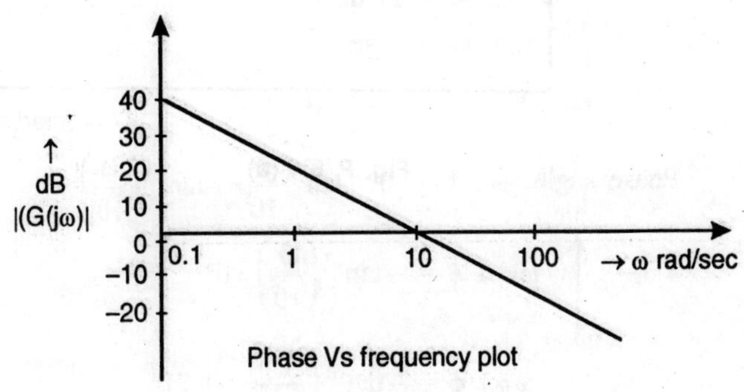

Fig. P 8.4 (c)

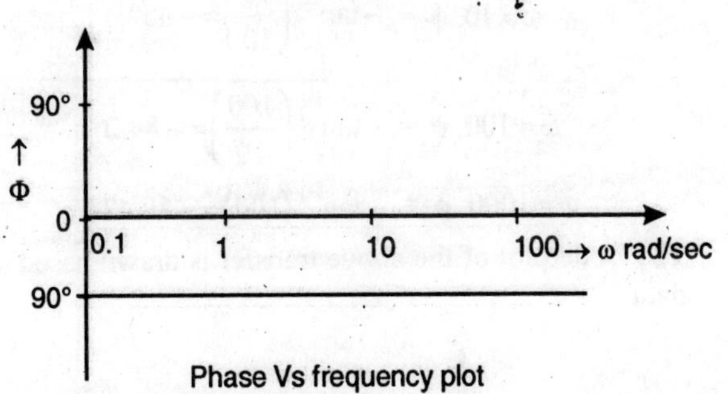

Fig. P 8.4 (d)

(c) $\qquad\qquad G(s) = \dfrac{100}{(10+s)}$

$$G(j\omega) = \dfrac{100}{10+(j\omega)}$$

Magnitude, $M = |G(j\omega)| = \dfrac{100}{\sqrt{10^2 + \omega^2}}$

$$M = 20\log|G(j\omega)| = 20\log 100 - 10\log\left(10^2 + \omega^2\right)$$

$$\Rightarrow \qquad M = 40 - 10\log\left(100 + \omega^2\right)$$

Now, $\qquad \omega = 0,\ M = 20$ dB $\left(40 - 10\log\left(100 + 0^2\right)\right)$

$\qquad \omega = 1,\ M = 19.96$ dB $[40 - 10\log(100 + 1)]$

$\qquad \omega = 10,\ M = 17$ dB

$\qquad \omega = 100,\ M = 0$ dB

$\qquad \omega = 1000,\ M = -20$ dB

Phase angle, $\qquad \phi = 0° - \tan^{-1}\dfrac{\omega}{10} = -\tan^{-1}\left(\dfrac{\omega}{10}\right)$

$$\omega = 0,\ \phi = -\tan^{-1}\left(\frac{0}{10}\right) = 0°$$

$$\omega = 1,\ \phi = -\tan^{-1}\left(\frac{1}{10}\right) = -5.71°$$

$$\omega = 10,\ \phi = -\tan^{-1}\left(\frac{10}{10}\right) = -45°$$

$$\omega = 100,\ \phi = -\tan^{-1}\left(\frac{100}{10}\right) = -84.2°$$

$$\omega = 1000,\ \phi = -\tan^{-1}\left(100\right) = -89.43°$$

The Bode plot of the above transfer is drawn based on the above data.

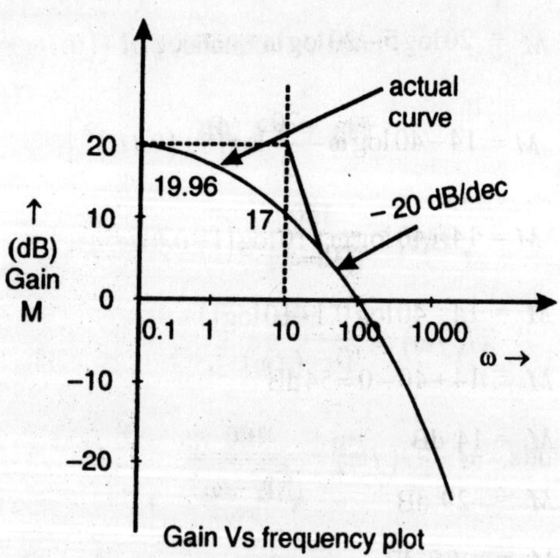

Fig. 8.4 (e)

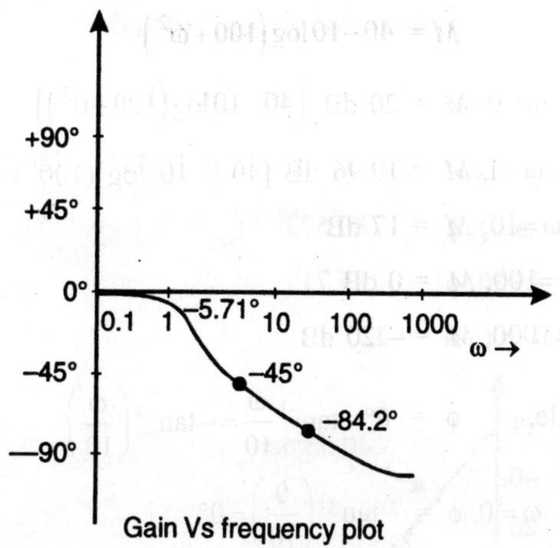

Fig. 8.4 (f)

Problem 8.5.   Plot Bode diagram for the given transfer function

$$G(s) = \frac{5}{(j\omega)^2 (1+j0.1\omega)}$$

Solution:

Magnitude,

$$M = 20\log|G(j\omega)|$$

$\Rightarrow$

$$M = 20\log 5 - 20\log\omega^2 - 20\log\sqrt{1+\left(0.1\omega^2\right)}$$

$\Rightarrow$

$$M = 14 - 40\log\omega - \frac{20}{2}\log\left[1+(0.1\omega)^2\right]$$

$\Rightarrow$

$$M = 14 - 40\log\omega - 10\log\left(1+0.01\omega^2\right)$$

Now, $\omega = 0.1$, $M = 14 - 40\log 0.1 - 10\log(1+0)$

$$M = 14 + 40 - 0 = 54\,\text{dB}$$

$\omega = 1$, $M = 14$ dB

$\omega = 10$, $M = -29$ dB

$\omega = 100$, $M = -86$ dB

Phase angle, $\qquad \phi = 0 - 90 \times 2 - \tan^{-1}(0.1\omega)$

$$\phi = -180° - \tan^{-1}(0.1\omega)$$

Now, $\omega = 0.1$, $\phi = -180° - \tan^{-1}(0.01) = -180.0573° = -180.0573° \simeq -180°$

$$\omega = 1, \quad \phi = -180.573°$$

$$\omega = 10, \quad \phi = -185.71°$$

$$\omega = 100, \quad \phi = -225°$$

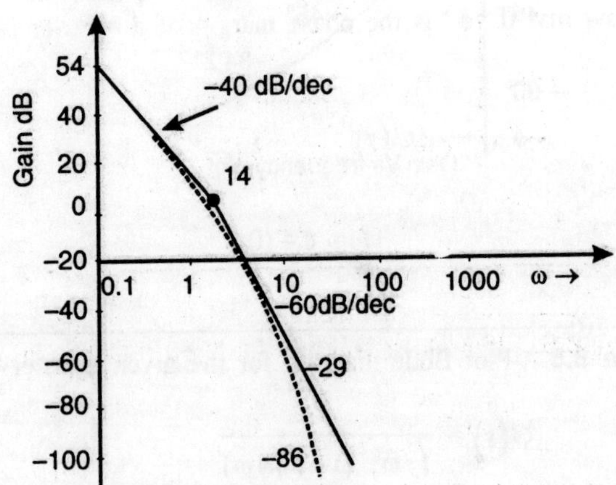

Gain Vs frequency plot

Fig. 8.5 (a)

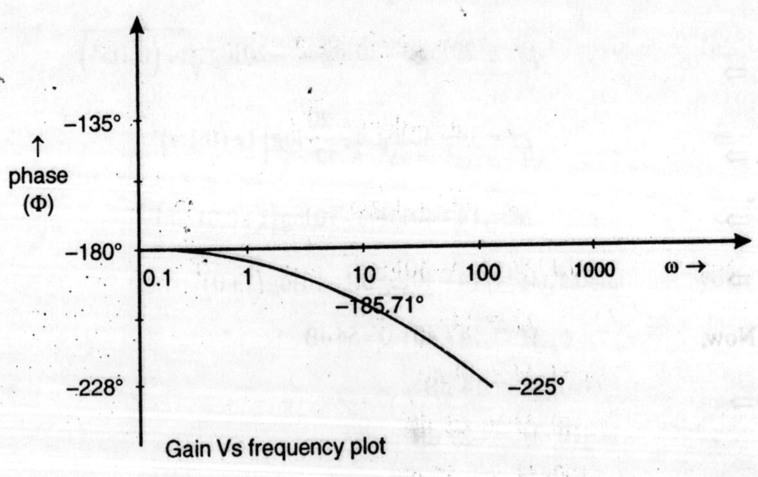

Gain Vs frequency plot

Fig. 8.5 (b)

> **Problem 8.6.** The open loop transfer function of a unity feedback control system is given as
>
> $$G\ (s) = \frac{as+1}{s^2}$$
>
> Find the value of 'a' to give a phase margin of 45° (Gate-2004)

Solution:

As we know that if 'ϕ' is the phase margin of a transfer function G (s) then

$$\phi_{pM} = \angle G(s)\Big|_{\omega=\omega_g}^{+180°}$$

where ω_g = gain cross over frequency.

To determine gain crossover frequency, we have,

$$20\log|G(j\omega)|_{\omega=\omega_g} = 0 \text{ dB}$$

$$\Rightarrow \quad 20\log\frac{\sqrt{1+(a\omega)^2}}{\omega^2} = 0$$

$$\Rightarrow \quad \log\frac{\sqrt{1+(a\omega)^2}}{\omega^2} = 0$$

$$\Rightarrow \quad \frac{\sqrt{1+a^2\omega^2}}{\omega^2} = 1$$

$$\Rightarrow \quad \sqrt{1+a^2\omega^2} = \omega^2$$

$$\Rightarrow \quad 1+a^2\omega^2 = \omega^4$$

$$\Rightarrow \quad \omega^4 - a^2\omega^2 - 1 = 0 \qquad (1)$$

Now, $\quad \phi_{pM} = \angle G(j\omega) + 180°$

$$\Rightarrow \quad 45° = \tan^{-1}(a\omega) - 180° + 180°$$

$$\Rightarrow \quad 45° = \tan^{-1}(a\omega)$$

$$\Rightarrow \quad a\omega = \tan 45° = 1 \qquad (2)$$

Putting this value in equation (1)

$$\omega^4 - a^2 \omega^2 - 1 = 0$$

$$\Rightarrow \qquad \omega^4 - 1 - 1 = 0$$

$$\Rightarrow \qquad \omega^4 = 2$$

$$\Rightarrow \qquad \omega^2 = \pm \sqrt{2} = 2^{1/2}$$

$$\Rightarrow \qquad \omega = \sqrt{(2)^{1/2}} = (2)^{1/4} = 1.189$$

From (2) Hence, $\quad a = \dfrac{1}{\omega} = \dfrac{1}{1.189} = 0.8408$

Problem 8.7. A unity feedback system has an open loop transfer function of

$$G\,(s) = \frac{10000}{s\left(s+10\right)^2}$$

(a) Determine the magnitude of $G(j\omega)$ in dB at an angular frequency of $\omega = 20$ rad/sec.

(b) Determine the phase margin in degrees.

(c) Determine the gain margin in dB.

(d) Is the system stable or unstable.   **(Gate-2001)**

Solution:

(a)   Magnitude $= 20\log|G(j\omega)|_{\omega=20}$

$$\Rightarrow \qquad M = 20\log 10^4 - 20\log\omega - 20\log\left(\sqrt{\omega^2 + 10^2}\right)^2$$

$$\Rightarrow \qquad M = 20\log 10^4 - 20\log\omega - 20\log\left(\omega^2 + 100\right)$$

Put $\omega = 20$

$$\Rightarrow \qquad M = 20\log 10^4 - 20\log 20 - 20\log\left(400 + 100\right)$$

$$\Rightarrow \qquad M = 80 - 26 - 54$$

$$\Rightarrow \qquad M = 0 \text{ dB.}$$

Hence, $|G(j\omega)|_{\omega=20} = 0$ dB

(b) Phase Margin ϕ_{pM} is given by

$$\phi_{pM} = \angle G(j\omega)\Big|_{\omega=\omega_g} + 180°$$

Where ω_g = gain cross over frequency which is 20 rad/sec in this case.

$$\Rightarrow \qquad \phi_{pM} = -90° - 2\tan^{-1}(0.1\omega) + 180°$$

Put $\qquad \omega = 20$

$$\Rightarrow \qquad \phi_{pM} = -2\tan^{-1}(0.1\times 20) + 90°$$

$$\Rightarrow \qquad \phi_{pM} = -126.87° + 90°$$

$$\Rightarrow \qquad \phi_{pM} = -36.87°$$

(c) For determining gain Margin we must know phase cross over frequency (ω_p) as below. At phase cross over frequency the angle of the transfer function is $-180°$

i.e. $\angle G(j\omega)_{\omega=\omega_p} = -180°$

$$\Rightarrow \quad -90° - 2\tan^{-1}(0.1\omega) = -180°$$

$$\Rightarrow -2\tan^{-1}(0.1\omega) = -90°$$

$$\Rightarrow \quad 2\tan^{-1}(0.1\omega) = 90°$$

$$\Rightarrow \qquad 0.1\omega = \tan 45°$$

$$\Rightarrow \qquad 0.1\omega = 1$$

$$\Rightarrow \qquad \omega = 10 \text{ rad/sec.}$$

$$\text{Gain Margin} = \frac{1}{|G(j\omega)|_{\omega=10}}$$

$$\Rightarrow \qquad M = \frac{10^4}{\omega\left(\sqrt{\omega^2 + 100}\right)^2}$$

$$\Rightarrow \qquad M = \frac{10^4}{10(100+100)}$$

$$\Rightarrow \qquad M = \frac{10^4}{10\times 200} = \frac{100}{2} = 50$$

$$\Rightarrow \qquad G\,M = 20\log\left(\frac{1}{50}\right) = -33.979\,\text{dB} \simeq -34 \text{ dB}$$

(d) Since the gain margin $(G\,M)$ and phase margin $(P\,M)$ are negative hence the system is unstable.

Problem 8.8. Draw the Bode plots for the following transfer functions and determine in each case, the system gain K for the gain cross-over frequency ω_c to be 5 rad/sec.

(a) $G(s) = \dfrac{Ks^2}{(1+0.2s)(1+0.02s)}$

(b) $G(s) = \dfrac{Ke^{-0.1S}}{s(1+s)(1+0.1s)}$

At gain cross-over frequency

$$|G(j\omega)|_{db} = 0$$

Solution:

(a) At gain cross-over frequency. $|G(j\omega)|_{db} = 0$

$$\Rightarrow \quad 20\log\frac{K|j\omega_c|^2}{|(1+j0.2\omega_c)|\cdot|(1+j0.02\omega_c)|} = 0$$

$$\Rightarrow \quad K = \frac{|(1+j)|\cdot|(1+0.1j)|}{25} = \frac{\sqrt{2}\times1.004}{25} = 0.056$$

The Bode plot is drawn in the semilog paper.

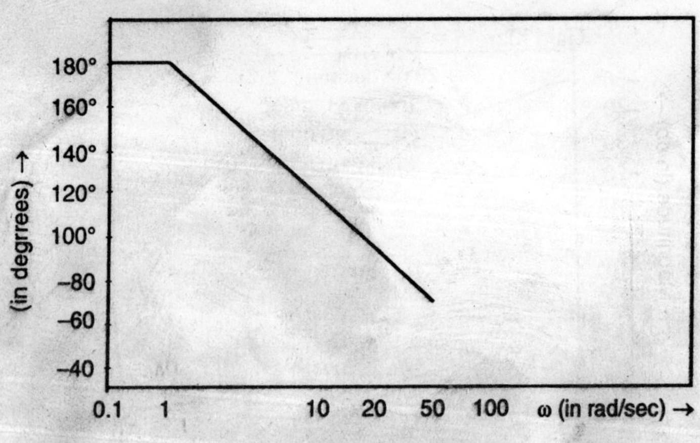

Fig. 8.8 (a)

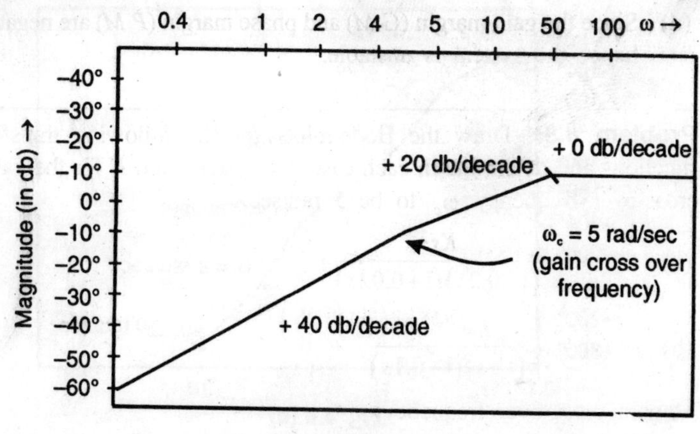

Fig. 8.8 (b)

(b) $G(s) = \dfrac{Ke^{-0.1s}}{s(s+1)(1+0.15)}$

At gain cross-over frequency. $\left|G(j\omega)\right|_{db} = 0$

$\Rightarrow \quad 20\log\dfrac{K}{\left|(j\omega_c)\right|\cdot\left|(1+j\omega_c)\right|\cdot\left|1+0.1\,j\omega_c\right|} = 0$

But Gain $\omega_C = 5$ rad/sec.

$$K = 5\left|(1+0.5\,j)\right|\cdot\left|(1+5\,j)\right| = 28.5$$

The Bode plot is drawn in the semilog paper.

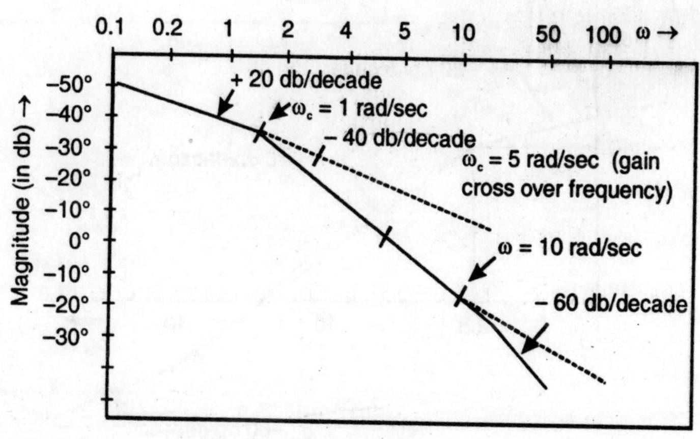

Fig 8.8 (c)

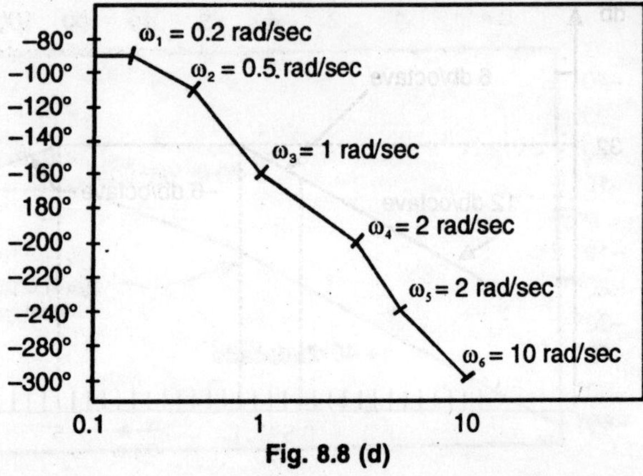

Fig. 8.8 (d)

Problem 8.9. The frequency response test data of certain elements plotted on Bode diagrams and asymptotically approximated are shown in the figure below. Find the transfer function of each element. (Elements are known to have minimum-phase characteristics).

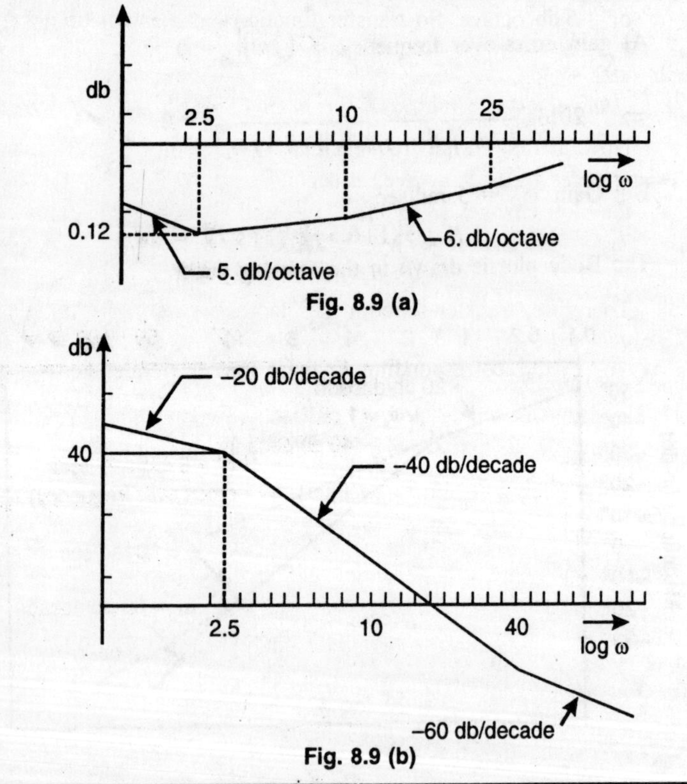

Fig. 8.9 (a)

Fig. 8.9 (b)

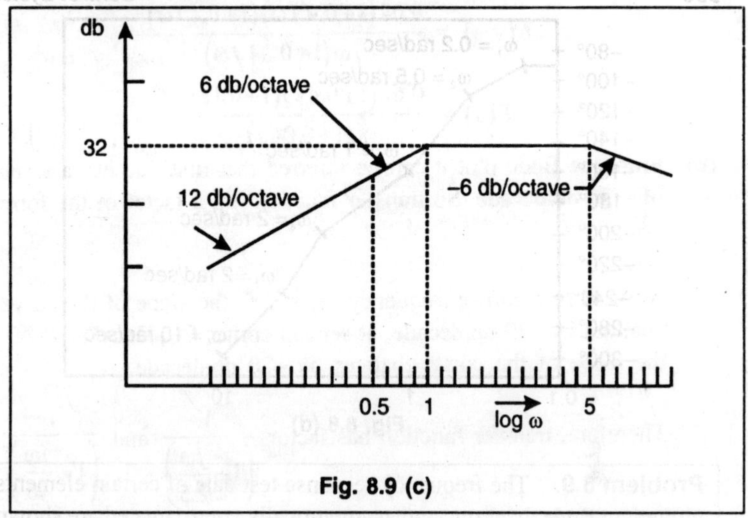

Fig. 8.9 (c)

Solution:

(a) From the Bode plot it can be inferred that, First line has a slope of – 6 db/octave. So transfer function has a factor of the form

$$\frac{K}{j\omega}.$$

Now, at first corner frequency ω_1 = 2.5, the slope of the curve changes by 6 db/octave, at second corner frequency ω_2 = 10, the slop changes by 6 db/octave, and at third corner frequency ω_3 = 25, it changes by– 6 db/octave.

Therefore, transfer function has factor $\left(1+\dfrac{j\omega}{2.5}\right), \left(1+\dfrac{j\omega}{10}\right)$ and $\dfrac{1}{1+j\omega/25}$ corresponding to these corner frequencies.

$\therefore$ Transfer function, $\quad T(j\omega) = \dfrac{K\left(1+\dfrac{j\omega}{2.5}\right)\left(1+\dfrac{j\omega}{10}\right)}{j\omega\left(1+\dfrac{j\omega}{25}\right)}$

$$= \frac{K(1+0.4\,j\omega)(1+0.1\,j\omega)}{j\omega(1+0.04\,j\omega)}$$

Now, from the plot, 20 log K = – 4.2 db

or $\qquad\qquad K = 0.62$ (approx).

$$T(j\omega) = \frac{0.62(1+0.4\,j\omega)(1+0.1\,j\omega)}{j\omega(1+0.04\,j\omega)}$$

$$T(s) = \frac{0.62(1+0.4s)(1+0.1s)}{s(1+0.04s)}$$

(b) From the Bode plot it can be inferred that first line has a slope of -20 db/decade. So transfer function has a factor of the form $\dfrac{K}{j\omega}$.

Now, at first corner frequency $\omega_1 = 2.5$, the slope of the curve changes by -20 db/decade, at second corner frequency $\omega_2 = 40$, the slope of the curve changes by -20 db/decade.

Therefore, transfer function has factors $\dfrac{1}{\left(1+\dfrac{j\omega}{2.5}\right)}$ and $\dfrac{1}{\left(1+\dfrac{j\omega}{40}\right)}$

corresponding to this corner frequencies.

Hence, the transfer function have the form

$$T(j\omega) = \frac{1}{j\omega\left(1+\dfrac{j\omega}{2.5}\right)\left(1+\dfrac{j\omega}{40}\right)}$$

Now, from the plot. 20 log $K = 48$ db (approx)

$\Rightarrow \qquad\qquad K = 251$

then, $\qquad T\,(r) = \dfrac{251}{s(1+0.4s)(1+0.025s)}$

(c) From the Bode plot, it can be inferred that, low-frequency asymptote has a slope $+12$ db/octave. Therefore, transfer function has the form $K(j\omega)^2$

Now, at first corner frequency $\omega_1 = 0.5$, the slope of the corner changes by -6 db/octave, at $\omega_2 = 1$, it changes by again -6 db/octave at $\omega_3 = 5$ and the slope changes by -6 db/octave.

Therefore, transfer function has factor

$$\left(\frac{1}{1+\dfrac{j\omega}{0.5}}\right),\ \frac{1}{\left(1+\dfrac{j\omega}{1}\right)} \text{and} \left(\frac{1}{1+\dfrac{j\omega}{5}}\right)$$

corresponding to these corner frequencies.

Hence, the transfer function has the form.

$$T(j\omega) = \frac{K(j\omega)^2}{(1+2j\omega)(1+j\omega)(1+0.2j\omega)}$$

Also from the plot.

$$\Rightarrow \qquad 20\log K + 20\log(10)^2 - 20\log 2 = 32 \text{ db}$$

$$\Rightarrow \qquad 20\log K + 0 - 20\log 2 = 32$$

$$\Rightarrow \qquad K = 79.43.$$

Thus, $$T(s) = \frac{(79.43)\cdot s^2}{(1+2s)(1+s)(1+0.2s)}$$

Problem 8.10. Draw the Bode plot for a unity feedback system characterised by the open-loop trasfer function.

$$G(s) = \frac{K(1+0.2s)(1+0.025s)}{s^3(1+0.001s)(1+0.005s)}$$

Show that the system is conditionally stable. Find the range of values of K for which the system is stable.

Solution:

Let $K = 1$

Magnitude plot

S. No.	Factor	Corner Frequency	Asymptotic log-magnitude characteristic
1.	$1/s^3$	None	Straight line of constant slope -60 db/decade passing through $\omega = 1$
2.	$(1 + 0.2\,s)$	5	Straight line of constant slope of $+20$ db/decade originating from $\omega = 5$
3.	$(1 + 0.025\,s)$	40	Straight line of constant slope of $+20$ db/decade originating from $\omega = 40$
4.	$1/(1 + 0.005\,s)$	200	Straight line of constant slope of -20 db/decade originating from $\omega = 200$
5.	$1/(1 + 0.001\,s)$	1000	Straight line of constant slope of -20 db/decade originating from $\omega = 1000$

Phase Plot

$$\phi = -270° + \tan^{-1} 0.2\,\omega + \tan^{-1} 0.025\,\omega - \tan^{-1} 0.005\,\omega - \tan^{-1} 0.001\,\omega.$$

S. No.	ω	ϕ
1.	0.1	$- 268°$
2.	0.8	$- 260°$
3.	1	$- 257°$
4.	3	$- 236°$
5.	10	$- 198°$
6.	15	$- 183°$
7.	30	$- 163°$
8.	60	$- 148°$
9.	100	$- 147°$
10.	300	$- 172°$
11.	400	$- 182°$

Resulting magnitude and phase plots are shown in the figure below.

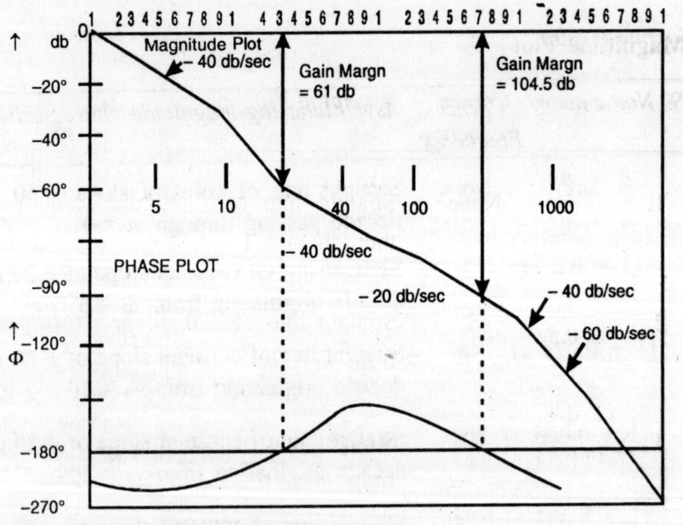

Fig. P 8.10

The phase plot crosses the $-180°$ line twice indicating that the system under consideration is conditionally stable.

Gain margins are 61 db and 104.5 db.

Now $\qquad$ $20 \log K = 61$ $\quad \therefore K = 1122$

Also $\qquad$ $20 \log K = 104.5 \therefore K = 167880$

$\therefore$ Condition for stability is $1122 \angle K \angle 167880$

Problem 8.12. The open-loop transfer function of a unity feedback system is

$$G\ (s) = \frac{1}{s(1+0.5s)(1+0.1s)}$$

Find gain and phase margin. If a phase-lag element with transfer function of $\dfrac{(1+2s)}{(1+5s)}$ is added in the forward path, find by how much the gain must be changed to keep the margin same.

Solution:

Magnitude Plot

S. No.	Factor	Corner frequency	Asymptotic log-magnitude characteristic
1.	$\dfrac{1}{s}$	None	Straight line of $-$ 20 db/decade passing thorugh $\omega = 1$
2.	$\dfrac{1}{1+0.5s}$	2	Straight line of -20 db/decade originating from $\omega = 2$
3.	$\dfrac{1}{1+0.1s}$	10	Straight line of -20 db/decade originating from $\omega = 10$

With phase lag element $G\ (s) = \dfrac{(1+2s)}{s(1+0.5s)(1+0.1s)(1+5s)}$

S. No.	Factor	Corner frequency	Asymptotic log-magnitude characteristic
1.	$\dfrac{1}{s}$	None	Similar to previous one.
2.	$\dfrac{1}{1+5s}$	0.2	Straight line of –20 db/decade originating from $\omega = 0.2$
3.	$\dfrac{1}{1+2s}$	0.5	Straight line of +20 db/decade originating from $\omega = 0.5$
4.	$\dfrac{1}{1+0.5s}$	2	Similar to previous one,
5.	$\dfrac{1}{1+0.5s}$	10	Similar to previous one.

Phase Plot

(a) Without phase lag, $\phi_1 = -90° - \tan^{-1} 0.5\omega - \tan^{-1} 0.1\omega$

(b) With phase lag, $\phi_2 = -90° - \tan^{-1} 0.5\omega - \tan^{-1} 0.1\omega - \tan^{-1} 2\omega - \tan^{-1} 5\omega$

S. No.	ω *rad/sec*	ϕ_1	ϕ_2
1.	0.1	– 93.4°	– 105°
2.	0.2	– 96.8°	– 120°
3.	0.5	– 107°	– 130°
4.	1	– 122.3°	– 137.5°
5.	2	– 146.31°	– 154.6°
6.	5	– 184.76°	– 188°
7.	10	– 228.7°	– 230.4°

Bode plots are shown in the figure below.

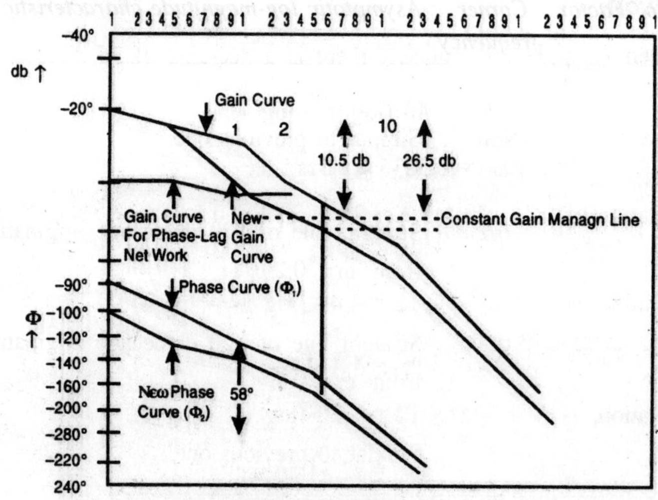

Fig. P. 8.11

(a) Gain margin = 19.5 db

(b) Phase margin = 58° with phase-lag network introduced the gain-margin is 26.5 db. But since *GM* is required to be kept constant at 19.5 db, the gain must be changed by 26.5 – 19.5 = 7 db 20

$\Rightarrow \qquad \log k = 7$

$\therefore \qquad K = 2.2$

Problem 8.12. Derive the transfer fuction of the system from the data given on the Bode diagram shown in the figure below.

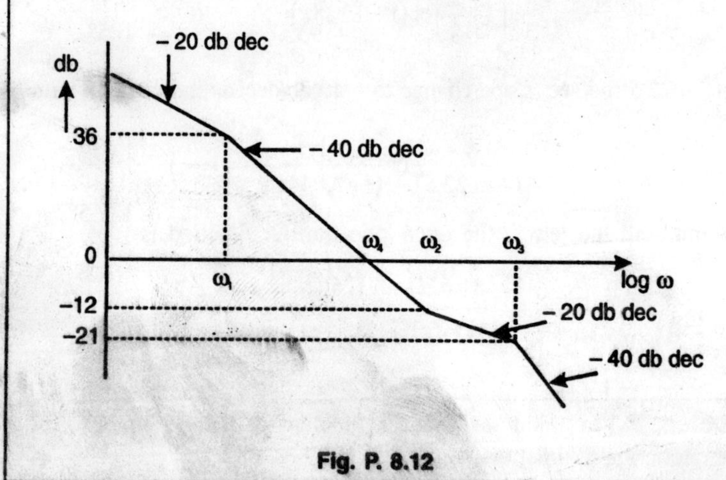

Fig. P. 8.12

Solution:

Between ω_1 and $\omega = 4$ rad/sec, there is a decrease of 35 db

$$\therefore \qquad -36 = -40 \,(\log 4 - \log \omega_1)$$

or $\qquad\qquad \omega_1 = 0.5036 \approx 0.5 \,\text{rad/sec}$

Calculation of 'K' $20 \log K = 36 + 20 \log 0.5$

or $\qquad\qquad\qquad K = 31.62$

Calculation 'ω_2' $\qquad -12 = -40 \,(\log \omega_2 - \log 4)$

or $\qquad\qquad\qquad \omega_2 = 8$ rad/sec

Calculation 'ω_3' $\qquad -21 + 12 = -20 \,(\log \omega_3 - \log 8)$

or $\qquad\qquad\qquad \omega_3 = 22.5$ rad/sec

First line has a slope of -20 db/decade indicating term $\dfrac{1}{s}$ and since it

is not passing through $\omega = 1$ rad/sec. the term is $\dfrac{K}{s}$ or $\dfrac{31.62}{s}$

At $\omega_1 = 0.5$ rad/sec. slope change to -40 db/decade indicating a term

$$\frac{1}{(1+s/0.5)} \,\text{or}\, \frac{1}{(1+2s)}$$

At $\omega_2 = 8$ rad/sec. slope change to -40 db/decade indicating a term

$$\left(1+\frac{s}{8}\right) \text{or} \,(1+0.125s)$$

At $\omega_3 = 22.5$ rad/sec. slope change to -40 db/decade indicating a term

$$\frac{1}{(1+s/22.5)} \,\text{or}\, \frac{1}{(1+0.044s)}$$

Combining all the terms, the open-loop transfer function is

$$G\,(s) = \frac{31.62\,(1+0.125s)}{s\,(1+2s)(1+0.044s)}$$

Problem 8.13. Find the transfer function of the system whose asymptotic approximation is given in the Fig. P.8.13.

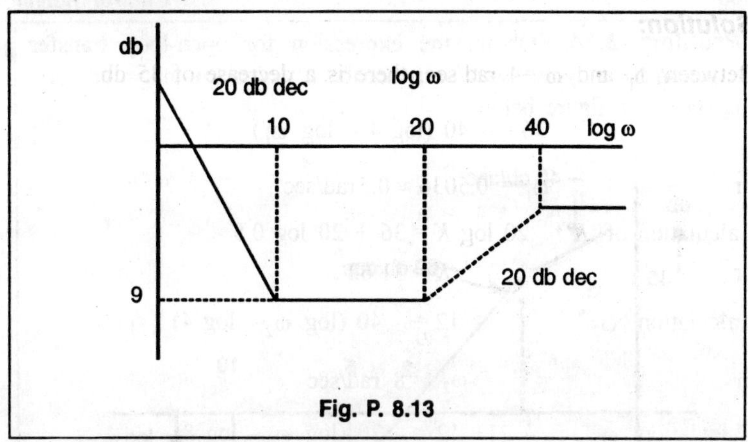

Fig. P. 8.13

Solution:

First line has a slope of −20 db/decade and is not passing through $\omega = 1$ rad/sec. Therefore, it indicates a term $\dfrac{K}{s}$

$$\therefore \qquad 20 \log K = -9$$

or $\qquad K = 0.35$

Therefore the term is $\dfrac{0.35}{s}$

At $\omega = 1$ rad/sec, slope changes to 0 db/decade indicating a term $(1 + s)$.

At $\omega = 20$ rad/sec, slope change to $+20$ db/decade indicating a term

$$\left(1+\frac{s}{20}\right) \text{or} \left(1+0.05 s\right)$$

At $\omega = 40$ rad/sec. slope changes to 0 db/decade indicating a term

$$\frac{1}{\left(1+\dfrac{s}{20}\right)} \text{or} \frac{1}{\left(1+0.025 s\right)}$$

Combining all the terms, we get

$$G\,(s) = \frac{0.35\left(1+s\right)\left(1+0.05 s\right)}{s\left(1+0.025 s\right)}$$

Problem 8.14. Obtain the expression for open-loop transfer function for a system with unity feedback whose log-magnitude plot is shown in figure below: ·

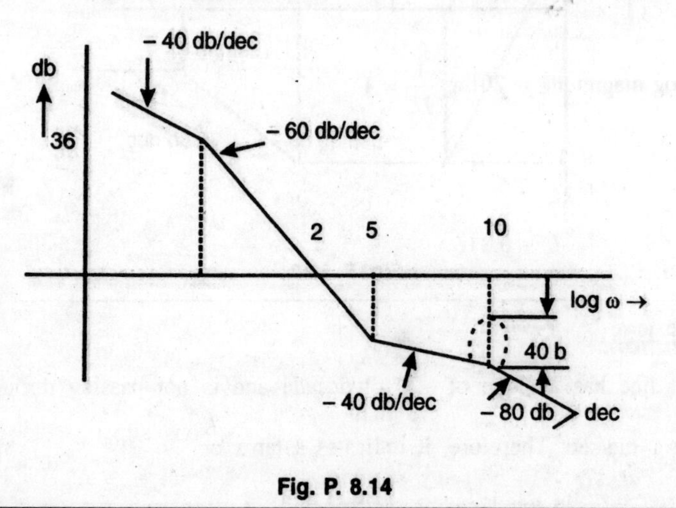

Fig. P. 8.14

Solution:

First line is having a slope of − 40 db/decade and since it is not passing

through ω = 1 rad/sec, it indicates a term $\dfrac{K}{s^2}$

Magnitude at ω = 1 rad/sec of initial part = 60 log 2 = 18.06

∴   20 log K = 18.06

or $K = 8$

At ω = 1rad/sec, the slope changes to − 40 db/decade to − 60 db/decade

indicating a term $\dfrac{1}{1+s}$.

At ω = 5 rad/sec, the slope changes from − 60 db/decade to − 40 db/decade indicating a term $\left(1+\dfrac{s}{5}\right)$ or $(1+0.2s)$

At ω = 10, there is a term of the form $\left\{\left(1+\dfrac{2\zeta s}{\omega_n}+\dfrac{s^2}{\omega_n^2}\right)\right\}^{-1}$ because the

slope changes from − 40 db/decade to − 80 db/decade and also a peak of 4 db is shown at ω_n = 10 rad/sec.

$$\text{At } \omega = \omega_n, \left\{1+\frac{2\zeta s}{\omega_n}+\left(\frac{s}{\omega_n^2}\right)^2\right\}^{-1} = \left\{\sqrt{\left(1-\frac{10}{10}\right)^2+\left(\frac{2\times\zeta\times10}{10}\right)^2}\right\}$$

$$\therefore \quad \log \text{ magnitude} = 20\log\frac{1}{2\zeta} = 4$$

$$\text{or} \qquad \frac{1}{2\zeta} = e^{1/5}$$

$$\text{or} \qquad \zeta = 0.316$$

Hence the term is $\left(1+\dfrac{2\times0.316\,s}{10}+\dfrac{s^2}{100}\right)^{-1}$

$$\therefore \quad G(s) = \frac{8(1+0.2s)}{s^2(1+s)\left(1+0.0632s+\dfrac{s^2}{100}\right)}$$

$$\text{or} \qquad G(s) = \frac{800(1+0.2s)}{s^2(1+s)\left(s^3+6.32s+100\right)}$$

Problem 8.15. Determine the value of K in the transfer function given below such that

(a) the gain margin is 20 dB.

(b) the phase margin is 30°

$$G(j\omega)\,H(j\omega) = \frac{K}{j\omega(1+j0.1\omega)(1+j0.05\omega)}$$

Solution:

(a) The phase cross over frequency occurs at

$$\angle G(j\omega)\,H(j\omega) = -180°$$

$$\Rightarrow \quad -90° - \tan^{-1}(0.1\omega) - \tan^{-1}(0.05\omega) = -180°$$

$$\Rightarrow \quad -\tan^{-1}(0.1\omega) - \tan^{-1}(0.05\omega) = -90°$$

$$\Rightarrow \quad \tan^{-1}\left(\frac{0.1\omega+0.05\omega}{1-0.1\omega\times0.05\omega}\right)=90°$$

$$\Rightarrow \quad \frac{0.15\omega}{1-0.005\omega^2} = \tan 90°=\infty$$

$$\Rightarrow \quad 1-0.005\omega^2 = 0$$

$$\Rightarrow \quad \omega^2 = \frac{1}{0.005}$$

$$\Rightarrow \quad \omega = \sqrt{200} =14.14\,\text{rad/sec}\,.$$

$$\text{Magnitude} = \left|G(j\omega)\,H(j\omega)\right|$$

$$M= \frac{K}{\omega\sqrt{1+(0.1\omega)^2}\,\sqrt{1+(0.05\omega)^2}}$$

Now, Gain Margin = 20 dB.

$$\Rightarrow \quad 20\log\frac{1}{M} = 20$$

$$\Rightarrow 20\log\frac{\omega\sqrt{1+(0.1\omega)^2}\,\sqrt{1+(0.05\omega)^2}}{K}\Bigg|_{\omega=14.14} = 20$$

$$\Rightarrow \quad \log\left[\frac{\omega\sqrt{1+(0.1\omega)^2}\,\sqrt{1+(0.05\omega)^2}}{K}\right] = 1$$

$$\Rightarrow \quad \frac{\omega\sqrt{1+(0.1\omega)^2}\,\sqrt{1+(0.05\omega)^2}}{K} = 10$$

$$\Rightarrow \quad K= \frac{\omega\sqrt{1+(0.1\omega)^2}\,\sqrt{1+(0.05\omega)^2}}{10}\Bigg|_{\omega=14.14}$$

$$\Rightarrow \quad K= \frac{14.14\sqrt{1+(1.414)^2}\,\sqrt{1+(0.707)^2}}{10}$$

$$\Rightarrow \qquad K = \frac{14.14 \times 1.732 \times 1.2246}{10}$$

$$\Rightarrow \qquad K = \frac{29.988}{10} = 2.9988$$

$$\Rightarrow \qquad K = 3$$

(b) The phase margin occurs at gain cross over frequency $\left(\omega_g\right)$ which is given as.

Magnitude of $G\left(j\omega\right)H\left(j\omega\right)\big|_{\omega=\omega_g} = 1.$

$$M = \frac{K}{\omega\sqrt{1+\left(0.1\omega\right)^2}\,\sqrt{1+\left(0.05\omega\right)^2}} = 1$$

$$K = \omega\sqrt{1+0.01\omega^2}\,\sqrt{1+0.0025\omega^2}$$

Now we find ω from the phase margin of $30°$

$$\angle G\left(j\omega\right)H\left(j\omega\right)\big|_{\omega=\omega_g} = -180° + 30°$$

$$\Rightarrow \qquad -90° - \tan^{-1}\left(0.1\omega\right) - \tan^{-1}\left(0.05\omega\right) = -150°$$

$$\Rightarrow \qquad -\tan^{-1} 0.1\omega - \tan^{-1} 0.05\omega = -60°$$

$$\Rightarrow \qquad \tan^{-1}\frac{0.1\omega + 0.05\omega}{1 - 0.005\omega^2} = 60°$$

$$\Rightarrow \qquad \frac{0.1\omega + 0.05\omega}{1 - 0.005\omega^2} = 1.732$$

$$\Rightarrow \qquad 0.15\omega = 1.732 - 0.00866\omega^2$$

$$\Rightarrow \qquad \omega^2 + 17.32\omega - 200 = 0$$

$$\Rightarrow \qquad \omega = \frac{-17.32 \pm \sqrt{\left(17.32\right)^2 + 4 \times 200 \times 1}}{2} = 7.922$$

Put $\omega = 7.922$ in the equation to find K

$$K = \omega\sqrt{1+\left(0.1\omega\right)^2}\,\sqrt{1+\left(0.05\omega\right)^2}$$

$$\Rightarrow \qquad K = 7.922 \sqrt{1+(0.7922)^2} \; \sqrt{1+(0.3961)^2}$$

$$\Rightarrow \qquad K = 7.922 \times 1.2758 \times 1.0756$$

$$\Rightarrow \qquad K = 10.87$$

Problem 8.16. Determine the gain margin and phase margin of the given transfer function

$$G(s)H(s) = \frac{2.6}{s+(s+1)(s+4)}$$

Solution:

Gain Margin occurs at phase crossover frequency and phase Margin occurs at gain crossover frequency. To determine phase crossover frequency,

$$\angle G(j\omega)H(j\omega)\Big|_{\omega=\omega_p} = -180°$$

$$\Rightarrow \qquad -90° - \tan^{-1}\omega - \tan^{-1}\left(\frac{\omega}{4}\right) = -180°$$

$$\Rightarrow \qquad \tan^{-1}\omega + \tan^{-1}(0.25\omega) = 90°$$

$$\Rightarrow \qquad \frac{\omega + 0.25\omega}{1 - 0.25\omega^2} = \tan 90° = \infty$$

$$\Rightarrow \qquad 1 - 0.25\omega^2 = \frac{0.35\omega}{\infty} = 0$$

$$\Rightarrow \qquad \omega^2 = \frac{1}{0.25} = 4$$

$$\Rightarrow \qquad \omega = 2 \, \text{rad/sec.}$$

$$\text{Gain Margin} = 20 \log \frac{1}{|GH(j\omega)|_{\omega=2}}$$

$$\Rightarrow \qquad GM = 20 \log \left[\frac{\omega \sqrt{1+(\omega)^2} \; \sqrt{(4)^2+\omega^2}}{2.6} \right]_{\omega=2}$$

$$\Rightarrow \quad GM = 20 \log \left[\frac{\omega\sqrt{1+(2)^2}\,\sqrt{(4)^2+(2)^2}}{2.6} \right]$$

$$\Rightarrow \quad GM = 20 \log 7.6923$$

$$\Rightarrow \quad GM = 17.72 \text{ dB}.$$

To determine gain crossover frequency

$$|G(j\omega) + H(j\omega)| = 1$$

$$\Rightarrow \quad \frac{2.6}{\omega\sqrt{1+\omega^2}\,\sqrt{4+\omega^2}} = 1$$

$$\Rightarrow \quad 2.6 = \omega\sqrt{1+\omega^2}\,\sqrt{16+\omega^2}$$

Squaring both sides, we get,

$$\Rightarrow \quad (2.6)^2 = \omega^2\left(1+\omega^2\right)\left(16+\omega^2\right)$$

$$\Rightarrow \quad 6.76 = \left(\omega^4+\omega^2\right)\left(16+\omega^2\right)$$

$$\Rightarrow \quad 6.76 = 16\omega^4+16\omega^2+\omega^6+\omega^4$$

$$\Rightarrow \quad \omega^6+17\omega^4+16\omega^2-6.76 = 0$$

By hit and trial we get,

$$\omega = 0.55 \text{ rad/sec.}$$

Phase Margin $\quad \phi = \angle GH(j\omega)\big|_{\omega=\omega g} +180°$

$$\Rightarrow \quad \phi = -90° - \tan^{-1}\omega - \tan^{-1}(0.25\omega) + 180°$$

$$\Rightarrow \quad \phi = -90° - \tan^{-1}(0.55) - \tan^{-1}(0.1375) + 180°$$

$$\Rightarrow \quad \phi = 90° - 28.81° - 7.83°$$

$$\Rightarrow \quad \phi = 90° - 36.64°$$

$$\Rightarrow \quad \phi = 53.34°$$

Problem 8.17.   Draw the Bode plot from the data given below:

ω rad/sec	0.2	0.5	1	2	2.5	5	10
$20 \log_{10} \|G(s)\|$ (dB)	30	21	14.5	5.0	2.0	−10	−26
$\angle G(s)$ (degree)	−98	−110	−130	−161	−173	−209	−236

(a)   Determine gain Margin

(b)   Determine Phase Margin

(c)   Determine the factor by which the gain is to be multiplied to have Gain Margin = 10 dB.

Solution:

(a)   Graphically we can find the GM and PM. The Gain margin is the gain in dB at the point where the phase plot cuts −180° line. The PM is the angle ϕ where the gain plot cuts the 0 dB line. The graphs are plotted to determine the phase crossover frequency and gain crossover frequency.

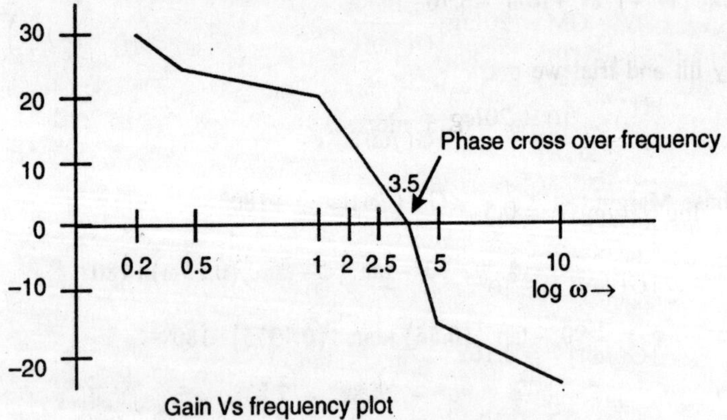

Fig. P. 8.18

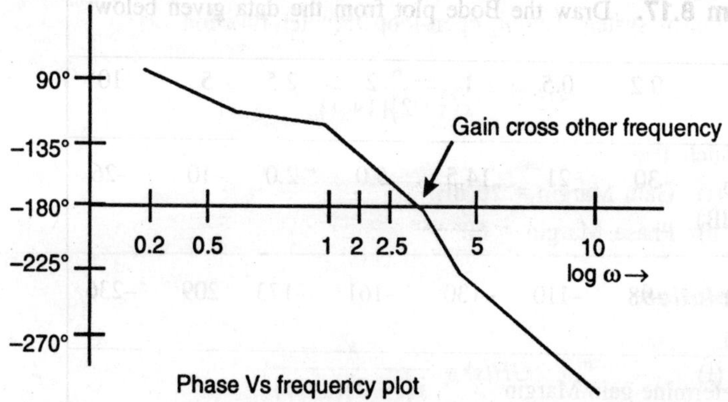

Fig. P. 8.18 (a)

From the figures it is clear that the phase crossover frequency
and gain crossover frequency coincides i.e., they are equal.
Hence, the Gain Margin and Phase Margin both are zero.

i.e., $\qquad$ GM = 0 dB.

$\qquad$ & PM = 0°

Hence, the given system is marginally stable.

(c) For Gain Margin = 10 dB

$\Rightarrow \qquad GM = 20\log\dfrac{1}{|G(j\omega)|}$

$\Rightarrow \qquad 10 = 20\log\dfrac{1}{|G(j\omega)|}$

$\Rightarrow \quad \log|G(j\omega)|^{-1} = 0.5$

$\Rightarrow \quad |G(j\omega)|^{-1} = 10^{0.5}$

$\Rightarrow \quad |G(j\omega)|^{-1} = 3.162$

$\Rightarrow \quad |G(j\omega)| = \dfrac{1}{3.162} = 0.316$

Therefore, gain must be multiplied by a factor of 0.316

Problem 8.18. Determine the value of K for a unity feedback control system having open-loop transfer function

$$GH(s) = \frac{K}{s(s+2)(s+4)}$$

Such that

 (i) Gain Margin = 20 dB.

 (ii) Phase Margin = 60°

Solution:

(i)
$$GH(s) = \frac{K}{s(s+2)(s+4)}$$

$$\Rightarrow \quad GH(j\omega) = \frac{K}{j\omega(j\omega+2)(j\omega+4)}$$

$$\Rightarrow \quad |GH(j\omega)| = \frac{K}{\omega\sqrt{\omega^2+(2)^2}\,\sqrt{\omega^2+(4)^2}}$$

$$\Rightarrow \quad |GH(j\omega)| = \frac{K}{\omega\sqrt{\omega^2+4}\,\sqrt{\omega^2+16}}$$

Now,
$$GM = 20.\log\frac{1}{|GH(j\omega)|}$$

$$\Rightarrow \quad 20 = 20\log\left[\frac{\omega\sqrt{4+\omega^2}\,\sqrt{16+\omega^2}}{K}\right]$$

$$\Rightarrow \quad \log\frac{\omega\sqrt{4+\omega^2}\,\sqrt{16+\omega^2}}{K} = 1 \tag{1}$$

Now to get phase cross over frequency,

$$\angle G(j\omega) = -180°$$

$$\Rightarrow \quad -90° - \tan^{-1}\left(\frac{\omega}{2}\right) - \tan^{-1}\left(\frac{\omega}{4}\right) = -180°$$

$$\Rightarrow \quad \tan^{-1}(0.5\omega) + \tan^{-1}0.25\omega = +90°$$

$$\Rightarrow \tan^{-1}\left(\frac{0.5\omega+0.25\omega}{1-0.125\omega^2}\right) = 90°$$

$$\Rightarrow \quad 1-0.125\omega^2 = 0$$

$$\Rightarrow \quad \omega^2 = \frac{1}{0.125}$$

$$\Rightarrow \quad \omega^2 = 8$$

$$\Rightarrow \quad \omega^2 = 2\sqrt{2} = 2.828 \text{ rad/sec}$$

Put in Equation (1) we get the value of 'K' as

$$\log\frac{2.828\sqrt{4+(2.828)^2}\;\sqrt{16+(2.828)^2}}{K} = 1$$

$$\Rightarrow \quad \log\frac{2.828\times3.463\times4.9}{K} = 1$$

$$\Rightarrow \quad \log\frac{47.975}{K} = 1$$

$$\Rightarrow \quad \frac{47.975}{K} = 10$$

$$\Rightarrow \quad K = 4.797 \simeq 4.8$$

(iii)   Phase Margin = 60°

Phase Margin = $\angle GH(j\omega)+180°$

$$\Rightarrow 60° = -90°-\tan^{-1}\left(\frac{\omega}{2}\right)-\tan^{-1}\left(\frac{\omega}{4}\right)+180°$$

$$\Rightarrow 60° = 90°-\tan^{-1}(0.5\omega)-\tan^{-1}(0.25\omega)$$

$$\Rightarrow \tan^{-1}\frac{0.5\omega+0.25\omega}{1-0.125\omega^2} = 30°$$

$$\Rightarrow \frac{0.5\omega+0.25\omega}{1-0.125\omega^2} = \frac{1}{\sqrt{3}}$$

$$\Rightarrow \quad 1.3\omega = 1-0.125\omega^2$$

$\Rightarrow \quad \omega^2 + 10.4\omega - 8 = 0$

$\Rightarrow \qquad \omega = \dfrac{-10.4 \pm \sqrt{(10.4)^2 + 4 \times 8}}{2}$

$\Rightarrow \qquad \omega = 0.72 \ \text{rad/sec.}$

At $\qquad \omega = 0.72, \ \left| GH(j\omega) \right|_{\omega = 0.72} = 1$

$\Rightarrow \quad \dfrac{K}{\omega\sqrt{4 + \omega^2}\ \sqrt{16 + \omega^2}} = 1$

$\Rightarrow \quad K = 0.72\sqrt{4 + (0.72)^2}\ \sqrt{16 + (0.72)^2}$

$\Rightarrow \quad K = 0.72\ \sqrt{4.5184}\ \sqrt{16.5184}$

$\Rightarrow \quad K = 6.22$

CHAPTER 9

Compensation in Control System

Problem 9.1.   Determine the maximum value for the Bode gain K_B which will result in a gain margin of 6 db or more and a phase margin of 45° or more for the system with the open-loop frequency response function

$$GH(j\omega) = \frac{K_B}{j\omega\left(1+j\omega/5\right)^2}$$

Solution:

The Bode plots for this system with $K_B = 1$ are shown in Fig. 9.1 (a) and (b)

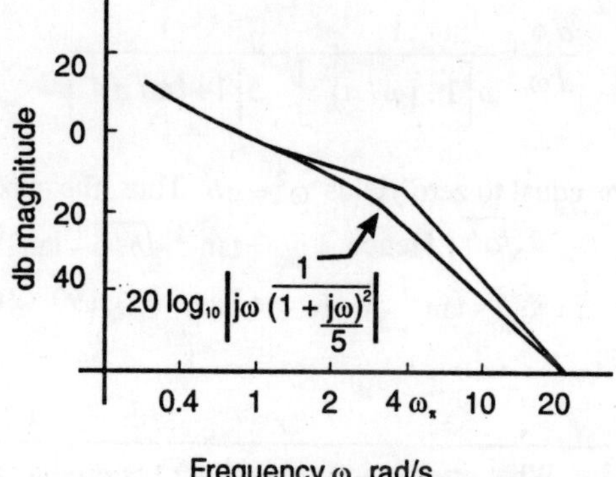

Fig. P. 9.1 (a)

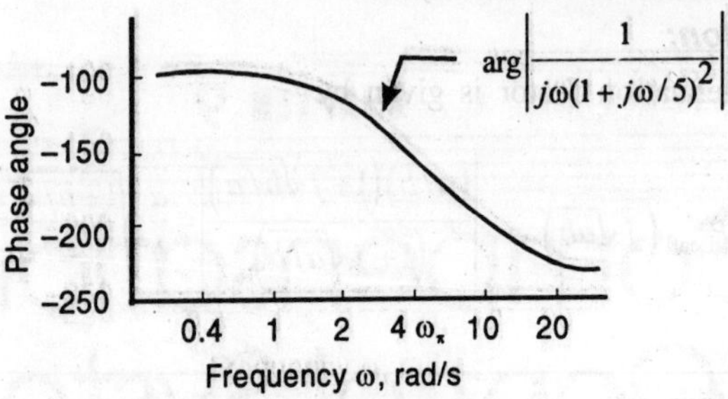

Fig. P. 9.1 (b)

The gain margin, measured at $\omega_\pi = 5$ rad/s, is 20 db. Thus, the Bode gain can be raised by as much as $20 - 6 = 14$ db and still satisfy the gain margin requirement. However, the Bode phase angle plot indicates that, for $\phi_{PM} \geq 45°$, the gain corssover frequency ω_1 must be less than about 2 rad/s. The magnitude curve can be raised by as much as 7.5 db before ω_1 exceeds 2 rad/s. Thus, the maximum value of K_B satisfying both specifications is 7.5 db, or 2.37.

Problem 9.2. Show that the maximum phase lead of the lead compensator occure at $\omega_m = \sqrt{ab}$ and prove equation $\phi_{max} = \left(90° - 2\tan^{-1}\sqrt{a/b}\right)$ degrees .

Solution:

The phase angle of the lead compensator is

$\phi = \arg P_{lead}(j\omega) = \tan^{-1}\omega/a - \tan^{-1}\omega/b.$ Then

$$\frac{d\phi}{d\omega} = \frac{1}{a\left[1+(\omega/a)^2\right]} - \frac{1}{b\left[1+(\omega/a)^2\right]}$$

Setting $d\phi/d\omega$ equal to zero yields $\omega^2 = ab$. Thus, the maximum phase lead occurs at $\omega_m = \sqrt{ab}$. Hence $\phi_{max} = \tan^{-1}\sqrt{b/a} - \tan^{-1}\sqrt{a/b}$. But since $\tan^{-1}\sqrt{b/a} = \pi/2 - \tan^{-1}\sqrt{a/b}$, we have $\phi_{max}\left(90 - 2\tan^{-1}\sqrt{a/b}\right)$ degrees.

Problem 9.3. What attenuation (magnitude) is produced by a lead compensator at the frequency of maximum phase lead $\omega_m = \sqrt{ab}$?

Solution:

The attenuation factor is given by

$$P_{\text{Lead}}\left(j\sqrt{ab}\right) = \left|\frac{(a/b)\left(1+j\sqrt{b/a}\right)}{\left(1+j\sqrt{a/b}\right)}\right| = \frac{a}{b}\sqrt{\frac{1+b/a}{1+a/b}} = \sqrt{\frac{a}{b}}$$

Problem 9.4. Design compensation for the system

$$GH(j\omega) = \frac{8}{(1+j\omega)(1+j\omega/3)^2}$$

which will yield an overall phase margin of 45° and the same gain crossover frequency ω_1 as the uncompensated system. The latter is essentially the same as designing of the same bandwidth.

Solution:

The Bode plots for the uncompensated system are shown in Fig. 9.4 (a) and (b).

The gain crossover frequency ω_1 is 3.4 rad/s and the phase margin is 10°. The specifications can be met with a cascade lead compensator and gain factor amplifier. Choosing a and b for the lead compensator is somewhat arbitrary, as long as the phase lead at $\omega_1 = 3.4$ is sufficient to raise the phase margin from 10° to 45°. However, it is often desirable, for economic reasons, to minimize the low frequency attenuation obtained from the lead network by choosing the largest lead ratio $a/b < 1$ that will supply the required amount of phase lead. Assuming this is the case, the maximum lead ratio that will yield $45° - 10° = 35°$ phase lead is about 0.3.

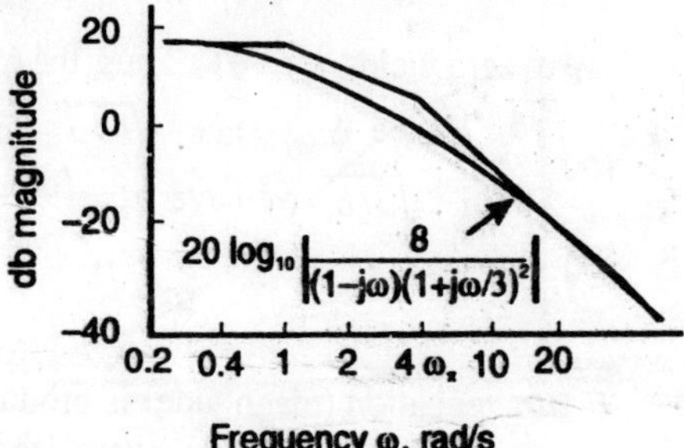

Fig. 9.4 (a)

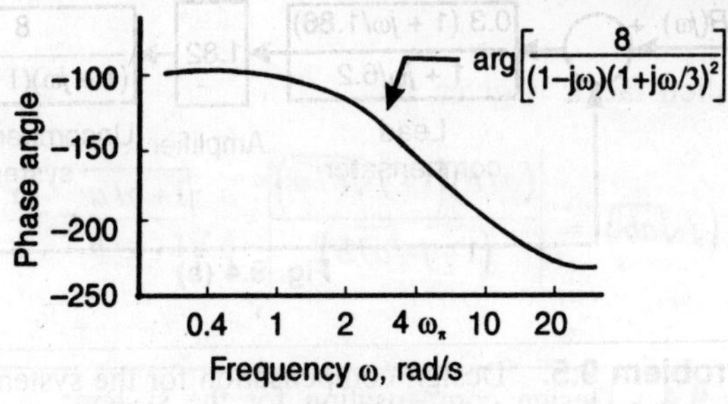

Fig. 9.4 (b)

From fig. 9.4 (a) & (b) yields a value of $a/b = 0.27$. But we shall use $a/b = 0.3$. We want to choose a and b such that the maximum phase lead, which occurs at $\omega_m = \sqrt{ab}$, is obtained at $\omega_1 = 3.4$ rad/s. Thus, $\sqrt{ab} = 3.4$. Substituting $a = 0.3\,b$ into this equation and solving for b, we find $b = 6.2$ and $a = 1.86$. But this compensator produces $20\log_{10}\sqrt{6.2/1.86} = 5.2$ db attenuation at $\omega_1 = 3.4$ rad/s (see Problem 9.3). Thus an amplifier with a gain of 5.2 db, or 1.82 is required, in addition to the lead compensator, to maintain ω_1 at 3.4 rad/s. The Bode plots for the compensated system are shown in Fig. 9.4 (c) & (d) and the block diagram in Fig. 9.4 (e) below.

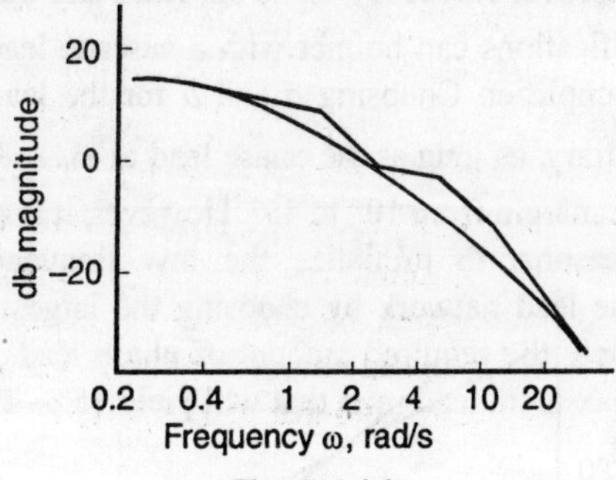

Fig. 9.4 (c)

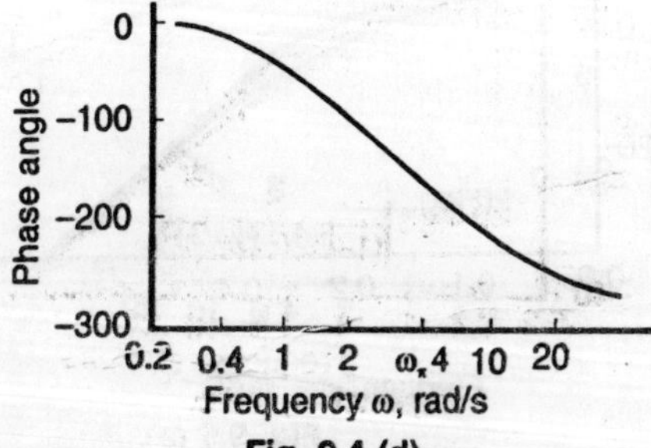

Fig. 9.4 (d)

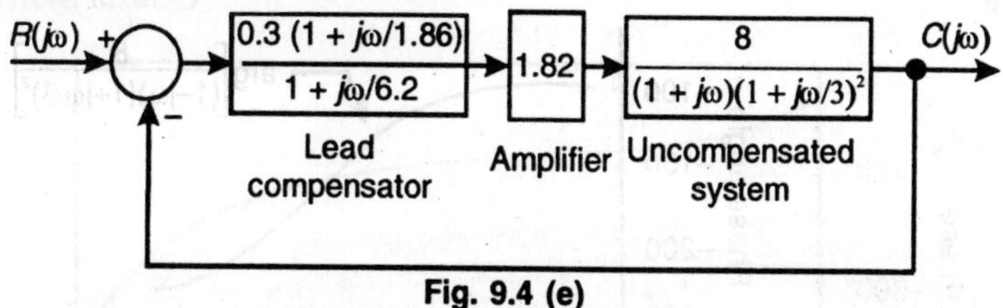

Fig. 9.4 (e)

Problem 9.5.   Design compensation for the system of Problem 9.1 to satisfy the same specifications and, in addition, to have a gain crossover frequency ω_1 less than or equal to 1.0 rad/s and a velocity error constant. $K_v > 5$.

Solution:

The Bode plots for this system, shown in Fig. 9.1 (a) & (b), indicate that $\omega_1 = 1$ rad/s for $K_B = 1$. Hence, $K_v = K_B = 1$ for $\omega_1 = 1$. The gain and phase margin requirements are easily met with any $K_B < 2.37$; but the steady-state specification requires $K_v = K_B > 5$. Therefore, a low frequency cascade lag compensator with $b/a = 5$ can be used to increase K_v to 5, while maintaining the crossover frequency and the gain and phase margins at their previous values. A lag compensator with $b = 0.5$ and $a = 0.1$ satisfies these requirements, as shown in Fig. 9.5 (a) and (b).

The compensated open-loop frequency response function is

$$\frac{5\left(1+j\omega/0.5\right)}{j\omega\left(1+j\omega/0.1\right)\left(1+j\omega/5\right)^2}$$

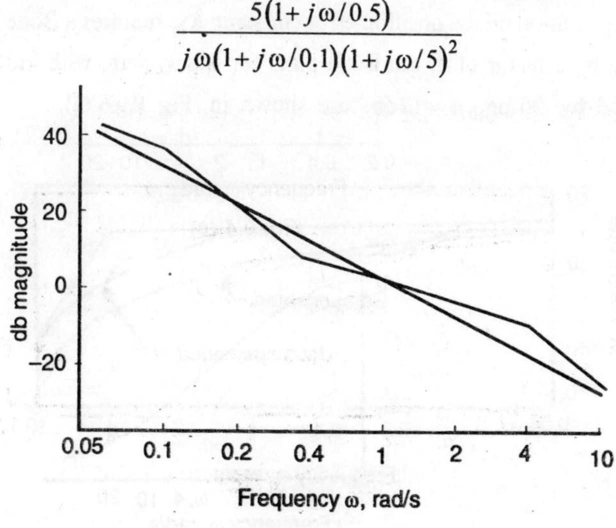

Fig. 9.5 (a)

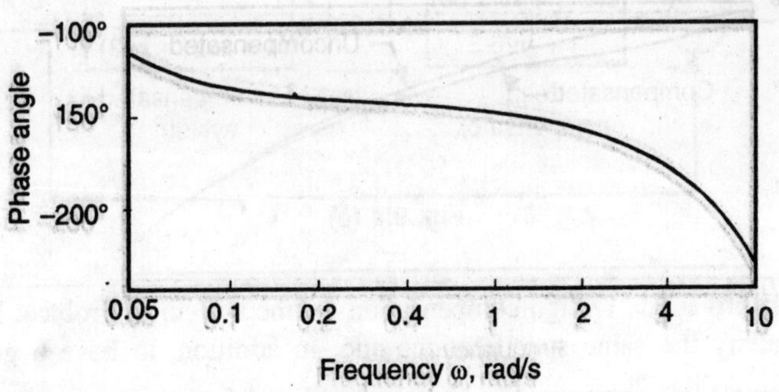

Fig. 9.5 (b)

Problem 9.6. Design a unity-feedback system, with the fixed plant

$$G_2(j\omega) = \frac{1}{\left(1+j\omega/3\right)^3}$$

satisfying the specifications: (1) $K_p \geq 4$, (2) gain margin ≥ 12 db, (3) phase margin $\geq 45°$.

Solution:

The specification on the position error constant K_p requires a Bode gain increase by a factor of 4. the Bode plots for this system, with the gain increased by $20\log_{10}4 = 12$ db, are shown in Fig. P.9.6 (a).

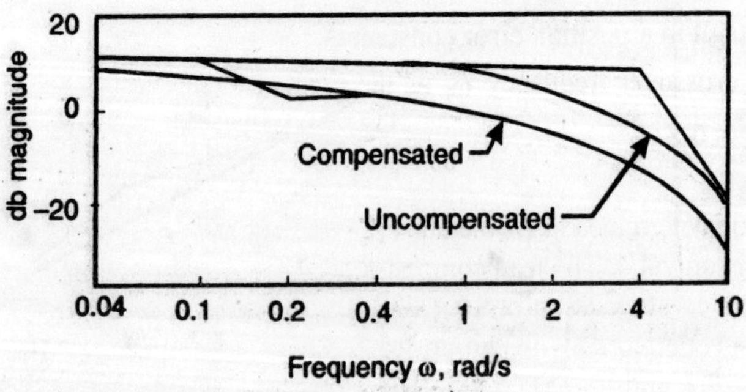

Fig. P. 9.6 (a)

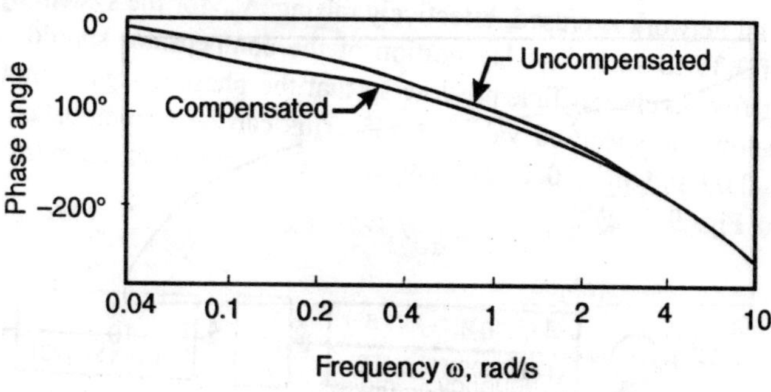

Fig. P. 9.6 (b)

The gain margin is 6 db and the phase margin is 30°. These margins can be increased by adding a lag compensator. To get the gain margin up to 12 db, the high frequency magnitude must be reduced by 6 db. To raise the phase margin to 45°, ω_1 must be lowered to 3.0 rad/s or less. This requires a magnitude attenuation of 3 db at that frequency. Therfore, let us choose a lag ratio $b/a = 2$ to yield a high frequency attenuation of $20\log_{10} 2 = 6$ db. For $a = 0.1$ and $b = 0.2$ the phase margin is 65° and the gain margin is 12 db, as shown in the compensated Bode plots of Fig. P. 9.6 (a) & (b).

The compensated open-loop frequency response function is.

$$\frac{4(1+j\omega/0.2)}{(1+j\omega/0.1)(1+j\omega/3)^3}$$

Problem 9.7. Determine compensation for the system of Problem 9.4 that will result in a position error constant $K_2 \geq 10$, $\phi_{PM} \geq 45°$ and the same gain crossover frequency ω_1 as the uncompensated system.

Solution:

The compensation determined in Problem 9.4 satisfies all the specifications except that K_p is only 4.4. The lead compensator chosen in that problem has a low frequency attenuation of 10.4 db, or a factor of 3.33. Let us replace the lead network with a lag-lead compensator, choosing $a_1 = 1.86$, $b_1 = 6.2$ and $a_2/b_2 = 0.3$. The low frequency magnitude becomes $a_1 b_2 / b_1 a_2 = 1$, or 0 db, and the attenuation produced by the lead network is erased, effectively raising K_p for the system by a factor of 3.33 to 14.5. The lag portion of the compensator should be placed at frequencies sufficiently low so that the phase margin is not reduced below the specified value of 45°. This can be accomplished with $a_2 = 0.09$ and $b_2 = 0.3$ The compensated system block diagram is shown in Fig. 9.7 (a).

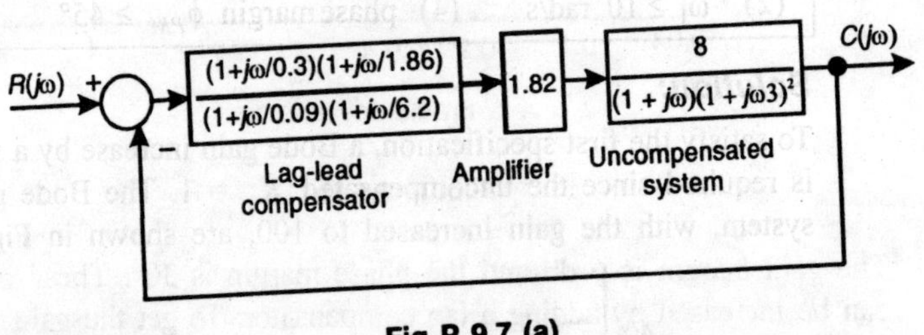

Fig P 9.7 (a)

The compensated Bode plots are shown in Fig. 9.7 (b) & (c).

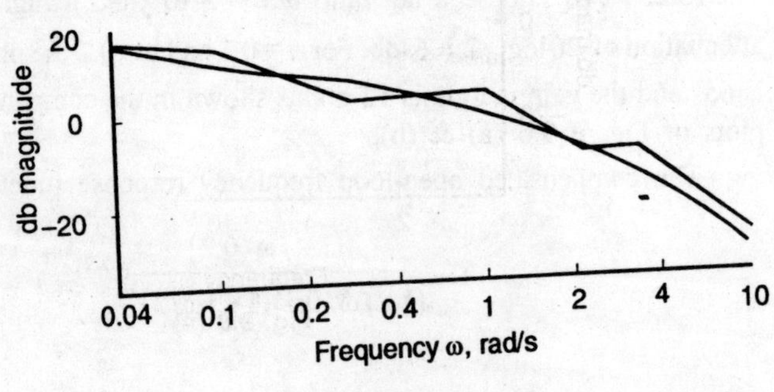

Fig. P 9.7 (b)

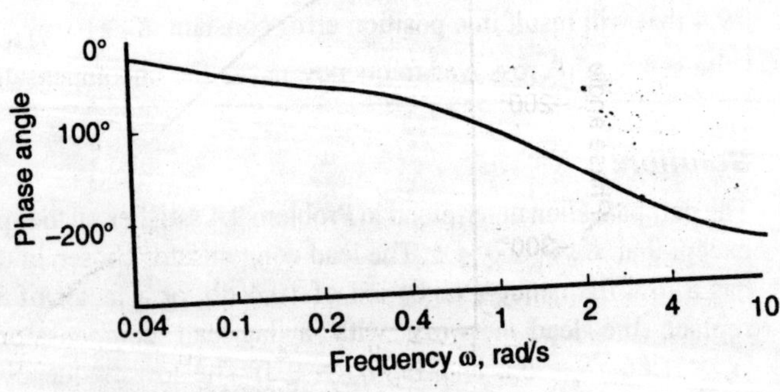

Fig. P 9.7 (c)

Problem 9.8. Design cascade compensation for a unity feedback control system, with the plant

$$G_2(j\omega) = \frac{1}{j\omega(1+j\omega/8)(1+j\omega/20)}$$

to meet the following specifications:

(1) $K_v \geq 100$ (3) gain margin ≥ 10 db

(2) $\omega_1 \geq 10$ rad/s   (4) phase margin $\phi_{PM} \geq 45°$

Solution:

To satisfy the first specification, a Bode gain increase by a factor of 100 is required since the uncompensated $K_v = 1$. The Bode plots for this system, with the gain increased to 100, are shown in Fig. 9.8 (a).

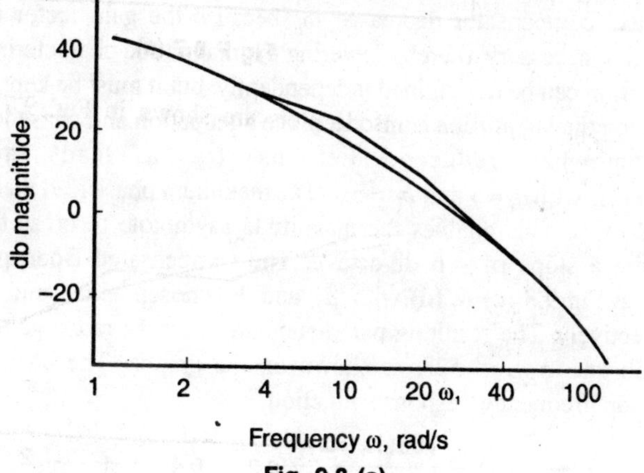

Fig. 9.8 (a)

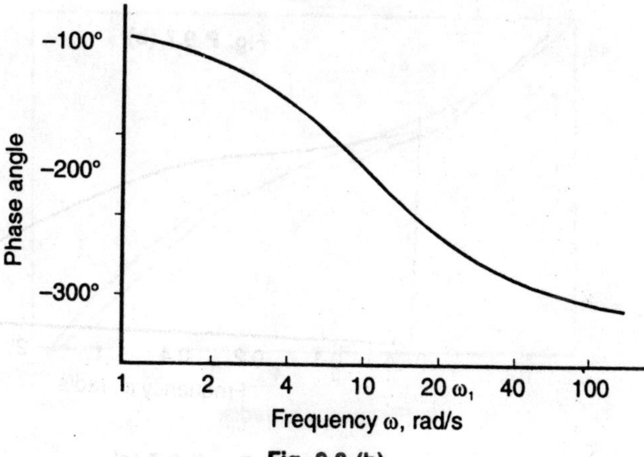

Fig. 9.8 (b)

The gain crossover frequency $\omega_1 = 23$ rad/s, the phase margin is $-30°$, and the gain margin is -12 db. Lag compensation could be used to increase the gain and phase margins by reducing ω_1. However, ω_1 would have to be lowered to less than 8 rad/s to achieve a 45° phase margin and to less than 6 rad/s for a 10 db gain margin. Consequently, we would not satisfy the second specification. With lead compensation, an additional Bode gain increase by a factor of b/a would be required and ω_1 would be increased, thus requiring substantially more than the 75° phase lead for $\omega_1 = 23$ rad/s. These disadvantages can be overcome using lag-lead compensation. The lead portion produces attenuation and phase lead. The frequencies at which these effects occur must be positioened near ω_1 so that ω_1 is slightly reduced and the phase margin is increased. Note that, although pure lead compensation increases ω_1, the lead portion of a lag-lead compensator decreases ω_1 because the gain factor increase of b/a is unnecessary, thereby lowering the magnitude characteristic. The lead portion can be determined independently but it must be kept in mind that, when the lag portion is included, the attenuation and phase lead may be somewhat reduced. Let us try a lead ratio of $a_1/b_1 = 0.1$, with $a_1 = 5$ and $b_1 = 50$. The maximum phase lead then occurs at 15.8 rad/s. This enables the magnitude asymptote to cross the 0 db line with a slope of -6 db/octave. The compensated Bode plots are shown in Fig 9.8 (c) & (d) with a_2 and b_2 chosen as 0.1 and 1.0 rad/s, respectively. The resulting parameters are $\omega_1 = 12$ rad/s, gain margin $= 14$ db and $\phi_{PM} = 52°$, as shown on the graphs. The compensated open-loop frequency response function is

$$\frac{100(1+j\omega)(1+j\omega/5)}{j\omega(1+j\omega/0.1)(1+j\omega/8)(1+j\omega/20)(1+j\omega/50)}$$

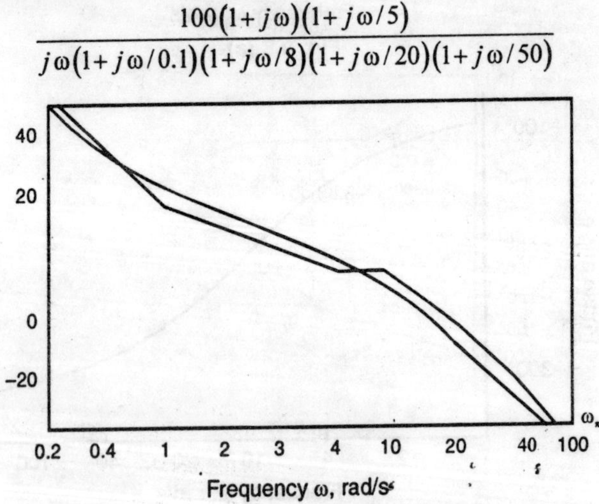

Fig. P. 9.8 (c)

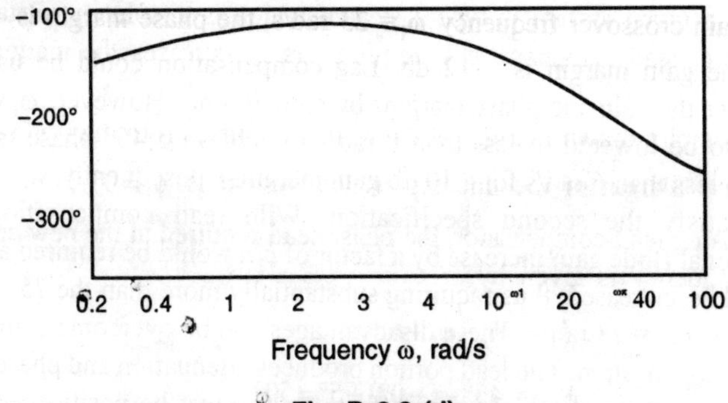

Fig. P. 9.8 (d)

Problem 9.9. A unity feedback control system has an open-loop transfer function of

$$G(s) = \frac{1}{s^2}$$

Design a suitable compensating network such that a phase margin of 45° is achieved without sacrificing system velocity error constant. Sketch the Bode plot of the uncompensated and compensated systems.

Solution:

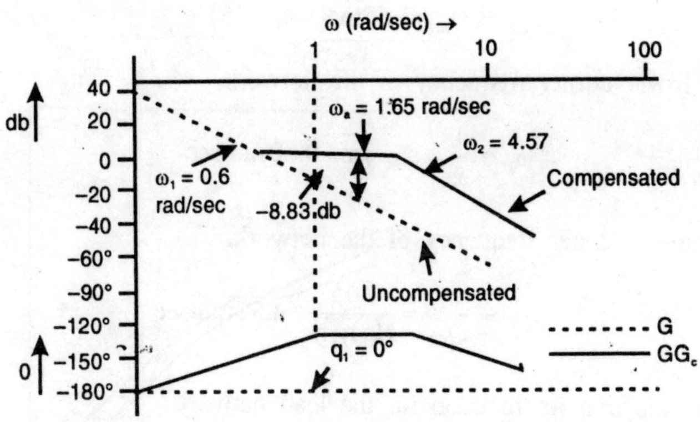

Fig. P. 9.9

Given $G(s) = \dfrac{1}{s^2}$

Since the given system is type-2 system which is absolutely unstable. Therefore, lead compensator is required as it increases the margin of stability.

The Bode plot drawn in Fig. P 9.9 the above figure. From the Bode plot, the phase margin of the uncompensated system is $\phi_1 = 0°$.

When using lead-compensator, the phase-lead required at the new cross-over frequency is given by

$$\phi_1 = \phi_s - \phi_1 + t$$

or

$$\phi_L = 45° - (-0°) + 5° = 50°$$

then

$$\alpha = \frac{1 - \sin 50°}{1 + \sin 50°} = 0.13.$$

The magnitude contribution of the compensated network at ω_m is,

$$10 \log\left(\frac{1}{\alpha}\right) = 10 \log\left(\frac{1}{0.13}\right) = -8.83\,\text{db}$$

Hence, the frequency at which the uncompensated system has a magnitude of -8.83 db becomes the new cross-over frequency $\omega_{c2} = \omega_m$, when the lead network is added.

From the Bode plot,

$$\omega_{c2} = \omega_m = 1.65\,\text{rad/sec.}$$

Then lower corner frequency of the network,

$$\omega_1 = \frac{1}{\tau} = \omega_m \sqrt{\alpha} = 0.60\,\text{rad/sec}$$

and, upper corner frequency of the network,

$$\omega_2 = \frac{\omega_m}{\sqrt{\alpha}} = \frac{1.65}{\sqrt{0.013}} = 4.57\,\text{rad/sec}$$

Hence, the transfer function for the lead network

$$G_c(s) = \frac{1.67\,s + 1}{0.218\,s + 1}$$

with amplification, $A = \dfrac{1}{\alpha} = 7.69.$

Problem 9.10. Consider the system shown in the figure below. Design a lead compensator for this stem to meet the following specifications:

Damping ratio $\zeta = 0.7$

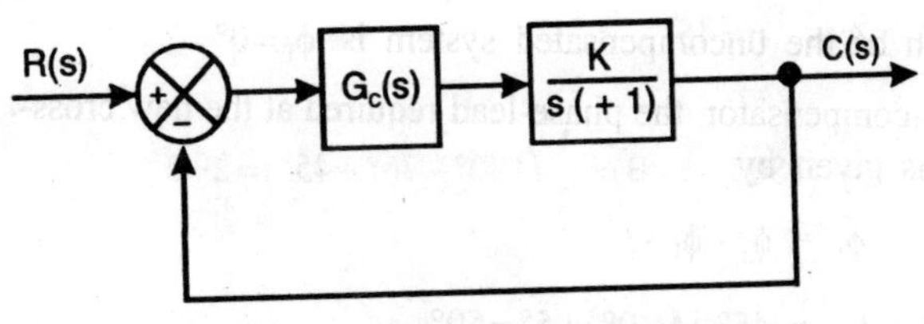

Fig. P. 9.10

Settling time $t_s = 1.4$ sec

Velocity error constant $K_v = 2$ sec^{-1}

Solution:

As
$$t_s = \frac{4}{r\,\omega_n}$$

then,
$$\omega_n = \frac{4}{1.4\times 0.7} = 4.08 \text{ rad/sec}$$

Then dominant closed loop poles are at

$$s_d = -r\,\omega_n \pm j\sqrt{1-r^2}\,\omega_n$$
$$= -2.85 \pm 2.91\,j$$

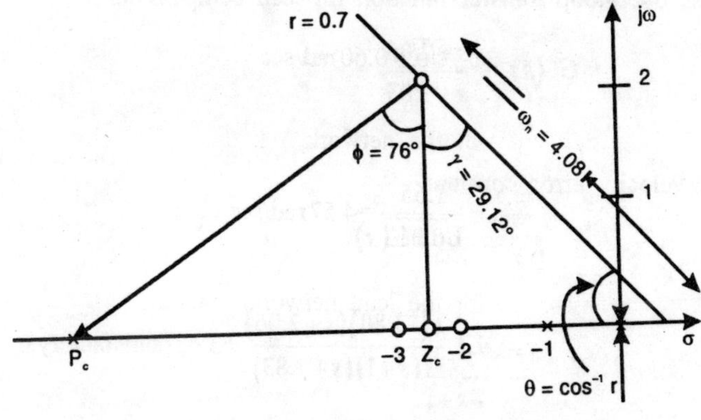

Fig. P. 9.10 (a)

The angle contribution at s_d is

$$\phi = -180° - (-134° - 122°) = 76°$$

$$\theta = \cos^{-1} r = 45°$$

then, $$\gamma = \frac{1}{2}(180° - \phi - \theta) = \frac{1}{2}(180° - 76° - 45°) = 29.5°$$

Now, $$Z_c = \omega_n \left[\frac{\sin\gamma}{\sin(\pi - \theta - \gamma)}\right]$$

$$= 4.08\left[\frac{\sin 29.12°}{\sin(180° - 45° - 29.12°)}\right] = 2.06$$

$$P_c = \omega_n \left[\frac{\sin(\gamma + \phi)}{\sin(\pi - \gamma - \theta - \phi)}\right]$$

$$= 4.08\frac{\sin(29.12° + 76°)}{\sin(180° - 29.12° - 76° - 45°)} = 7.83$$

At $$s_d K = |s_d| \cdot |s_d + 1| = 13.80$$

Hence, open-loop transfer function for lead compensator is

$$G_c(s) = \frac{s + 2.06}{s + 7.83}$$

Now, velocity error constant,

$$K_p = \underset{s \to 0}{\text{Lt}}\, sG(s)$$

$$= \underset{s \to 0}{\text{Lt}}\, \frac{s(13.80)(s + 2.06)}{s(s + 1)(s + 7.83)} = 3.63\,(\text{satisfactory})$$

Since K_p should be greater than 2 sec^{-1} as specified obtained value of 3.63 sec^{-1} shows satisfactory steady state performance.

Problem 9.11. A unity feedback system has an open-loop transfer function of

$$G(s) = \frac{4}{s(2s+1)}$$

It is desired to obtain a phase margin of 40° without sacrificing the K_v of the system. Design a suitable lag-network and compute the value of network components assuming any suitable impedance level.

Solution:

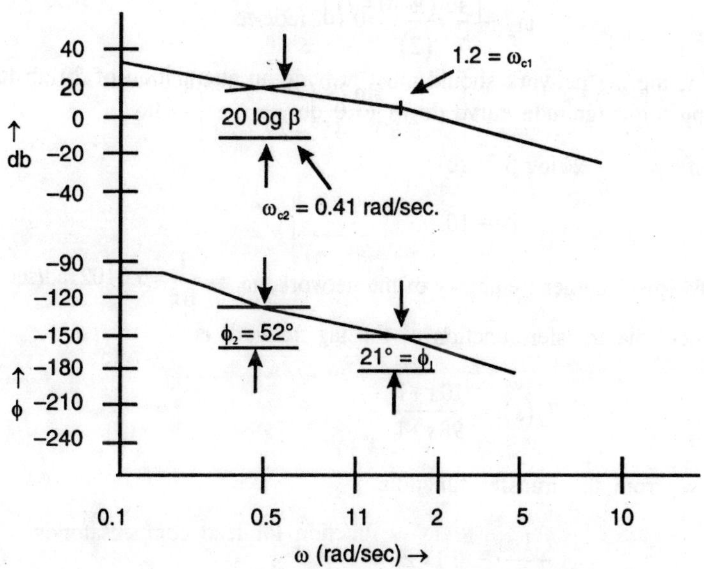

Fig. P. 9.11

Given, $\qquad G\,(s) = \dfrac{4}{s(2s+1)}$

or $\qquad G(j\omega) = \dfrac{4}{j\omega(2j\omega+1)}$

The Bode plot of $G(j\omega)$ is shown in the Fig. P.9.11.

From the plot the cross-over frequency $\omega_{c1} = 1.2$ rad/sec and phase-margin, $\phi_1 = +21°$.

Given, $\phi_s = 40°$

then uncompensated system must make a phase contribution of

$$\phi_2 = \phi_s + \epsilon = 40° + 12° = 52°$$

From the Bode plot,

At $\omega_{c2} = 0.41$ rad/sec.

Then, upper corner frequency which is two octaves below ω_{c2} is

$$\omega_2 = \frac{1}{\tau} = \frac{\omega_{c2}}{(2)^2} = 0.102 \, \text{rad/sec.}$$

Now, the lag network should must provide an attenuation of 20 db to bring log-magnitude curve down to 0 db

then, $20 \log \beta = 20$

or $\beta = 10.$

Now lower corner frequency of the network, $\omega_1 = \dfrac{1}{\beta\tau} = 0.0102 \, \text{rad/sec.}$

Hence, the transfer function of the lag network is,

$$G_c(s) = \frac{10s+1}{98s+1}$$

Now, from the transfer function,

$$\frac{1}{R_2 C} = 0.1 = Z_c$$

and $\dfrac{1}{\left[\dfrac{R_1+R_2}{R_2}\right] \cdot R_2 C} = P_c = \dfrac{1}{98}$

or $\left(\dfrac{R_1+R_2}{R_2}\right) R_2 C = 98$

Choosing $C = 10\,\mu F$

we have, $R_2 = 1\,\text{M}\Omega$ and $R_1 = 8.3\,\text{M}\Omega$

Problem 9.12. Design a phase-lead compensator for the system shown in the figure, to satisfy the following specifications:

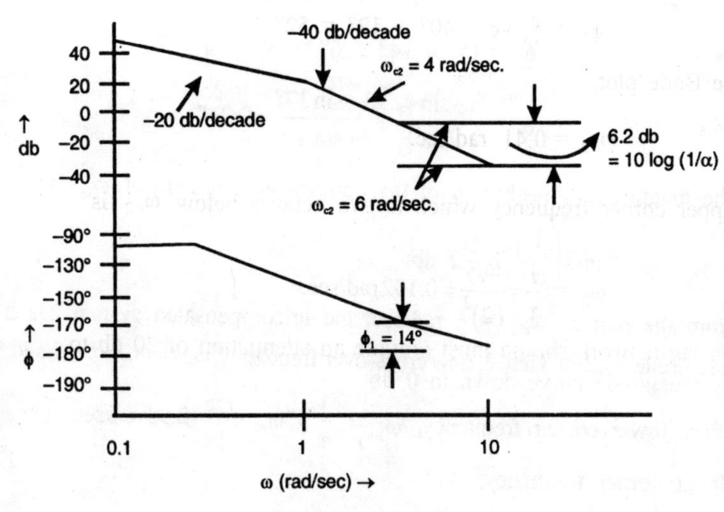

Fig. P. 9.12

(i) The phase margin of the system must be greater than 45°.

(ii) Steady-state error for a unit step input should be less than 1/15 deg per deg/sec of the final output velocity.

(iii) The gain cross-over frequency of the system must be less than 7.5 rad/sec.

Solution:

Given, e_{ss} (steady-state error) $\geq 1/15$

then K_v, (velocity-error constant) ≤ 15.

Now, as
$$K_v = \underset{s \to 0}{\text{Lt}}\ s\,G(s)$$

then.
$$15 = \underset{s \to 0}{\text{Lt}}\ \frac{s \cdot K}{s(s+1)} = K.$$

Thus, open-loop transfer function becomes,

$$G(j\omega) = \frac{15}{j\omega(j\omega+1)}$$

The Bode plots of the system drawn in the above figure.

From the plot phase-margin for the uncompensated system, $\phi_1 = 14°$.

If a lead compensator have to be used then phase-lead requirement at new cross-over frequency is

$$\phi_1 = 45° - 14° + 6° = 37° = \phi_m$$

then

$$\alpha = \frac{1 - \sin\phi_1}{1 + \sin\phi_1} = \frac{1 - \sin 37°}{1 + \sin 37°} = 0.24°$$

The magnitude contribution of the compensated system at ω_m is

$$10\log\left(\frac{1}{\alpha}\right) = 6.2 \text{ db.}$$

From the plot at $\omega_{c2} = 6$ rad/sec, the uncompensated system has a magnitude -6.2db. Hence, new cross-over frequency $\omega_{c2} = \omega_m = 6$ rad/sec.

Then, lower corner frequency, $\omega_1 = \dfrac{1}{\tau} = \omega_m \sqrt{\alpha} = 2.93$ rad/sec

Upper corner frequency,

$$\omega_2 = \frac{1}{\alpha\tau} = \frac{\omega_m}{\sqrt{\alpha}} = 12.24 \text{ rad/sec.}$$

Hence, the lead compensator transfer function

$$G_c(s) = \frac{(0.34s + 1)}{(0.081s + 1)}$$

and amplification, $A = \dfrac{1}{\alpha} = 4.16$

Problem 9.13. Design cascade compensation for a unity feedback control system, with the plant

$$G_2(j\omega) = \frac{1}{j\omega\left(1 + \dfrac{j\omega}{8}\right)\left(1 + \dfrac{j\omega}{20}\right)}$$

to meet the following specifications:

(i) $K_v \geq 100$

(ii) gain margin ≥ 10 db

(iii) $\omega_1 \geq 10$ rad/s

(iv) phase margin $\phi_{PM} \geq 45°$

Solution:

To satisfy the first specification, a Bode gain increase by a factor of 100 is required since the uncompensated $K_v = 1$. The Bode plots for this system, with the gain increased to 100, are shown in Fig. 9.13 (a).

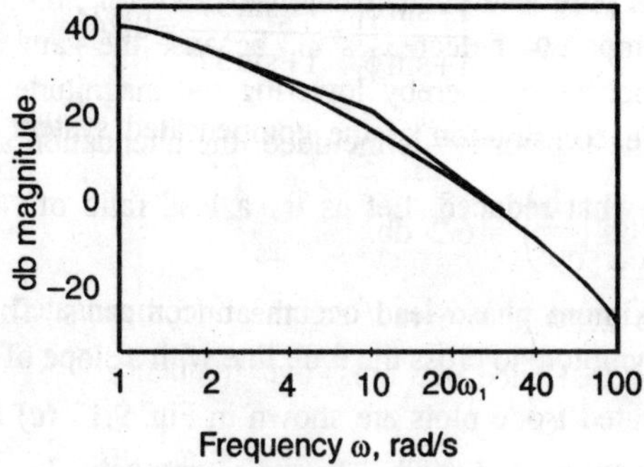

Fig. P. 9.13   (a)

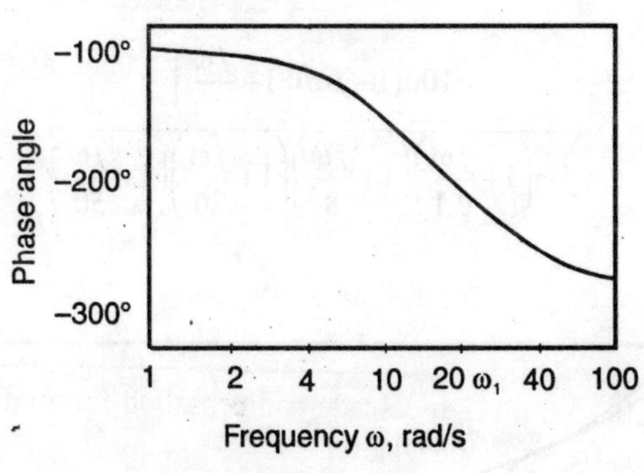

Fig. P. 9.13(b)

Gain crossover frequency, $\omega_1 = 23$ rad/s, the phase margin is $-30°$, and the gain margin is -12 db.

Lag compensation could be used to increase the gain and phase margins by reducing ω_1. However, ω_1 would have to be lowered to less than 8 rad/s to a 45° phase margin and to less than 6 rad/s for a 10 db gain margin. Consequently, we would not satisfy the second specification.

With lead compensation, an additional Bode gain increase by a factor of *b/a* would be required and ω_1 would be increased, thus, requiring substantially more than the 75° phase lead for $\omega_1 = 23$ rad/s.

These disadvantages can be overcome using lag-lead compensation. The lead portion produces attenuation and phase lead. The frequencies at which these effects occur must be positioned near ω_1 so that ω_1 is slightly reduced and the phase margin is increased.

Although pure lead compensation increases ω_1, the lead portion of a lag-lead compensator decreases ω_1 because the gain factor increase of *b/a* is unnecessary, thereby lowering the magnitude characteristic.

When the lag portion is included, the attenuation and phase lead may be somewhat reduced. Let us try a lead ratio of $\dfrac{a_1}{b_1} = 0.1$, with $a_1 = 5$ and $b_1 = 50$.

The maximum phase lead occurs at 15.8 rad/s. This enables the magnitude asymptote to cross the 0 db line with a slope of -6 db/octave.

The compensated Bode plots are shown in Fig. 9.13 (c) & (d) with a_2 and b_2 chosen as 0.1 and 1.0 rad/s, respectively. The resulting parameters are $\omega_1 = 12$ rad/s, gain margin = 14 db and $\phi_{PM} = 52°$, as shown on the graphs. The compensated open-looped frequency response function is

$$\dfrac{100(1-j\omega)\left(1+\dfrac{j\omega}{5}\right)}{j\omega\left(1+\dfrac{j\omega}{0.1}\right)\left(1+\dfrac{j\omega}{8}\right)\left(1+\dfrac{j\omega}{20}\right)\left(1+\dfrac{j\omega}{50}\right)}$$

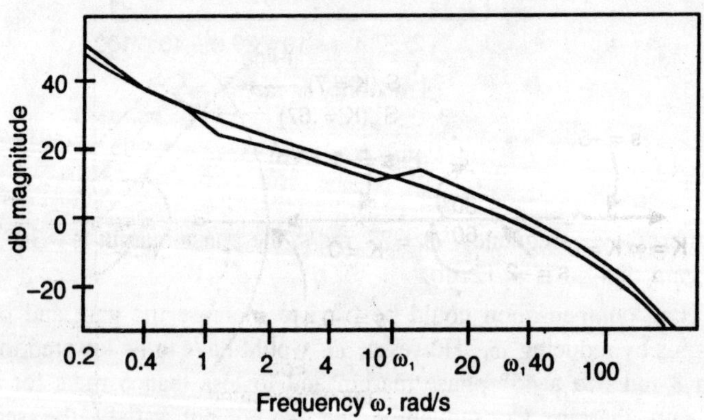

Fig. P. 9.13

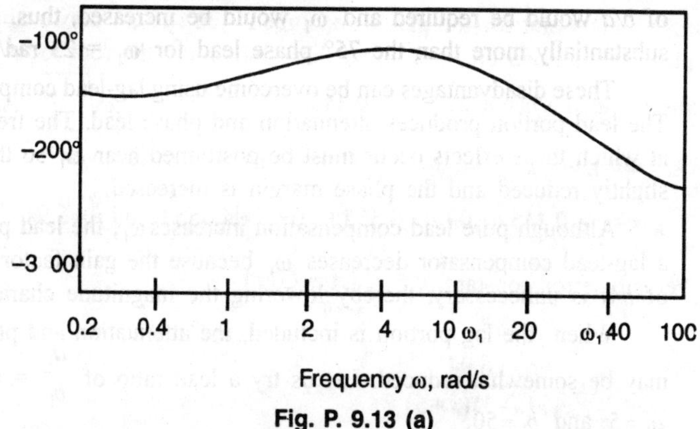

Fig. P. 9.13 (a)

Problem 9.14. A unity feedback system has an open-loop transfer function of $G(s) = \dfrac{K}{s(s+1)(s+5)}$

Draw the root locus plot and determine the value of K to give a damping ratio of 0.3. A network having a transfer function of $\dfrac{10(1+10s)}{(1+100s)}$ is now introduced in tandem. Find the new value of K which gives the same damping ratio for the closed-loop response. Compare the velocity error constant and settling time of the original and the compensated systems.

Solution:

Given, open-loop transfer function $G\,(s) = \dfrac{K}{s(s+1)(s+5)}$

Fig. P. 9.14

The root-locus plot for the given uncompensated system drawn above by hit and trial procedure, it is found that *r*-line cut the root locus for uncompensated system at $s_d = -0.345 + 1.09\,j = -r\omega_n \pm j\sqrt{1-r^2}\,\omega_n$

At s_d, the gain K^{uc} of the uncompensated system is given by

$$K^{uc} = \left|-0.345+1.09\,j\right| \times \left|0.655+1.09\,j\right| \times \left|4.655+1.09\,j\right| = 6.94$$

$\therefore$ Velocity error constant, $K_v^{uc} = \underset{s\to 0}{\mathrm{Lt}}\; s \cdot G(s)$

$$\underset{s\to 0}{\mathrm{Lt}}\; \frac{s\,6.94}{s(s+1)(s+5)} = 1.38\,j$$

Settling time, $\qquad t_s = \dfrac{4}{r\,\omega_n} = 11.59\,\mathrm{sec.} \qquad\qquad \left[\because r\omega_n = 0.345\right]$

Now, redrawing root-locus for the compensated system in the above figure, we get *r*-line cut the root-locus at

$$s_d' = -0.315 \pm 1\,j$$

$$= -r\omega_n \pm j\sqrt{1-r^2}\,\omega_n$$

At, $s_s'(-0.315+1\,j)$, the gain of the compensated system is given by,

$$K^c = \left|-0.345+j\right| \times \left|0.685+j\right| \times \left|4.685+j\right| = 6.02$$

$\therefore$ Velocity error constant, $K^c = \underset{s\to 0}{\mathrm{Lt}}\; s\,G(s) = \underset{s\to 0}{\mathrm{Lt}}\; \dfrac{s\,6.02}{(s+1)(s+5)} = 1.20$

Settling time, $\qquad t_s = \dfrac{4}{r\,\omega_n} = 12.69 \qquad\qquad \left[\because r\omega_n = 0.315\right]$

Problem 9.15. A servomechanism has an open-loop transfer function of

$$G\,(s) = \frac{10}{s(1+0.5s)(1+0.1s)}$$

Draw the Bode plot and determine the phase and gain margins. A network having the transfer function $\dfrac{1+0.23s}{1+0.023s}$ is now introduced in tandem. Determine the new gain and phase margins. Comment upon the improvement in system response caused by the network.

Solution:

The Bode plots for compensated and uncompensated system is drawn in the figure below. For the uncompensated system.

$$GM \text{ (gain-margin)} = 1.7 \, db \text{ at } \omega_c$$

$$= 4.3 \, rad/sec \, (\text{phase cross-over frequency})$$

$$\text{phase-margin, } |\phi_{uc}| = 2.68°$$

For the compensated system

$$GM = 17 \, db \text{ at } \omega_c = 16 \, rad/sec$$

$$\phi_c \text{ (phase-margin)} = 37°$$

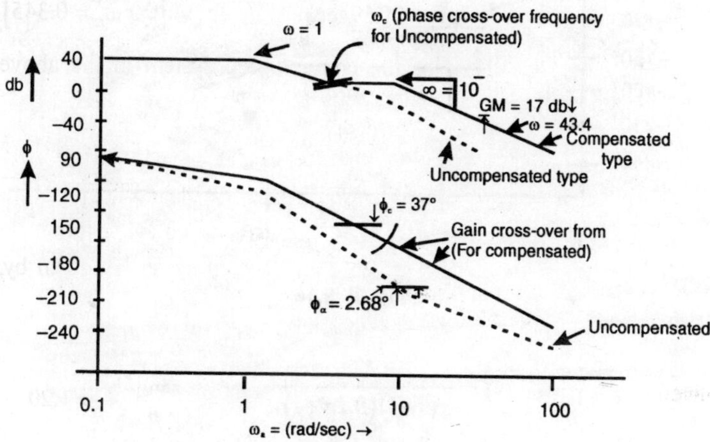

Fig. P. 9.15

Problem 9.16. A unity feedback system has an open-loop transfer function

$$G(s) = \frac{K}{s(s+1)(0.2s+1)}$$

Design phase-lag compensation for the system to achieve the following specifications:

Velocity error constant, $K_v = 8$

Phase margin $\approx 40°$.

Also compare the cross-over frequency of the uncompensated and compensated systems.

Solution:

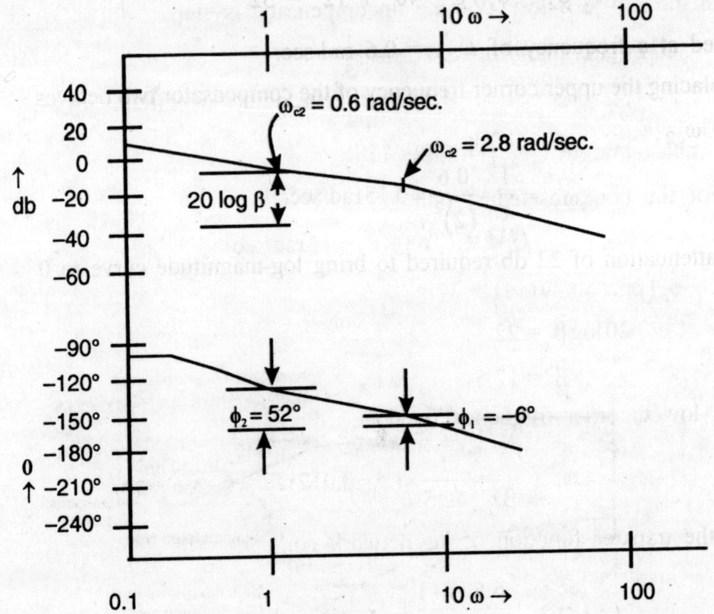

Fig. P. 9.16

Given
$$G(s) = \frac{K}{s(s+1)(0.2s+1)}$$

and\
$$\phi_s = 40°, \quad K_v = 8$$

$$K_v = \underset{s \to 0}{\text{Lt}}\, sG(s) = \underset{s \to 0}{\text{Lt}}\, s\frac{K}{s(s+1)(0.2s+1)} = K = 8$$

Thus,
$$G_f(j\omega) = \frac{8}{j\omega(j\omega+1)(0.2\,j\omega+1)}$$

From, the Bode plot of $G_f(j\omega)$, it is found that corss-over-frequency for uncompensated system is

$$\omega_{c1} = 2.8 \ \text{rad/sec}$$

and phase-margin, $\phi_1 = -6°$

so the system is initially unstable.

Now, phase-margin required for compensated system,

$$\phi_2 = \phi_s + \epsilon = 40° + 12° = 52°$$

obtained at a frequency of $\omega_{c2} = 0.6$ rad/sec.

Now placing the upper corner frequency of the compensator two octaves below ω_{c2},

$$\omega_2 = \frac{1}{\tau} = \frac{0.6}{(2)^2} = 0.15 \text{ rad/sec.}$$

Now, attenuation of 22 db required to bring log-magnitude curve to 0 db.

$$20 \log \beta = 22$$

or $$\beta = 12.5$$

Hence, lower corner frequency fixed at,

$$\omega_1 = \frac{1}{\beta\tau} = \frac{1}{12.5} \times 1.5 = 0.012 \text{ rad/sec.}$$

Then, the transfer function of lag-network compensation is

$$G_c(s) = \left(\frac{6.67\,s + 1}{83.3\,s + 1} \right)$$

Now, ω_c (uncompensated) $= 2.8$ rad/sec.

$$\omega_c \text{ (compensated)} = 0.6 \text{ rad/sec.}$$

Problem 9.17. Figure shows a unity feedback system with a forward path transfer function

$$G(s) = \frac{10}{s(s+1)}$$

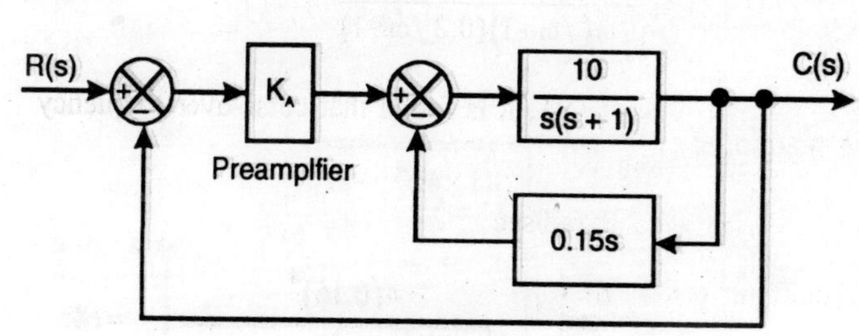

Fig. P. 9.17

The system is compensated by means of rate feedback $H(s) = 0.15$ s, in the minor feedback loop. Determine the effect of compensation by comparing the phase margins and bandwidths of the compensated and uncompensated schemes. In each case adjust K_A to a value which gives a velocity error constant of $K_v = 10$ for the purpose of comparison.

Solution:

Without compensation, $\dfrac{C(s)}{R(s)} = \dfrac{10\,K_A}{s^2+s+10\,K_A}$

(1)

Given, $K_v = 0$ (velocity error constant)

then, $10 = \underset{s \to 0}{Lt}\, s\,G(s) = \underset{s \to 0}{Lt}\, s \cdot \dfrac{10\,K_A}{s^2+s}$ [Using eq^n (1)]

$$= \underset{s \to 0}{Lt} \cdot \dfrac{10\,K_A}{s+1} = 10\,K_A$$

which gives, $K_A = 1$

From equation (1) $\dfrac{C(s)}{R(s)} = \dfrac{10}{s^2+s+10}$

By definition, $\omega_n = \sqrt{10} = 3.16\,\text{rad/sec.}$

and $2r\,\omega_n = 1$

which gives, $r = 0.16$

Now, Bandwidth,

$$\omega_b = \omega_n\left[1+2r^2+\sqrt{2-4r^2+4r^4}\right]^{1/2}$$

$$= 3.16\left[1+2\times0.16^2+\sqrt{2-4\times0.16^2+4\times0.16^4}\right]^{1/2} = 4.9\,\text{rad/sec.}$$

Phase margin, $\phi_{pm} = \tan^{-1}\left[\dfrac{4r^4}{\sqrt{(1+4r^4)}-2r^2}\right]^{1/2}$

$$= \tan^{-1}\left[\dfrac{4(0.16)^4}{\sqrt{(1+4\times0.16^4)}-2\times0.16^2}\right]^{1/2} = 18°$$

With Compensation: When $H(s)$ introduced, then

$$\frac{C(s)}{R(s)} = \frac{10\,K_A}{s^2+2.5s+10\,K_A} = \frac{\dfrac{10\,K_A}{\left(s^2+2.5s\right)}}{1+\dfrac{10\,K_A}{s^2+2.5s}} \qquad (1)$$

Given, Velocity error constant $K_v = 0$

$$\therefore \qquad 10 = \underset{s\to0}{\mathrm{Lt}}\; s\,G(s) = \underset{s\to0}{\mathrm{Lt}}\; s\cdot\frac{10\,K_A}{s^2+2.5s} = \frac{10\,K_A}{2.5}$$

or $\qquad K_A = 2.5$

From equation (1) $\quad \dfrac{C(s)}{R(s)} = \dfrac{25}{s^2+2.5s+25}$

By definition, $\qquad \omega_n = \sqrt{25} = 5$ rad/sec

$$\therefore \qquad r = \frac{2.5}{2\times5} = 0.25$$

then, bandwidth, $\omega_b = \omega_n\left[\left(1+2r^2\right)+\sqrt{2-4r^2+4r^4}\right]^{1/2}$

$$= 5\left[0.125+\sqrt{2-4\times0.25^2+4\times0.25^4}\right]^{1/2}$$

$$= 7.30 \ \text{rad/sec.}$$

ϕ_{pm} (phase margin)

$$= \tan^{-1}\left\{\left[\frac{4r^4}{\sqrt{\left(1+4r^4\right)}-2r^2}\right]^{1/2}\right\}$$

$$= \tan^{-1}\left\{\left[\frac{4(0.25)^4}{\sqrt{\left(1+4\times(0.25)^4\right)}-2\times(0.25)^4}\right]^{1/2}\right\} = 28°$$

Problem 9.18. A system with derivative control is shown in the figure below.

(a) Determine the transfer function and the characteristic equation of the system.

(b) Given that $K > 0$. Find the range of T for which the system is unconditionally stable.

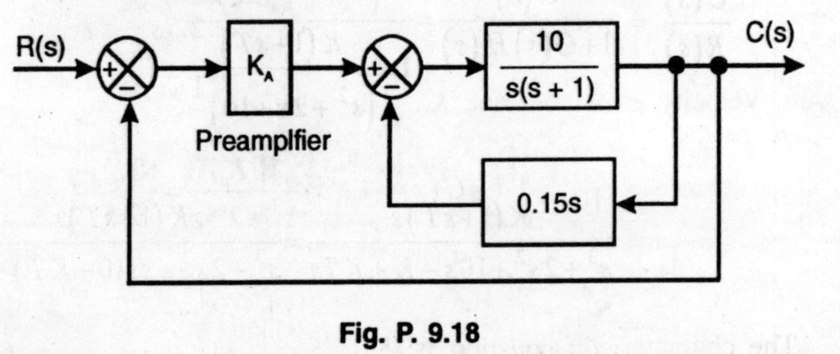

Fig. P. 9.18

Solution:

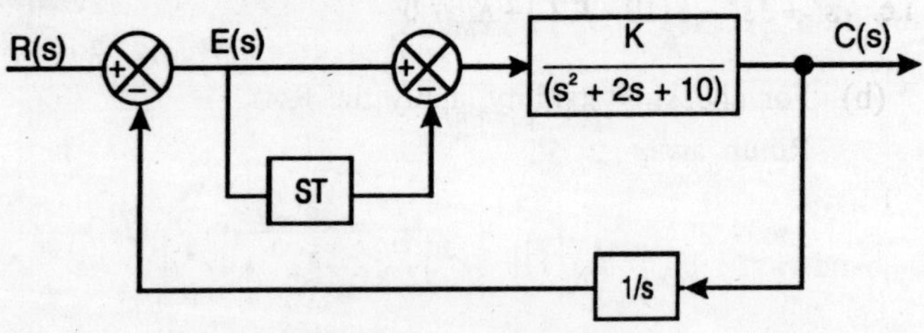

Fig. P. 9.18 (a)

This can be simplified as:

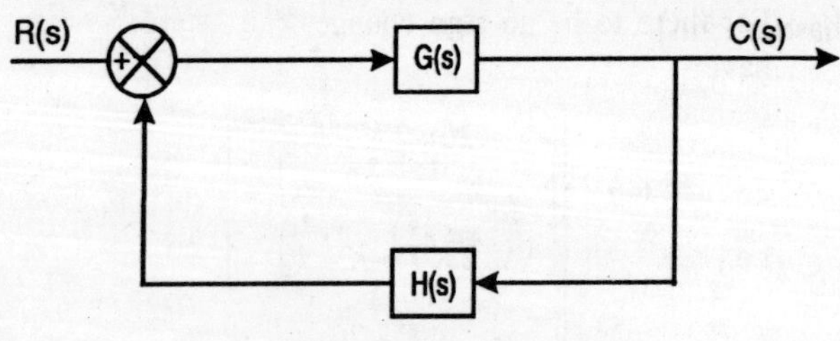

Fig. P. 9.18 (b)

where
$$G\,(s) = \frac{K(1+sT)}{\left(s^2+2s+10\right)}$$

and $$H(s) = \frac{1}{s}$$

The system transfer function is

$$\frac{C(s)}{R(s)} = \frac{G(s)}{1+G(s)H(s)} = \frac{\dfrac{K(1+sT)}{\left(s^2+2s+10\right)}}{1+\dfrac{K(1+sT)}{\left(s^2+2s+10\right)\dfrac{1}{s}}}$$

$$= \frac{K(1+sT)s}{s^3+2s^2+10s+K+KTs} = \frac{K(1+sT)s}{s^3+2s^2+s(10+KT)+K}.$$

The characteristic equation is

$$1 + G(s)H(s) = 0$$

i.e. $\quad s^3 + 2s^2 + s(10+KT) + K = 0$

(b) For checking stability, apply the Routh-Hurwitz rule.
Routh array

$$
\begin{array}{c|cc}
s^3 & 1 & (10+KT) \\
s^2 & 2 & K \\
s^1 & \dfrac{20+2KT-K}{2} & 0 \\
s^0 & K & 0
\end{array}
$$

For there to be no sign changes (i.e., stable system), we should have

$$\frac{20+2KT-K}{2} > 0$$

i.e., $\qquad 2KT > K - 20$

or $\qquad T > \dfrac{K-20}{2K}$

i.e., $\qquad T > \left(\dfrac{1}{2} - \dfrac{10}{K}\right)$

Problem 9.19. (a) Consider the system shown below, where $G_c(s)$ is the transfer function of a compensator. Find the steady-state error of the system when the compensator is

(i) P-type

(ii) PI-type

Assume $r(t)$ to be a unit step.

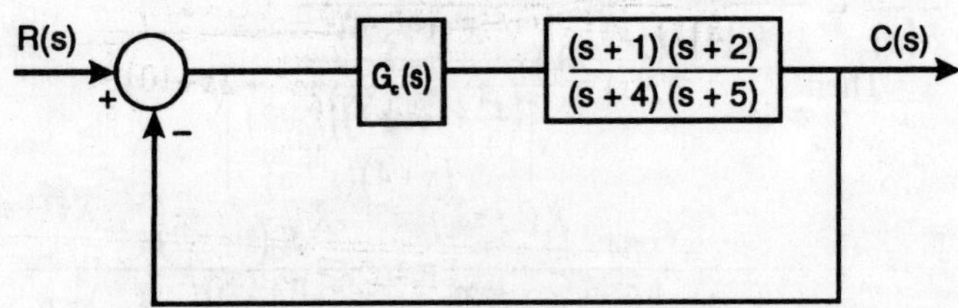

Fig. P. 9.19

(b) The output of a system is given in difference equation form as
$$y(k) = a\,y(k-1) + x(k)$$
where $x(k)$ is the input.

If $x(k) = 0$ for $k = 0$, $x(0) = 1$ and $y(0) = 0$, find $y(k)$ for all k. Determine the range of 'a' for which $y(k)$ is bounded.

Solution:

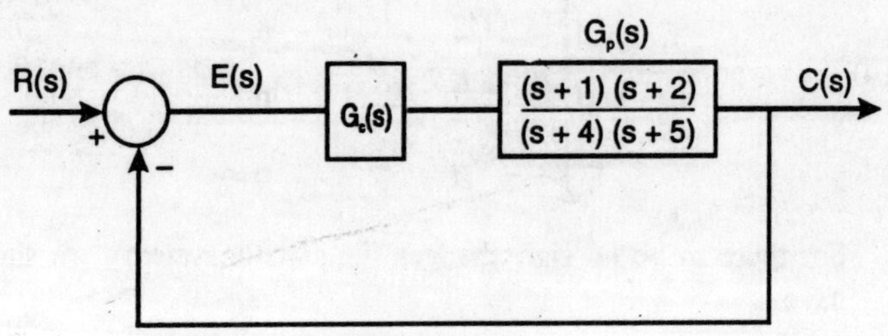

Fig. P. 9.19 (a)

(a) Steady state error. $\quad e_{ss} = \lim_{s \to 0} s\,E(s)$

where $\quad E(s) = \dfrac{R(s)}{1 + G_c(s)\,G_p(s)}$

For a unit step input. $R(s) = \dfrac{1}{s}$

and
$$G_p(s) = \frac{(s+1)(s+2)}{(s+4)(s+5)}$$

(i) P-type controller

$$G_p(s) = K_p \quad \text{where } K_p \text{ is a constant}$$

Then
$$E(s) = \frac{1}{s\left[1+\dfrac{k_p(s+1)(s+2)}{(s+4)(s+5)}\right]}$$

$$= \frac{(s+4)(s+5)}{s\left[(s+4)(s+5)+K_p(s+1)(s+2)\right]}$$

Steady state error, $e_{ss} = \lim_{s \to 0} s\,E(s)$

$$= \lim_{s \to 0}\frac{(s+4)(s+5)}{s\left[(s+4)(s+5)+K_p(s+1)(s+2)\right]}$$

$$= \frac{20}{20+2K_p} = \frac{1}{1+0.1K_p}$$

(ii) PI-type controller

$$G_c(s) = \frac{K_p s + K_1}{s}$$

where K_p and K_1 are constants.

Then
$$E(s) = \frac{1}{s\left[1+\left(\dfrac{K_p s + K_1}{s}\right)\dfrac{(s+1)(s+2)}{(s+4)(s+5)}\right]}$$

$$= \frac{s(s+4)(s+5)}{s\left[s(s+4)(s+5)+(K_p s + K_1)(s+1)(s+2)\right]}$$

Then, steady state error, $e_{ss} = \lim_{s \to 0} s\,E(s)$

$$= \lim_{s \to 0} \frac{s(s+4)(s+5)}{\left[s(s+4)(s+5) + (K_p s + K_1)(s+1)(s+2) \right]}$$

$$= \lim_{s \to 0} \frac{(s+4)(s+5)}{(s+4)(s+5) + \left(K_p + \dfrac{K_1}{s} \right)(s+1)(s+2)} = \frac{20}{\infty} = 0$$

(b) Difference equation is

$$y(k) = ay\,(k-1) + x\,(k) \quad \text{(i)}$$

Given, $x(0) = 1$, $y(0)$ and $x(k) = 0$ for $K \neq 0$.

Value of $k > 0$

$$y\,(1) = ay\,(0) + x\,(1)$$
$$= 0 + 0 = 0$$
$$y\,(2) = ay\,(1) + x\,(2)$$
$$y\,(3) = ay\,(2) + x\,(3) = 0 \text{ and so on.}$$

Hence $y\,(k) = 0 \text{ for } k \geq 0$.

Value of $k < 0$

From equation (1). $ay\,(k-1) = y\,(k) - x\,(k)$

For $k = 0$. $ay\,(-1) = y\,(0) - x\,(0)$
$$= 0 - 1 = -1$$

or $$y\,(-1) = -\frac{1}{a}$$

For $k = -1$. $a\,y\,(-2) = y\,(-1) - x\,(-1) = \dfrac{-1}{a} - 0$

or $$y\,(-2) = \frac{-1}{a^2}$$

$$k = -2,\; ay\,(-3) = y\,(-2) - x\,(-2) = \frac{-1}{a^2} - 0 = \frac{-1}{a^2}$$

or $$y\,(-3) = \frac{-1}{a^3}$$

Hence for $k < 0$. $y\,(k) = \dfrac{-1}{a^{-k}} = -a^k$

For $y\,(k)$ to be bounded, we should have $0 < a < 1$.

Problem 9.20. Consider a feedback system where the open loop transfer function is

$$G(s) = \frac{K}{s(1+sT_1)(1+sT_2)(1+sT_3)}$$

Determine the asymptote, which the Nyquist plot approaches as $\infty \to 0$.

Find also the range for K in terms of the crossover frequency ω_p for stability.

Solution:

The Open Loop Transfer Function is

$$H(s)\ G(s) = \frac{K}{s(1+sT_1)(1+sT_2)(1+sT_3)}$$

$$H(j\omega)G(j\omega) = \frac{K}{j\omega(1+j\omega T_1)(1+j\omega T_2)(1+j\omega T_3)}$$

$$\lim_{\omega \to 0} G(j\omega)H(j\omega) = \lim_{\omega \to 0}\frac{K}{j\omega} = \lim_{\omega \to 0}\frac{K}{\omega}\underline{|-90°}$$

Hence, the asymptote of the Nyquist plot tends to an angle of $-90°$ as $\omega \to 0$.

Value of K for stability

The characteristic equation is, $1 + G(s)\ H(s) = 0$

i.e.
$$1 + \frac{K}{s(1+sT_1)(1+sT_2)(1+sT_3)} = 0$$

$$s(1+sT_1)(1+sT_2)(1+sT_3) + K = 0$$

$$s\left[s^2 T_1 T_2 + s(T_1+T_2)+1\right](1+sT_3) + K = 0$$

Expanding we get

$$s^4 T_1 T_2 T_3 + s^3(T_1 T_2 + T_1 T_3 + T_2 T_3) + s^2(T_1+T_2+T_3) + s + K = 0$$

The Routh array is

s^4	$T_1 T_2 T_3$	$T_1+T_2+T_3$	K
s^3	$T_1 T_2 + T_1 T_3 + T_2 T_3$	1	0
s^2	C_1	K	0
s^1	$\dfrac{\left[K(T_1 T_2 + T_1 T_3 + T_2 T_3) - C_1\right]}{-C_1}$	0	0
s^0	K	0	0

where $\quad C_1 = \dfrac{(T_1+T_2+T_3)+(T_1T_2+T_1T_3+T_2T_3)-T_1T_2T_3}{T_1T_2+T_2T_3+T_3T_1}$

For $K = 0$ we have the auxiliary equation $s_1 = 0$ or $\omega_1 = 0$

For $\quad K = \dfrac{C_1}{T_1T_2+T_2T_3+T_1T_3}$

The auxiliary equation is $C_1 s^2 + \dfrac{C_1}{T_1T_2+T_2T_3+T_1T_3} = 0$

or $\quad s^2 = \dfrac{1}{T_1T_2+T_2T_3+T_1T_3}$

or $\quad s = \pm j \sqrt{\dfrac{1}{T_1T_2+T_2T_3+T_1T_3}}$

The crossover point is at, $\omega_1 = \sqrt{\dfrac{1}{T_1T_2+T_2T_3+T_1T_3}}$

Problem 9.21. A system has an OLTF given by

$$G(s) = \dfrac{K}{s\left(1+\dfrac{s}{10}\right)\left(1+\dfrac{s}{100}\right)}$$

Find the maxm. value of K for which the system is stable under unity feedback conditions. Compute the velocity error co-efficient and phase margin at $K = 10$. Give the design of an equalizer to provide a velocity error coefficient of 100 and an adequate phase-margin. Draw the Bode diagram of the system designed.

Solution:

$$G(s)\,H(s) = \dfrac{K}{s\left(1+\dfrac{s}{10}\right)\left(1+\dfrac{s}{100}\right)} = \dfrac{K\times 1000}{s(s+10)(s+100)}$$

$$= \dfrac{1000\,K}{s\left(s^2+110s+1000\right)}$$

The characteristic equation is

$$1 + G(s)\,H(s) = 0$$

$$\Rightarrow \quad s^3 + 110\,s^2 + 1000\,s + 1000\,K = 0$$

Constructing the Routh's array,

$$
\begin{array}{c|cc}
s^3 & 1 & 1000 \\
s^2 & 110 & 1000\,K \\
s^1 & \dfrac{110\times10^3 - 10^3\,K}{110} & - \\
s^0 & 1000\,K & -
\end{array}
$$

Thus, for the system to be stable

$$110\times10^3 - 10^3\,K > 0$$

$$\Rightarrow \quad K < 110$$

and $\qquad 1000\,K > 0 \Rightarrow K > 0$

The maxm. value of K for which the system is stable is $K = 110$

Velocity error coefficient (K_v)

K_v is defined as $\quad K_v = \underset{s\to0}{\mathrm{Lt}}\; s\,G(s)\,H(s)$

$$= \underset{s\to0}{\mathrm{Lt}}\; \frac{K}{\left(1+\dfrac{s}{10}\right)\left(1+\dfrac{s}{100}\right)} = K = 10 \text{ (given)}$$

Hence, $\qquad K_v = 10$

Phase-margin at $K = 10$

Gain cross-over freq. ω_g is found to be equal to 10 i.e. $\omega_g = 10$ (from Graph-I).

Now $\qquad GH(j\omega) = \dfrac{10^4}{j\omega(j\omega+10)(j\omega+100)}$

$$GH(j10) = \dfrac{10^4}{j10(j10+10)(j10+100)}$$

$$\angle GH(j10) = 0 - 90^\circ - \tan^{-1}1 - \tan^{-1}0.1$$

$$= -90^\circ - 45^\circ - 5.7^\circ = -140.7^\circ$$

Hence, phase margin $= -140.7° + 180° = 39.3°$

Design of equalizer to provide $K_v = 100$

$$K_v = \underset{s \to 0}{Lt}\, s\, G(s) = 100$$

$$\Rightarrow \quad \underset{s \to 0}{Lt}\, \frac{K}{\left(1+\dfrac{s}{10}\right)\left(1+\dfrac{s}{100}\right)} = K = 100$$

Hence, $\quad G(s) = \dfrac{100}{s\left(1+\dfrac{s}{10}\right)\left(1+\dfrac{s}{100}\right)} = \dfrac{10^5}{s(s+10)(s+100)}$

$$|GH(j0.1)| = \frac{10^5}{0.1 \times 10 \times 100} = 10^3$$

$$|GH(j0.1)|\,dB = 20\log_{10} 10^3 = 60\,dB.$$

The gain cross-over freq. $\omega_g = 30$ (from Graph II)

$$\angle GH(j\omega_g) = 0 - 90° - \tan^{-1}(3) - \tan^{-1}(0.3) = -178.27°$$

The phase margin $= -178.27° + 180° = 1.73°$

A phase margin of $1.73°$ implies that the system is quite oscillatory. The original system with $K_v = 10$ has a phase margin of $\approx 40°$. Hence, the additional phase lead required is $\approx 38°$.

In order to achieve a phase margin of $50°$ without decreasing the value of K, it is necessary to insert a suitable lead compensator into the system.

$$\sin\phi_m = \sin 38° = \frac{1-\alpha}{1+\alpha} \Rightarrow \alpha = 0.24$$

From the plot-II, $|G(j\omega)| = -6.2$ dB corresponds to $\omega = 44.7$ radiaus/ sec.

We select this as the new gain crossover frequency ω_c and

$$\frac{1}{T} = \sqrt{\alpha}\,\omega_c = 21.9 \text{ and } \frac{1}{\alpha T} = \frac{\omega_c}{\sqrt{\alpha}} = 91.24.$$

The lead networking is given by $\dfrac{s+21.9}{s+91.24} = \dfrac{0.24(1+0.046s)}{(1+0.011s)}$

To compensate for the attenuation due to the lead network, we increase the amplifier gain by a factor of $\dfrac{1}{0.24} = 4.17$.

Then the T. F. of the compensator which consists of the lead network and amplifier becomes

$$G_c(s) = \frac{(4.17)(s+21.9)}{(s+91.4)} = \frac{(1+0.046s)}{(1+0.011s)}$$

The compensated system has the following OLTF.

$$G_1(s) = G_c(s) \cdot G(s) = \frac{4.17(s+21.9) \times 10^5}{(s+91.4)s(s+10)(s+100)}$$

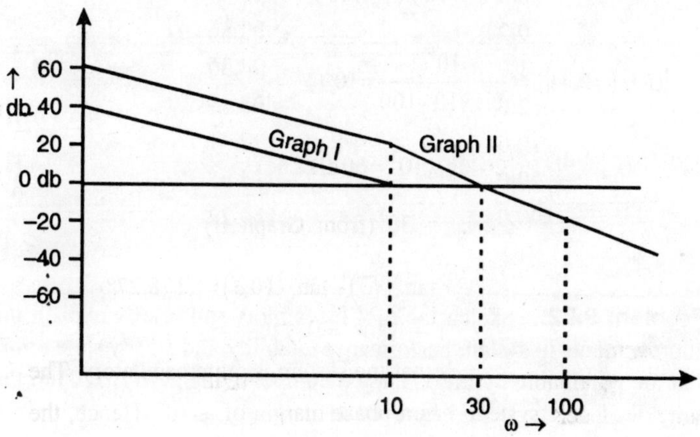

Fig. P. 9.21 (a)

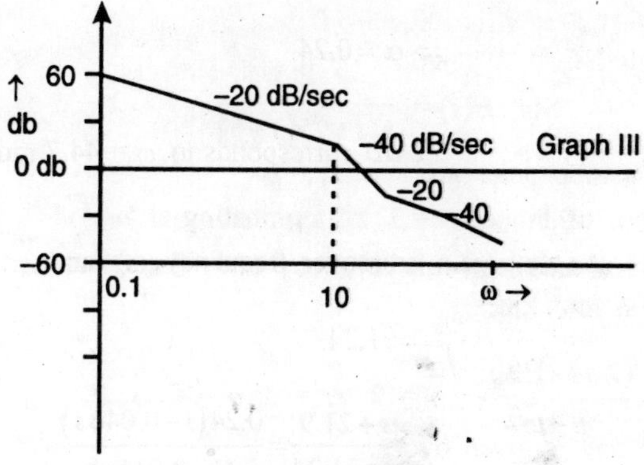

Fig. P. 9.21 (b)

The Bode plot is given in Graph-III

$$|G_1(j0.1)| \approx \frac{4.17\times10^5 \times 21.9}{91.4\times0.1\times10^3} \approx 999.16$$

$$|G_1(j0.1)|\,\text{dB} \approx 60 \text{ dB}.$$

$$\angle G_1(j\omega) = \tan^{-1}\left(\frac{\omega}{21.9}\right) - \tan^{-1}\left(\frac{\omega}{91.4}\right) - 90° - \tan^{-1}\left(\frac{\omega}{10}\right) - \tan^{-1}\left(\frac{\omega}{100}\right)$$

ω	$\angle G_1(j\omega)$
0.1	-90.43
0.2	-90.86
1.0	-94.30
2.0	-98.49
10.0	-122.41
20.0	-134.68
100.0	-189.22

Problem 9.22.   Sketch the root locus plots and briefly explain the improvements in system performance (stability and steady state error) that are obtainable by introducing a compensating zero $(s + 3)$ to the unity feedback system where OLTF is

$$G(s) = \frac{K}{s(s+2)(s+6)}$$

Solution:

$$G(s)\,H(s) = \frac{K}{s(s+2)(s+6)}$$

(a)   Open loop poles are $0, -2, -6$.

$\therefore$ no. of branches $= 3$, all terminating at ∞.

(b)   The real axis segments between 0 and -2 and between -6 and ∞ lie on root loucs.

(c) $\phi = \dfrac{(2q+1)180°}{n-m}$, $n = 3$, $m = 0$, $q = 0, 1, 2$

Three asymptotic angles are $60°, 180°, -60°$

(d) Centroid $-\sigma = \dfrac{-2-6}{3} = \dfrac{-8}{3} = -2.67$

(e) Breakaway points

$$K = -\,s\,(s+2)\,(s+6)$$

$$= -\left[s^3 + 8s^2 + 12s\right]$$

$$\therefore \qquad \frac{dK}{ds} = -\left[3s^2 + 16s + 12\right] = 0,$$

$$\Rightarrow \quad 3s^2 + 16s + 12 = 0,$$

$$\Rightarrow \qquad s = \frac{-16 \pm \sqrt{16^2 - 4\times3\times12}}{2\times3}$$

$$= -\,0.9 \text{ or } -4.43$$

$\therefore$ actual breakaway point is -0.9

(f) Intersection with imaginary axis $s^3 + 8s^2 + 12s + K = 0$

$$
\begin{array}{ccc}
s^3 & 1 & 12 \\[4pt]
s^2 & 8 & K \\[4pt]
s^1 & \dfrac{96-K}{8} & \\[6pt]
s^0 & K &
\end{array}
$$

$$\frac{96-K}{8} > 0 \;\Rightarrow\; K < 96$$

also, $K > 0$ Thus, $K = 96$

$$8s^2 + 96 = 0 \Rightarrow s^2 = 12,$$

$$s = \pm j2\sqrt{3}$$

Root locus plot

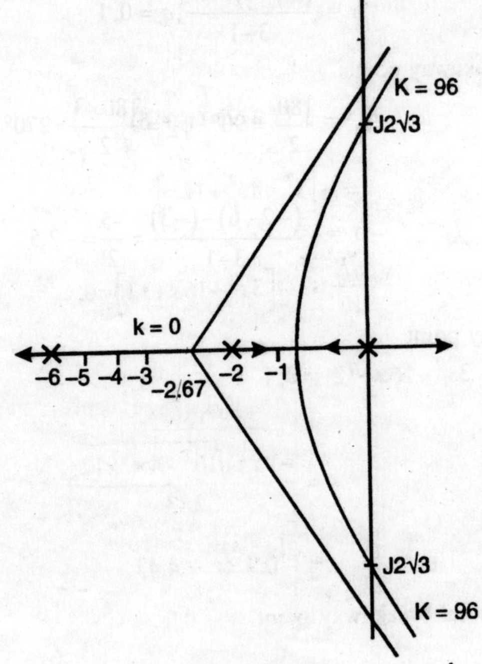

Fig. P. 9.22

Now, $H(s) = (s + 3),$

$$G(s)\,H(s) = \frac{K(s+3)}{s(s+2)(s+6)}$$

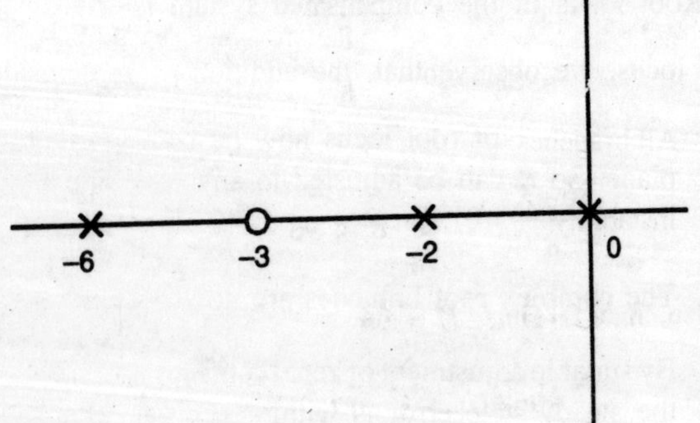

Fig. P. 9.22 (a)

The regions between 0 and -2; -3 and -6 lie on root locus.

$$\phi = \frac{(2q+1)180°}{3-1}, q = 0, 1$$

$$\phi_1 = \frac{180}{2} = 90°, \quad \phi_2 = \frac{180 \times 3}{2} = 270° = -90°$$

Centroid $\qquad -\sigma = \dfrac{(-2-6)-(-3)}{3-1} = \dfrac{-5}{2} = -2.5$

Breakaway point

$$K = -\left[\frac{s^3 + 8s^2 + 12s}{s+3}\right]$$

$$\frac{dK}{ds} = -\left[\frac{(s+3)(3s^2+16s+12)-(s^3+8s^2 12s)}{(s+3)^2}\right]$$

$\Rightarrow \quad 3s^3 + 25s^2 + 60s + 36 - s^3 - 8s^2 - 12s = 0$

$\Rightarrow \quad 2s^3 + 17s^2 + 48s + 36 = 0$

One value of the above eqn is $s \approx -1.17$.

The Root locus of the compensated system is shown in the figure.

From locus, we observe that, the addition of zero results in:

(i) All branches of root locus now lie completely in left half of s-plane. So K can be adjusted to any $+$ ve value without causing instability.

(ii) The complex root branches are now bent away from $j\omega$ axis.

(iii) By suitable adjustment of zero (compensating) along $-$ ve real axis, the steady state error is within specified limits.

Root Locus

The bold portion shows the root locus.

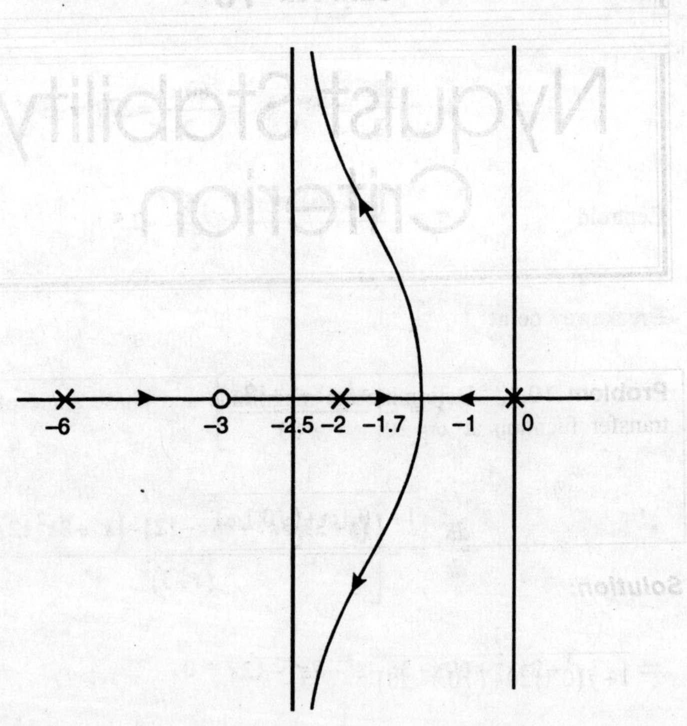

Fig. P. 9.22 (b)

CHAPTER 10

Nyquist Stability Criterion

Problem 10.1.   Determine the magnitude and phase angle of the transfer fucntion at $\omega = 20$

$$\frac{1}{1+j0.1\omega+(j0.1\omega)^2}$$

Solution:

$$\frac{1}{1+j(0.1)20+(j0.1\times20)^2} = \frac{1}{1+2j-4} = \frac{1}{-3+j2}$$

Magnitude

$$M = \frac{1}{\sqrt{(-3)^2+(2)^2}} = \frac{1}{3.67} = 0.277$$

Phase angle

$$\phi = 0° - \tan^{-1}\left(\frac{2}{-3}\right)$$

$$= 33.69° \quad \text{or} \ -146.31$$

Polar form of T.F. at $\omega = 20$

$$= 0.277 \angle -146.31 \text{ or } -0.2177 \angle 33.69$$

$$= -0.23 - j\,0.154$$

Problem 10.2.   Sketch the approximate polar plot of above problem.

Solution:

$$GH(j\omega) = \frac{1}{1+0.1j\omega+(j0.1\omega)^2} = \frac{1}{1-0.01\omega^2+j0.1\omega}$$

$$= \frac{1}{\sqrt{\left(1-0.01\omega^2\right)^2 +(0.1\omega)^2}} \angle \tan^{-1}\frac{0.1}{1-0.01\omega^2}$$

at $\qquad \omega = 0, \ M=1, \ \phi=0$

at $\qquad \omega = \infty, \ M=0, \ \phi=-180°$

Nyquist (Polar) Plot

at $\qquad \omega = 20$ (any arbitrary value)

$$G(j\omega) = (-\,0.23 - j\,0.154)$$

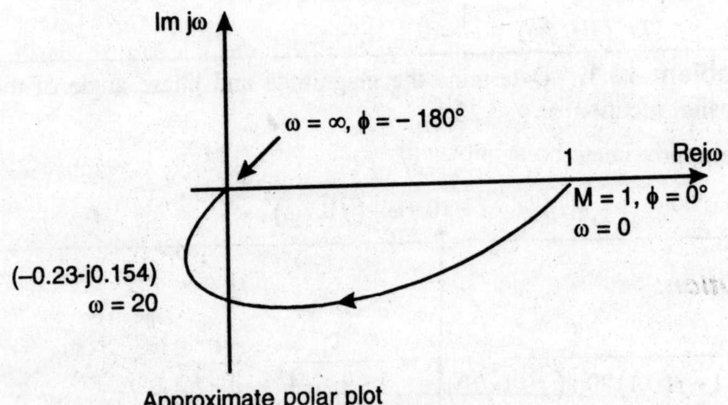

Fig. P. 10.2

Problem 10.3. Sketch the approximate polar plot of the frequency response for the transfer function

$$\frac{1}{(1+0.5s)(1+2s)}$$

Solution:

$$GH(s) = \frac{1}{(1+0.5s)(1+2s)}$$

This is a type-0 system. There are no poles on the $j\omega$ axis.

Putting $\qquad s = j\omega$ for $0<\omega<\infty$

$$GH(j\omega) = \frac{1}{(1+0.5\,j\omega)(1+2\,j\omega)}$$

This is of the general form

$$GH(j\omega) = \frac{1}{(j\omega + P_1)(j\omega + P_2)} = \frac{1}{\sqrt{(\omega^2 + P_1^2)(\omega^2 + P_2^2)}}$$

$$\phi = -\tan^{-1}\left(\frac{\omega}{P_1}\right) - \tan^{-1}\left(\frac{\omega}{P_2}\right)$$

Here $\qquad P_1 = 2, \quad P_2 = 0.5$

NNow $\qquad GH(j0) = \dfrac{1}{P_1 P_2} = \dfrac{1}{2 \times 0.5} = 1$

$$\phi = 0°$$

$$\underset{\omega \to \infty}{Lt}\ GH(j\omega) = \frac{1}{\infty} = 0$$

$$\phi = 180°$$

The approximate polar plot will be as

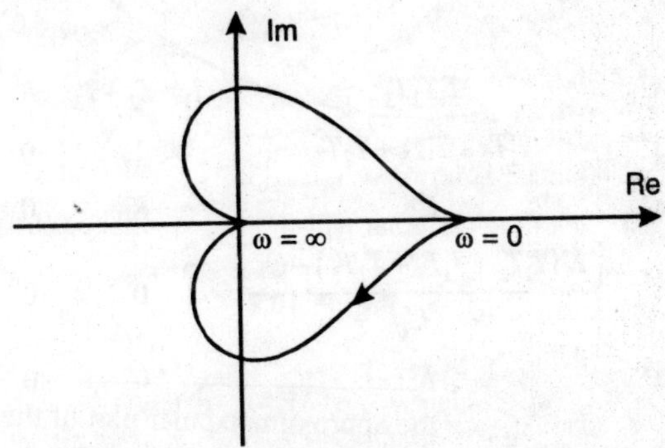

Fig. P. 10.3

Problem 10.4. Consider a feedback system where the OLTF is

$$G(s) = \frac{K}{s(1 + sT_1)(1 + sT_2)(1 + sT_3)}$$

Determine the asymptote, which the Nyquist plot approaches as $\omega \to 0$. Find also the range of k in terms of the crossover frequency ω_p for stability. (GATE)

Solution:

$$G(j\omega)H(j\omega) = \frac{K}{j\omega(1+j\omega T_1)(1+j\omega T_2)(1+j\omega T_3)}$$

$$\underset{\omega\to 0}{Lt}\; G(j\omega)H(j\omega) = \underset{\omega\to 0}{Lt}\frac{K}{j\omega} = \underset{\omega\to 0}{Lt}\frac{K}{\omega}\angle -90°$$

Hence the asymptote of the Nyquist plot tends to an angle of $-90°$ as $\omega\to 0$

Value of K for stability:-

The ch. equation is $1 + G(s)\;H(s) = 0$

$$\Rightarrow\quad 1+\frac{K}{s(1+sT_1)(1+sT_2)(1+sT_3)} = 0$$

$$\Rightarrow\quad s(1+sT_1)(1+sT_2)(1+sT_3)+K = 0$$

Expanding, we get

$$s^4 T_1 T_2 T_3 +s^3(T_1 T_2 +T_1 T_3 +T_2 T_3)+s^2(T_1+T_2+T_3)+s+k=0$$

The Routh array is

s^4	$T_1 T_2 T_3$	$T_1+T_2+T_3$	K
s^3	$T_1 T_2 +T_1 T_3 +T_2 T_3$	1	0
s^2	C_1	K	0
s^1	$\dfrac{\left[K(T_1 T_2 +T_1 T_3 +T_2 T_3)-C_1\right]}{-C_1}$	0	0
s^0	K	0	0

where
$$C_1 = \frac{(T_1+T_2+T_3)(T_1 T_2 +T_1 T_3 +T_2 T_3)-T_1 T_2 T_3}{T_1 T_2 +T_2 T_3 +T_3 T_1}$$

For $K = 0$, we have the auxiliary equation

$$s^1 = 0 \text{ or } \omega_1 =0$$

For
$$K = \frac{C_1}{T_1 T_2 +T_2 T_3 +T_1 T_3}$$

we have the auxiliary equation,

$$C_1 s^2 + \frac{C_1}{T_1 T_2 +T_2 T_3 +T_1 T_3} = 0$$

or
$$s^2 = \frac{-1}{T_1T_2 + T_2T_3 + T_1T_3}$$

or
$$s = \pm j\sqrt{\frac{1}{T_1T_2 + T_2T_3 + T_1T_3}}$$

The cross-over point is at

$$\omega_1 = \sqrt{\frac{1}{T_1T_2 + T_2T_3 + T_1T_3}}$$

Problem 10.5. The transfer function block diagram of a feedback control system is shown in figure where $T > 0$ is a scalar parameter.

On the s-plane sketch the loci of the poles of the closed loop system when T takes all the values from 0 to ∞. Mark all salient values in the diagram. (GATE)

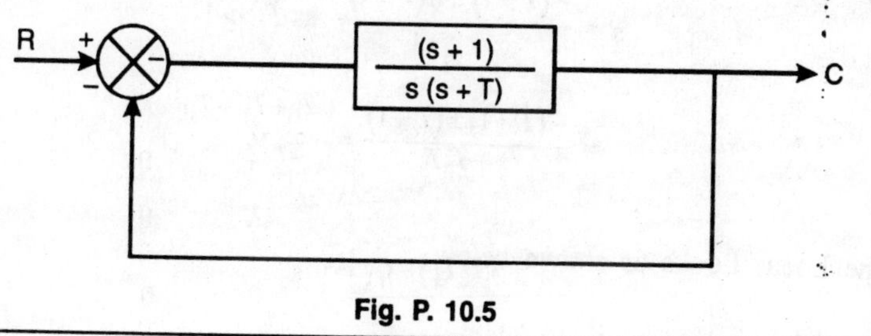

Fig. P. 10.5

Solution:

For given system, ch. equation will be

$$1 + \frac{(s+1)}{s(s+T)} = 0$$

$$\Rightarrow \quad s^2 + Ts + (s+1) = 0$$

$$\Rightarrow \quad s^2 + s(T+1) + 1 = 0$$

The poles of the CLTF are the roots of this equation

$$s = \frac{-(T+1) \pm \sqrt{(T+1)^2 - 4}}{2}$$

For $\qquad\qquad T = 0,$

The ch. equation is $s^2 + s + 1 = 0$

$$s = \frac{-1 \pm \sqrt{1-4}}{2} = \frac{-1 \pm j\sqrt{3}}{2} = -0.5 \pm j\,0.866$$

For $\qquad T = 1$

$$s^2 + 2s + 1 = 0$$

$\Rightarrow \qquad (s+1)^2 = 0 \Rightarrow s = -1, -1$

For $\qquad T = 2,$

$$s^2 + 3s + 1 = 0$$

$$s = \frac{-3 \pm \sqrt{9-4}}{2} = -0.382, \ -2.618$$

and at $\qquad T = \infty$

$$s = \frac{-(T+1) \pm \sqrt{(T+1)^2}}{2} \quad \text{as } T \gg 1$$

$$= \frac{-(T+1) \pm (T+1)}{2} = -\infty, \ 0 \ .$$

The Locus Looks as shown below

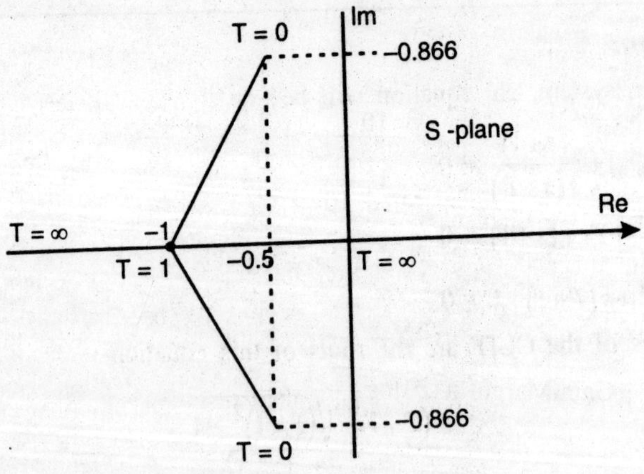

Fig. P. 10.5 (a)

Problem 10.6. Consider a unity feedback system having an OLTF of

$$G(s) = \frac{K}{s(1+0.1s)(1+s)}$$

Determine K for a gain margin of 20 dB and a phase margin of 60°.

Solution:

$$G(s) = \frac{K}{s(1+0.1s)(1+s)} = \frac{10K}{s(s+10)(s+1)}$$

$$|G(j\omega)| = \frac{10K}{\omega\sqrt{\omega^2+100}\sqrt{\omega^2+1}}$$

$$\angle G(j\omega) = -90° - \tan^{-1}\frac{\omega}{10} - \tan^{-1}\omega$$

To calculate gain margin

$$-90° - \tan^{-1}\frac{\omega}{10} - \tan^{-1}\omega = -180°$$

$$\Rightarrow \qquad \tan^{-1}\frac{\omega}{10} + \tan^{-1}\omega = 90°$$

$$\Rightarrow \qquad \tan^{-1}\frac{\dfrac{\omega}{10}+\omega}{1-\dfrac{\omega^2}{10}} = 90°$$

$$\Rightarrow \qquad \frac{\dfrac{\omega}{10}+\omega}{1-\dfrac{\omega^2}{10}} = \frac{1}{0} \Rightarrow 1-\frac{\omega^2}{10}=0$$

$$\Rightarrow \qquad \omega^2 = 10 \Rightarrow \omega_p = \sqrt{10}$$

(phase cross-over freq.)

$$\text{Gain Margin} = 20\log\frac{1}{|G(j\omega_p)|}$$

$$|G(j\omega)|_{\omega=\omega_p} = \frac{10K}{\sqrt{10}\sqrt{110}\sqrt{11}} = \frac{10K}{110} = \frac{K}{11}$$

$$\therefore \quad GM = 20\log\frac{11}{K}\,dB = 20$$

$$\Rightarrow \quad \log\frac{11}{K} = 1 \Rightarrow \frac{11}{K} = 10^1 = 10$$

hence, $\quad K = 1.1$

(ii) The phase margin point (i.e., Gain cross-over freq. ω_g) is given by

$$|GH|_{\omega=\omega_g} = 1$$

i.e., $$\frac{K}{\omega_g\sqrt{1+0.01\omega_g^2}\,\sqrt{1+\omega_g^2}} = 1 \qquad (1)$$

The PM is given by

$$\text{PM} = \angle GH\left(k\omega_g\right) + 180°$$

$$= -90° - \tan^{-1}0.1\omega_g - \tan^{-1}\omega_g + 180° = 60° \text{ (given)}$$

$$\Rightarrow \quad -90° - \tan^{-1}\left(\frac{0.1\omega_g + \omega_g}{1-0.1\omega_g^2}\right) + 180° = 60°$$

$$\Rightarrow \quad \frac{1.1\omega_g}{1-0.1\omega_g^2} = \tan 30° = 0.577$$

$$\Rightarrow \quad \omega_g^2 + 19.06\,\omega_g - 10 = 0 \Rightarrow \omega_g = -19.57,\ 0.51$$

But $\quad \omega_g \neq -ve, \quad \therefore \omega_g = 0.51\,\text{rad/sec.}$

Substituting this value in equation (1),

$$\frac{K}{0.51\times\sqrt{1+0.01\times(0.51)^2}\times\sqrt{1+(0.51)^2}} = 1$$

$$\Rightarrow \quad K = 0.5732$$

Problem 10.7. Sketch the polar Nyquist plot on a plain paper for the Transfer function

$$G(s) = \frac{10}{s(s+1)(1+0.5s)}$$

Solution:

Given

$$G(s) = \frac{10}{s(s+1)(1+0.5s)} = \frac{20}{s(s+1)(s+2)}$$

Putting

$$s = j\omega$$

$$G(j\omega) = \frac{20}{j\omega(1+j\omega)(2+j\omega)}$$

$$\underset{\omega\to 0}{\mathrm{Lim}}\, G(j\omega) = \frac{20}{j\omega\times 2} = \infty \angle -90^\circ$$

$$\underset{\omega\to 0}{\mathrm{Lim}}\, G(j\omega) = \frac{1}{(j\omega)^3} = 0 \angle -270^\circ$$

Intersection with real and Imaginary axes

$$G(j\omega) = \frac{20(-j\omega)(1-j\omega)(2-j\omega)}{\omega^2(1+\omega^2)(4+\omega^2)} = \frac{20\left[j(\omega^3-2\omega)-3\omega^2\right]}{\omega^2(1+\omega^2)(4+\omega^2)}$$

Equating the imaginary part to zero

$$\frac{20(\omega^3-2\omega)}{\omega^2(1+\omega^2)(4+\omega^2)} = 0 \text{ or}, \omega^3 = 2\omega$$

$$\Rightarrow \qquad \omega^2 = 2; \ \omega = \sqrt{2}$$

$$G\left(j\sqrt{2}\right) = \frac{20\times(-3\times 2)}{2\times 3\times 6} = -\frac{6\times 20}{6\times 6} = -\frac{10}{3}$$

The required polar plot will be as:

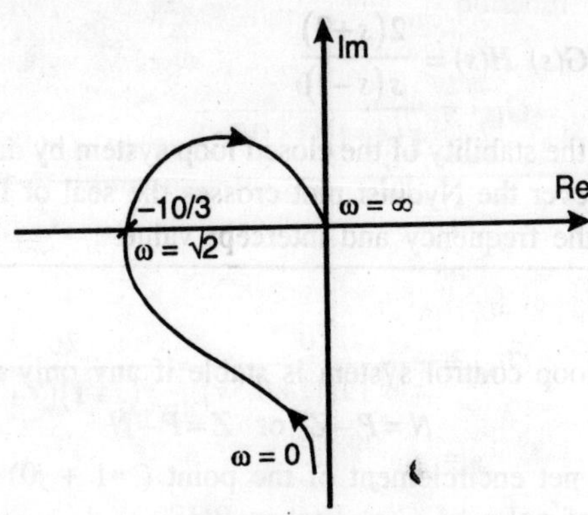

Fig. P. 10.7

Problem 10.8. For the open loop Transfer function $G(s) = \dfrac{14}{(s+1)(s+2)}$ find the resonant frequency (ω_r) and Resonant peak M_r.

Solution:

The ch. equation of given system

$$1+\frac{14}{(s+1)(s+2)} = 0 \Rightarrow s^2+3s+16=0$$

$$\therefore \qquad \omega_n^2 = 16 \Rightarrow \omega_n = 4$$

$$2\xi\omega_n = 3 \Rightarrow \xi = \frac{3}{2\times 4} = \frac{3}{8} = 0.375$$

Resonant frequency

$$\omega_r = \omega_n \sqrt{1-2\xi^2} = 4\sqrt{1-2\times(0.375)^2}$$

$$= 3.39 \text{ rad sec.}$$

Resonant peak $\qquad M_r = \dfrac{1}{2\xi\sqrt{1-\xi^2}} = \dfrac{1}{2\times 0.375\sqrt{1-0.1406}}$

$$= 1.438.$$

Problem 10.9. The OLTF of a system is

$$G(s)\ H(s) = \frac{2(s+3)}{s(s-1)}$$

Investigate the stability of the closed loop system by drawing Nyquist plot. Whenever the Nyquist plot crosses the seal or imaginary axis, determine the frequency and intercept value.

Solution:

The closed loop control system is stable if any only if

$$N = P - Z \quad \text{or} \quad Z = P - N$$

where N = net encirclement of the point $(-1 + j0)$ by the plot

P = no. of poles of $G(s)\ H(s)$ in RHS

Z = no. of zeroes of $1 + G(s)\ H(s)$ in RHS of s-plane

OLTF $\qquad G(s)\ H(s) = \dfrac{2(s+3)}{s(s-1)} = \dfrac{6(1+0.33s)}{s(-1+s)}$

$$M = \frac{6\sqrt{1+(0.33\omega)^2}}{\omega\sqrt{1+\omega^2}}$$

$$\phi = \tan^{-1}0.33\omega - 90° - \tan^{-1}\frac{\omega}{-1}$$

at $\qquad \omega = 0,\ M = \infty;\ \phi = 0° - 90° + 180° = 90°$

at $\qquad \omega = 0.5,\ M = \dfrac{6\times1.015}{0.5\times1.12} = 10.88$

$$\phi = 9.369° - 90° + 153.45° = 72.80°$$

at $\qquad \omega = \infty,\ M = 0,\ \phi = 90° - 90° - (-270°)$

$$= 270° = -90°$$

$$G(j\omega)H(j\omega) = \frac{6(1+j0.33\omega)}{-\omega^2 - j\omega}$$

$$= \frac{6\left(-\omega^2 + j\omega - j0.33\omega^3 - 0.33\omega^2\right)}{\omega^4 + \omega^2}$$

Equating imaginary part to zero

$$\frac{6\left(\omega - 0.33\omega^3\right)}{\omega^4 + \omega^2} = 0 \Rightarrow 6\left(1 - 0.33\omega^2\right) = 0$$

$$\Rightarrow \qquad \omega = \sqrt{\frac{1}{0.33}} = 1.741 \, \text{rad/sec.}$$

$$M \text{ at } \qquad 1.741 = \frac{6\sqrt{1+(0.33\,\omega)^2}}{\omega\sqrt{1+\omega^2}} = \frac{6\sqrt{1+(0.33\times1.741)^2}}{1.741\sqrt{1+(1.741)^2}}$$

$$= \frac{6\times1.153}{1.741\times2.008} = \frac{6.918}{3.498} \approx 2$$

Thus, the required value of frequency = 1.741 rad/sec. and value of magnitude = 2.

The Nyquist plot is shown in the figure. From the Fig. P. 10.9 (a), we observe that $(-1 + j0)$ point lies inside the plot.

Since, there is one encirclement of $(-1 + j0)$ in anticlockwise direction.

$$\therefore \qquad N = + 1$$

From transfer function,

No. of open loop pole in RHS plane = 1

$$\Rightarrow \qquad P = 1$$

Now $\qquad N = P - Z$

$$\Rightarrow \qquad + 1 = 1 - Z$$

$$\Rightarrow \qquad Z = 1 - 1 = 0$$

Thus, open loop system is unstable as $P = 1$. However, closed loop system is stable as $Z = 0$.

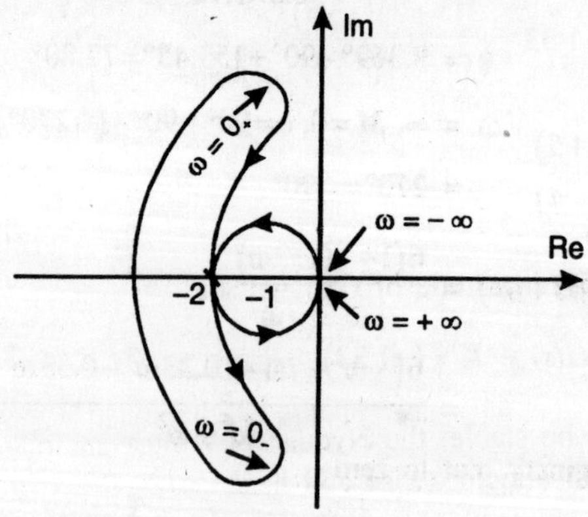

Fig. P. 10.9 (a)

Problem 10.10. The Nyquist plot of a unity feedback system having open loop transfer function

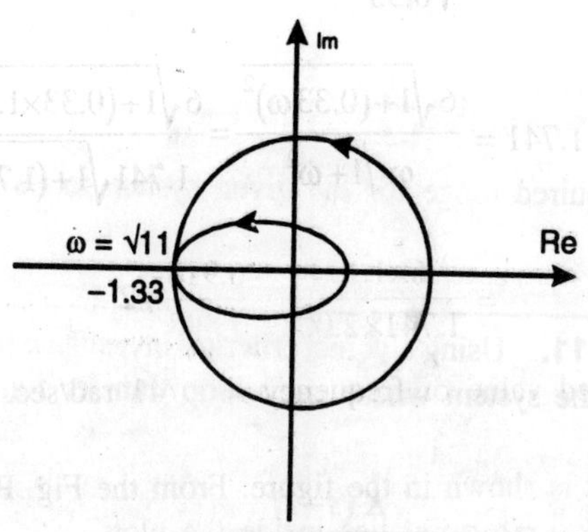

Fig. P. 10.10

$$G(s) = \frac{K(s+3)(s+5)}{(s-2)(s-4)}$$

for $K = 1$ is shown for the system to be stable, Find the range of K.

Solution:

From the diagram, we see that magnitude of $G(s) = -1.33$, given $K = 1$

So,

$$-1.33 = \frac{1(s+3)(s+5)}{(s-2)(s-4)}$$

$$\Rightarrow \quad \frac{(s+3)(s+5)}{(s-2)(s-4)} = -1.33$$

Now, any value K, magnitude of $G(s)$ will be

$$G(s) = K \times (-1.33)$$

For the system to be stable, the Nyquist plot must encircle the point $(-1 + j0)$ twice in anticlockwise and hence,

$$|K \times (-1.33)| > |-1|$$

$$\Rightarrow \qquad |-1.33\,K\,|>1$$

or $\qquad 1.33\,K>1$

$$\Rightarrow \qquad K>\frac{1}{1.33}$$

This is the required range for the given system to be stable.

Problem 10.11. Using Nyquist criterion investigate the closed-loop stability of the system whose open-loop transfer function is given below:

$$G(s)\ H(s) = \frac{K(s+1)}{(s+0.5)(s-2)}$$

Consider (i) $K = 1.25$ (ii) $K = 2.5$

Solution:

$$G(j\omega)H(j\omega) = \frac{K(j\omega+1)}{(j\omega+0.5)(j\omega-2)}$$

Rationalizing the above equation and separating into real and imaginary parts

$$G(j\omega)H(j\omega) = -K\left[\frac{1+2.5\omega^2}{\left(\omega^2+0.25\right)\left(\omega^2+4\right)} + \frac{j\omega\left(\omega^2-0.5\right)}{\left(\omega^2+0.25\right)\left(\omega^2+4\right)}\right]$$

Nyquist plot intersects the real axis at $\omega=\sqrt{0.5}$ where imaginary term is zero.

ω	$\mathrm{Re}\left[G(j\omega)H(j\omega)\right]$	$\mathrm{Im}\left[G(j\omega)H(j\omega)\right]$
0	$-K$	$j0$
0.5	$-0.76\,K$	$+j0.058\,K$
$\sqrt{0.5}$	$-0.66\,K$	$+j0$
5	$-0.044\,K$	$-j0.17\,K$
∞	0	$-j0$

with the help of above values, the shape of the Nyquist plot is drawn as shown.

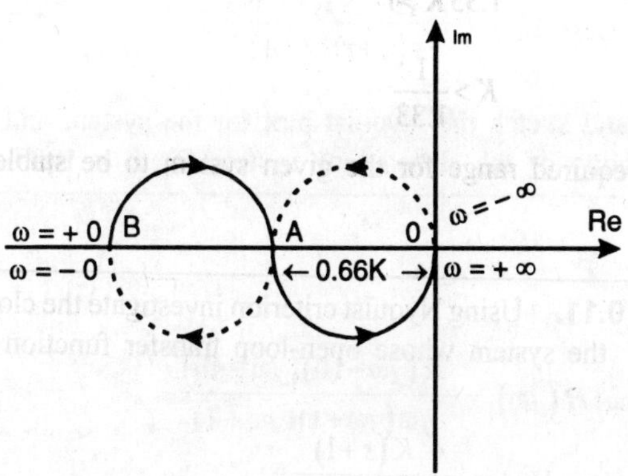

Fig. P. 10.11

As the type of the system is zero therefore the plot will make angle 0 i.e., $\omega=-0$ and $\omega=+0$ are coincident.

(i) For $K = 2.5$, $OA=0.66\times2.5=1.65$, thus, the point $(-1+j0)$ lies between O and A and no. of encirclements $N = +1$ (anti-clockwise).

Since $P_+ = 1$

from $N = P_+ - Z_+$

$$+1 = 1 - Z_+ \Rightarrow Z_+ = 0$$

Thus, the closed-loop characteristic equation has no roots with +ve real part and the system is stable.

(ii) For $K = 1.25$, $OA=0.66\times1.25=0.825$, thus, the point $(-1+j0)$ lies between A and B and no. of encirclements $N = -1$ (clockwise)

$$N = P_+ - Z_+ \Rightarrow -1 = 1 - Z_+$$

$\Rightarrow$ $Z_+ = 2$ and the system is unstable.

The limiting value of K for stability

$$0.66\,K \geq 1$$

$\Rightarrow$ $K \geq 1.5$

Problem 10.12. The open loop gain *G(s) H(s)* of a feedback control system is

$$G(s) \; H(s) = \frac{K(s+10)(s+40)}{s(s+1)(s+4)}$$

Work out and sketch the Nyquist plot for the system and comment of the stability of the closed loop system for $K = 100$.   (ES' 99)

Solution:

$$G(j\omega)H(j\omega) = \frac{K(j\omega+10)(j\omega+40)}{j\omega(j\omega+1)(j\omega+4)}$$

magnitude of OLTF

$$M = \frac{K\sqrt{\omega^2+100}.\sqrt{\omega^2+1600}}{\omega\sqrt{\omega^2+1}.\sqrt{\omega^2+4}}$$

Phase angle $\phi = \tan^{-1}\dfrac{\omega}{10}+\tan^{-1}\dfrac{\omega}{40}-90°-\tan^{-1}\omega-\tan^{-1}\dfrac{\omega}{4}$

At $\omega = 0, \; M = \infty, \; \phi = -90°$

and at $\omega = \infty, \; M = 0, \; \phi = -90°$

The Nyquist plot intersects the real axis at a point given by

$$I_m\big[G(j\omega)H(j\omega)\big]=0$$

which yields, $\omega = 11.2 \, \text{rad/sec}, \, 142.8 \, \text{rad/sec}$

Now, for $K = 100$

$|G(j\omega)H(j\omega)|$ at $\omega = 11.2$ is $= 41.6$ and,

$|G(j\omega)H(j\omega)|$ at $\omega = 142.8$ is $= 0.725$

and the plot is drawn as shown in Fig. P.10.12.

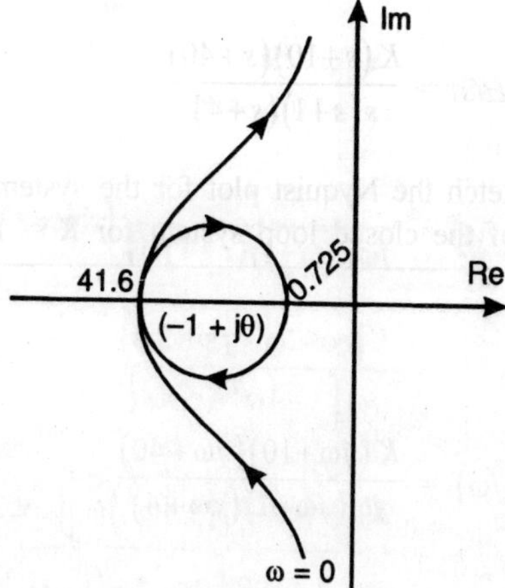

Fig. P. 10.12

Since there is one encirclement of the point $(-1 + j0)$ in clockwise direction,

So,
$$N = -1, \; P_+ = 0$$

∴
$$N = P_+ - Z_+$$

⇒
$$-1 = 0 - Z_+$$

⇒
$$Z_+ = 1$$

Thus, there is one root having positive real part and hence the system is unstable.

Problem 10.13. Consider the feedback system shown in Fig. P. 10.13 Sketch the Nyquist plot of this system when $G_c(s)=1$ and determine the maximum value of K for stability.

If $G_c(s)$ is modified to a controller having the transfer function $(1 + 1/s)$, what then is the maximum value of K for stability? Comment upon the result.

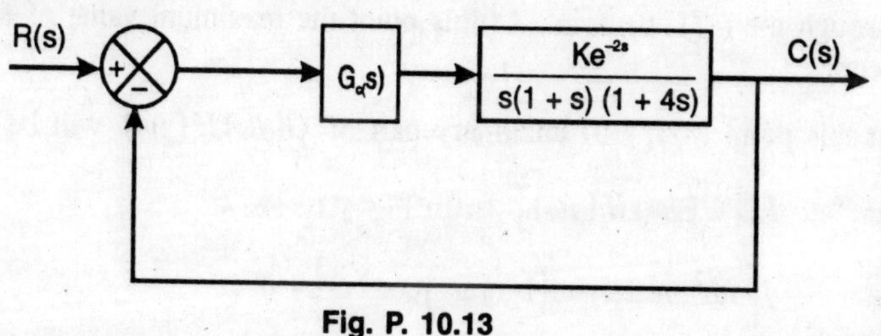

Fig. P. 10.13

Solution:

Given

$$G(s)\,H(s) = \frac{G_c(s)\,K e^{-2s}}{s(s+1)(1+4s)}$$

then,

$$G(j\omega)H(j\omega) = \frac{K e^{-2j\omega}}{j\omega(j\omega+1)(1+4j\omega)}\ \left[\text{since } G_c(s)=1\right]$$

$$= \frac{K\left[\cos 2\omega - j\sin 2\omega\right]}{j\omega\left[1-4\omega^2+5j\omega\right]}$$

$$= \frac{K\left[-5\omega^2 - j\left(-4\omega^2\right)\omega\right]\left[\cos 2\omega - j\sin 2\omega\right]}{\omega\left[25\omega^4 + \left(1-4\omega^2\right)^2\right]}$$

Nyquist plot for the above system is drawn below:

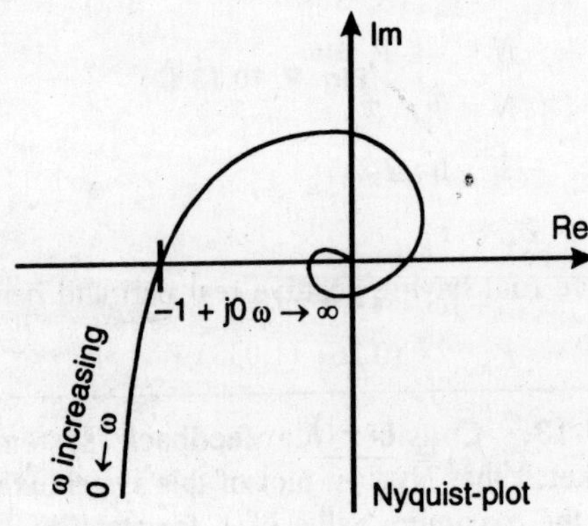

Fig. P. 10.13 (a)

The system will be marginally stable when $G(j\omega)H(j\omega)$ plot will pass through the (− 1, 0) point. At this point the maximum value of K will occur.

At this point (− 1, j 0) imaginary part of $G(j\omega)H(j\omega)$ will be zero.

$$\Rightarrow \qquad I_m\left[G(j\omega)H(j\omega)\right] = 0$$

$$\Rightarrow \qquad j\left[j\omega^2 \sin 2\omega - \omega\left(1-4\omega^2\right)\cos 2\omega\right] = 0$$

$$\Rightarrow \qquad 5\omega^2 \sin 2\omega = \omega\left(1-4\omega^2\right)\cos 2\omega$$

$$\Rightarrow \qquad \tan 2\omega = \frac{1-4\omega^2}{5\omega}$$

From the graph,

at $\omega_1 = 0.26$ rad/sec. both the curves cut each other

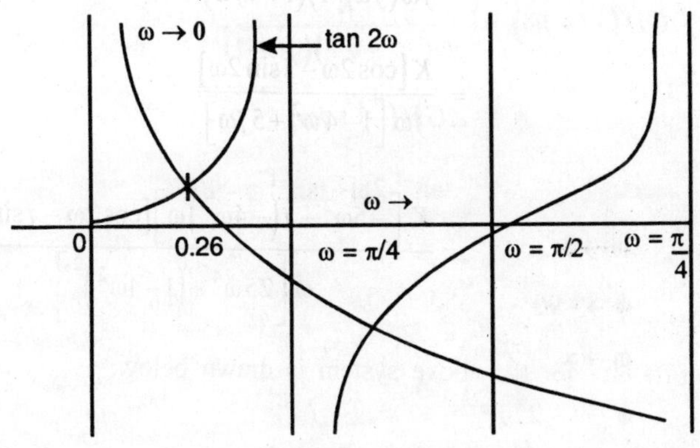

Fig. P. 10.13 (b)

At this point, $|G(j\omega)H(j\omega)|_{\omega=\omega_1} = 1$

$$\frac{K}{|j\omega_1|\,|1+j\omega_1|\,|1+4\,j\omega_1|} = 1$$

$$\Rightarrow \qquad K_{max} = (0.26)\,(1.033)\,(1.44) = 0.387$$

Now when $\qquad G_c(s) = \dfrac{(s+1)}{s}$

Then, $\quad G(j\omega)H(j\omega) = \dfrac{Ke^{-2j\omega}}{(j\omega)^2\,(1+4\,j\omega)}$

Problem 10.14. Consider a feedback system having the characteristic equation

$$1 + \frac{K}{(s+1)(s+1.5)(s+2)} = 0$$

It is desired that all the roots of the characteristic equation have real parts less than -1. Extend the Nyquist stability criterion to find the largest value of K, satisfying this condition.

Solution:

$$G(j\omega)H(j\omega) = \frac{K}{(j\omega+1)(j\omega+1.5)(j\omega+2)}$$

Now, for the roots having real parts less than -1,

$$GH(-1+j\omega) = \frac{K}{j\omega(j\omega+0.5)(j\omega+1)}$$

$$\phi = \angle GH(-1+j\omega)$$

$$= -\tan^{-1}2\omega - \tan^{-1}\omega - 90°$$

when, $\omega \to 0$

$$\phi \to -90°$$

when, $\omega \to \infty$

$$\phi = -270°$$

Now, the Nyquist plot will cut the real axis where $\phi = -180°$, i.e. imaginary parts are zero.

Then, $-90° - \tan^{-1}2\omega - \tan^{-1}\omega = -180°$

$$\Rightarrow \quad \tan^{-1}\left(\frac{3\omega}{1-2\omega^2}\right) = 90°$$

which gives $\omega = \dfrac{1}{\sqrt{2}}$ rad/sec

Given,

$$|G(-1+j\omega)H(-1+j\omega)|_{\omega=\frac{1}{\sqrt{2}}} = \frac{K}{\dfrac{1}{\sqrt{2}}\left|\dfrac{1}{2}+\dfrac{1}{\sqrt{2}}j\right|\left|1+\dfrac{1}{\sqrt{2}}j\right|} \leq 1$$

$$\Rightarrow \qquad\qquad \frac{4K}{3} \leq 1$$

$$\Rightarrow \qquad\qquad K \leq \frac{3}{4}$$

Hence, the maximum value of $K = 0.75$

Problem 10.15. Sketch the Nyquist plot of a closed-loop system which has the open-loop transfer function

$$G(s)\ H(s) = \frac{2e^{-sT}}{s(1+s)(1+0.5s)}$$

determine the maximum value of T for the system to be stable.

Solution:

The open-loop transfer function

$$G(s)\ H(s) = \frac{2e^{-sT}}{s(1+s)(1+0.5s)}$$

Now, Let $\quad L(j\omega) = 4(j\omega)e^{-j\omega T}$

where, $\quad (4\,j\omega) = \dfrac{4}{(j\omega)(j\omega+1)(j\omega+2)}$

Nyquist plot of $4(j\omega)e^{-j\omega T}$ is shown below.

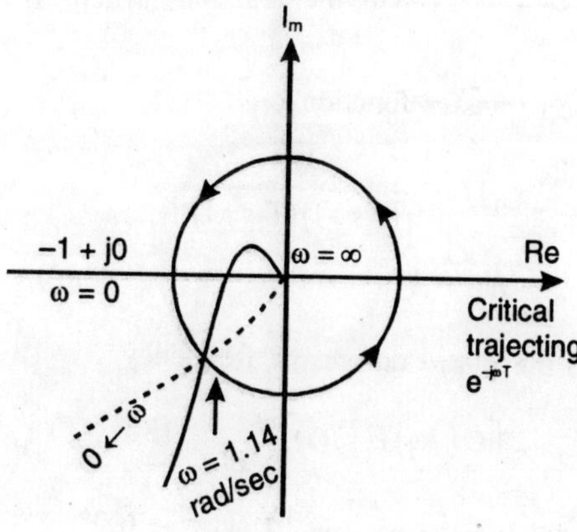

Fig. P. 10.15

The frequency at which $4(j\omega)$ will cut the trajectory.

$e^{-j\omega T}$ can be formed by

$$|4(j\omega)| = 1$$

$$\Rightarrow \qquad 4 = \sqrt{\omega(\omega^2+1)(\omega^2+4)}$$

Solving this equation, we get

$$\omega = 1.14 \text{ rad/sec.}$$

Now, at the point of intersection,

$$\angle 4(j\omega)\big|_{\omega=1.14} = 1.14T + \pi \text{ rad}$$

$$\Rightarrow \quad \pi + \tan^{-1}\left(\frac{2-\omega^2}{3\omega}\right)\bigg|_{\omega=1.14} = \pi + 1.14T \text{ rad}$$

$$\Rightarrow \quad T = 0.17 \text{ sec.}$$

Problem 10.16. The open-loop transfer function of a unity feedback system is given by

$$G(s) = \frac{K}{s(T_1 s + 1)(T_2 s + 1)}$$

Derive an expression for gain K in terms of $T_1 T_2$ and specified gain margin GM.

Solution:

Given open-loop transfer function,

$$G(s) = \frac{K}{s(T_1 s + 1)(T_2 s + 1)}$$

and

$$H(s) = 1$$

Let $\omega = \omega_1$ be the phase cross-over frequency

then,

$$\angle\left[G(j\omega)H(j\omega)\right]_{\omega=\omega_1} = 180°$$

$$\Rightarrow \quad -90° - \tan^{-1} T_1\omega_1 - \tan^{-1} T_2\omega_1 = -180°$$

$$\Rightarrow \quad \tan^{-1}\frac{(T_1+T_2)\omega_1}{1-T_1T_2\omega_1^2} = 90°$$

$$\Rightarrow \quad \omega_1 = \frac{1}{\sqrt{T_1T_2}}$$

Now gain-margine

$$GM = \frac{1}{|G(j\omega)H(j\omega)|_{\omega=\omega_1}}$$

$$GM = \cfrac{1}{K/\cfrac{1}{\sqrt{T_1 T_2}}\left[\sqrt{\dfrac{T_1}{T_2}}+\sqrt{\dfrac{T_2}{T_1}}\right]}$$

$$\frac{1}{GM} = \frac{K(T_1 T_2)}{T_1 + T_2}$$

$$\Rightarrow \qquad K = \frac{1}{GM}\cdot\frac{T_1+T_2}{T_1 T_2} = \frac{1}{GM}\left(\frac{1}{T_1}+\frac{1}{T_2}\right)$$

Problem 10.17. The straight line Bode plot of a feedback system is shown in Fig. P. 10.17 Derive an expression for the value of ω_c to yield maximum phase margin, in terms of system constants ω_1, ω_2. Determine the maximum phase margin when $m = 1$.

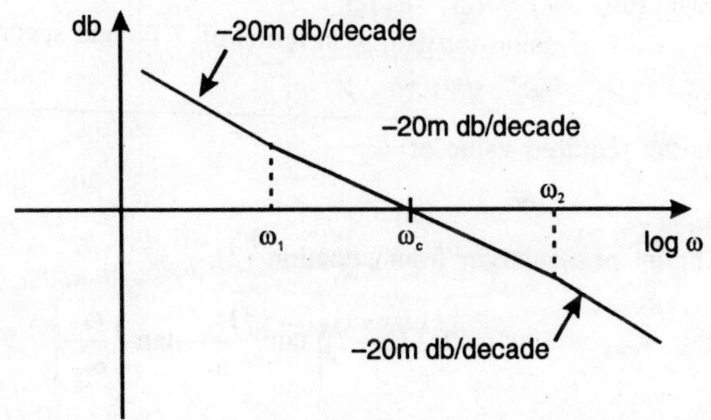

Fig. P. 10.17

Solution:

From the given figure:

At
$$\omega = \omega_c\,;$$

$$20\log G(j\omega)\big|_{\omega=\omega_c} = 0$$

or,
$$G(j\omega)\big|_{\omega=\omega_c} = 1$$

Again, the open loop T.F. from the given figure:

$$G(s) = \frac{K(s+\omega_1)^{m-1}}{s^m (s+\omega_2)^{m-1}}$$

At $\omega=\omega_c$, $\angle G(j\omega)=-m\,90°+(m-1)\tan^{-1}\left(\dfrac{\omega_c}{\omega_1}\right)-(m-1)\tan^{-1}\left(\dfrac{\omega_c}{\omega_2}\right)$

$$\tag{1}$$

then phase Margin,

$$\phi = \angle G(j\omega)+180°$$

$$= 180°-90°\,m(m-1)\left(\tan^{-1}\dfrac{\omega_c}{\omega_1}-\tan^{-1}\dfrac{\omega_c}{\omega_2}\right) \tag{2}$$

ϕ will be maximum when,

$$\frac{d\phi}{d\omega_c} \doteq 0=(m-1)\left[\frac{\omega_1}{\omega_1^2+\omega_c^2}-\frac{\omega_2}{\omega_2^2+\omega_c^2}\right]$$

or $\qquad \omega_1\omega_2\left(\omega_2-\omega_1\right) = \left(\omega_2-\omega_1\right)\omega_c^2$

$$\therefore \qquad \omega_c = \sqrt{\omega_2\,\omega_1}$$

Which is the required value of ω_c

when $\qquad\qquad m = 1$

the maximum phasemargin from equation (2),

$$\phi_{\max} = 180°-90°+(m-1)\left[\tan^{-1}\dfrac{\omega_c}{\omega_1}-\tan^{-1}\dfrac{\omega_c}{\omega_2}\right]= 90°$$

Problem 10.18.   A unity feedback system has the following open-loop frequency response.

ω	2	3	4	5	6	8	10
$\lvert G(j\omega)\rvert$	7.5	4.8	3.15	2.25	1.70	1.00	0.64
$\angle G(j\omega)$	$-118°$	$-130°$	$-140°$	$-150°$	$-157°$	$-170°$	$-180°$

(a) Evaluate the gain margin and phase margin of the system.

(b) Determine the change in gain required so that the phase margin of the system is 60 db.

(c) Determine the change in gain required so that the phase margin of the system is 60 deg.

Solution:

(a) From the frequency response

At $\omega_1 = 10$ rad/sec

$$\angle G(j\omega) = -180°$$

It means that $\omega_1 = 10$ rad/sec is phase cross-over frequency.

$$\text{gain margin} = 20 \log (1/a)$$

where $a = |G(j\omega)|_{\omega=10\,\text{rad/sec}} = 0.64$

Therefore $GM = 20 \log (1/0.64)$
$$= 3.87 \text{ dB}$$

Now Gain cross over frequency,

$$\omega_2 = 8 \text{ rad/sec}$$

At $\omega = \angle G(j\omega) = -170°$

Then phase margin,

$$\phi = +180° + \angle G(j\omega)|_{\omega=\omega_2}$$
$$= 180° - 170° = 10°$$

(b) Now, for a GM of 20 dB, the Nyquist plot should intersect the real axis where,

$$20 \log 1/a = 200$$

$$\therefore \qquad a = 0.1$$

This can be achived if the system gain is changed by a factor

$$\frac{0.1}{0.64} = 0.156$$

(c) Now, let at $\omega = \omega_3$, the phase margin, i.e.,

$$60° = \angle G(j\omega)|_{\omega=\omega_3} + 180°$$

or $\angle G(j\omega)|_{\omega=\omega_3} = -120°$

From frequency response table given,

At $\omega_3 = 2.1$ rad/sec

$$\angle G(j\omega) = -120° \text{ (approx.)}$$

and $|G(j\omega)| = 6.8$ (approx.)

Then, the change in gain should be decreased by a factor of

$$\frac{1}{6.8} = 0.147$$

Problem 10.19. Sketch the inverse Nyquist plot of a feedback system characterized by the open-loop transfer function

$$G(s) = \frac{K}{s(1+0.1s)(1+s)}$$

Find the value of M_r, for $K = 1$. By what factor should the gain K be changed so that M_r is 1.4? Determine the value of ω_r for the new setting of gain.

Solution:

Given O.L. T.F., $G(s) = \dfrac{K}{s(1+0.1s)(1+s)}$

Then, $\qquad \dfrac{1}{G(j\omega)} = \dfrac{1}{K} j\omega(1+0.1\,j\omega)(1+j\omega)$

$$= \frac{1}{K}\left[-1.1\omega^2 + j\omega\left(1-\omega^2\right)\right]$$

The, $\dfrac{1}{G(j\omega)}$; locus intersect the real axis,

where $\qquad\qquad\qquad \angle\dfrac{1}{G(j\omega)} = -180°$

or, $\qquad 90° + \tan^{-1}(\omega) + \tan^{-1}(0.1\omega) = -180°$

or, $\qquad\qquad \tan^{-1}\left(\dfrac{1.1\omega}{1-0.1\omega^2}\right) = -270°$

Which yields, $\qquad\qquad\qquad \omega = \sqrt{10}$ rad/sec.

Now, the mapping of Nyquist contour is shown in fig. 10.19 (a) is obtained as follows:

(i) Semicircular indent around the origin is given as $s = \lim\limits_{t \to 0} te^{jQ}$

(where Q varies from $-90°$ through $0°$ to $+90°$) is mappedinto,

$$\lim\limits_{t \to 0} te^{jQ}\left(1+0.1te^{jQ}\right)\left(1+te^{jQ}\right) = 0(\angle-90°+0° \text{ to} \angle 90°)$$

(ii) The infinite semicircle of Nyquist contours given as $s = \lim\limits_{R \to \infty} Re^{j\phi}$

(ϕ varies from $+90°$ through $0°$ to $-90°$) is Mapped into,

$$\lim_{R \to \infty} \mathrm{Re}^{j\phi}\left(1+0.1\mathrm{Re}^{j\phi}\right)\left(1+\mathrm{Re}^{j\phi}\right)$$

$$= \infty\left(\angle 270° \to \angle 0° \to \angle -270°\right)$$

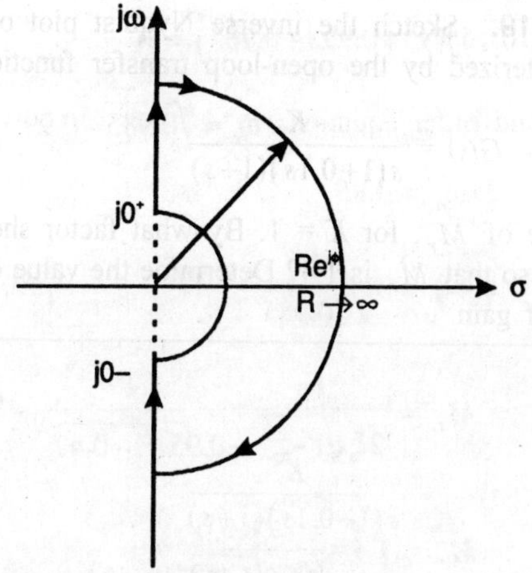

Fig. P. 10.19 (a)

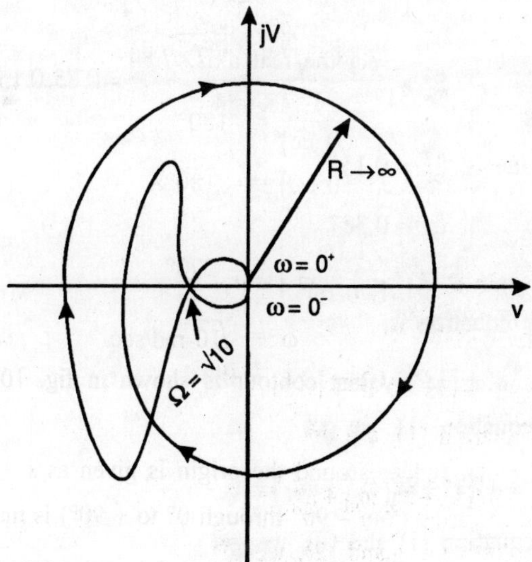

Fig. P. 10.19 (b)

The complete mapped plot corresponding to Nyquist contour is shown in Fig. 10.19 (b)

Characteristic equation is,

$$s(1+0.1s)(1+s)+1 = 0$$

$$\Rightarrow \quad s^3+11s^2+10s+10 = 0$$

$$(s+10.15)(s^2+0.95s+0.985) = 0$$

From the second order equation, $\omega_n = \sqrt{0.985} = 0.99$

$$\Rightarrow \quad 2r\omega_n = 0.95$$

$$\Rightarrow \quad \xi = \frac{0.95}{2\sqrt{0.985}} = 0.47$$

Then, $\quad M_r = \dfrac{1}{2\xi\sqrt{1-\xi^2}} = \dfrac{1}{0.95\sqrt{1-0.47}^2} = 1.2$

when $\quad M_r = 1.4 = \dfrac{1}{2\xi_1\sqrt{1-\xi_1^2}}$

$$\Rightarrow \quad 7.84\xi_1^4 - 7.84\xi_1^2 + 1 = 0$$

$$\therefore \quad \xi_1^2 = \frac{7.84 \pm \sqrt{7.84^2 - 7\times7.84}}{2\times7.84} = 0.85, 0.15$$

Accepted value is, $\xi_1^2 = 0.15$

$$\therefore \quad \xi_1 = 0.387$$

Let the new gain is K_1 and new natural frequency be ω_{n1}. Then, the characteristic equation is,

$$s^3+11s^2+10s+10K_1 = 0 \tag{1}$$

Factorising equation (1), we get

$$(s+a)(s^2+2\xi_1\omega_{n1}+\omega_{n1}^2) = 0 \tag{2}$$

Comparing equation (1) and (2), we get

$$a\omega_{n1}^2 = 10K_1 \tag{3}$$

$$2a\xi_1\omega_{n1} + \omega_{n1}^2 = 10 \tag{4}$$

$$a + 2\xi_1 \omega_{n1} = 11 \qquad (5)$$

From equations (4) and (5)

$$\omega_{n1}^2 + 2\xi_1 \omega_{n1}\left(11 - 2\xi_1 \omega_{n1}\right) = 10$$

Putting $\xi_1 = 0.38$, we get

$$0.4\omega_{n1}^2 + 8.5\omega_{n1} - 10 = 0$$

which gives, $\omega_{n1} = 1.10$ rad/sec

Then $K_1 = 1.3$

Thus, gain changed by a factor $\dfrac{1}{1.3} = 0.76\,(\text{approx.})$

and ω_n (resonant frequency) $\omega_n\sqrt{1 - 2\xi^2} = 1.05$ rad/sec (approx.)

Problem 10.20. Draw the polar plots of transfer functions given below. Determine whether these plots cross the real axis. If so, determine the frequency at which the plots cross the real axis and the corresponding magnitude $|G(j\omega)|$

(a) $$G(s) = \frac{1}{(1+s)(1+2s)}$$

(b) $$G(s) = \frac{1}{s(1+s)(1+2s)}$$

(c) $$G(s) = \frac{1}{s^2(1+s)(1+2s)}$$

Solution:

(a) $$G(s) = \frac{1}{(1+s)(1+2s)}$$

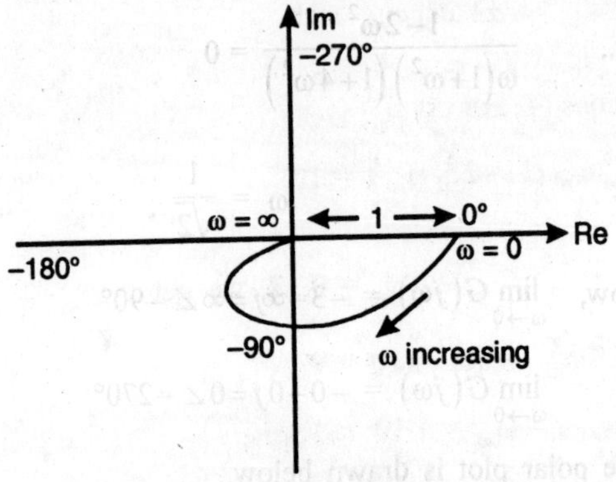

Fig. P. 10.20

$$G(j\omega) = \frac{1}{(1+j\omega)(1+2j\omega)} = \frac{1-2\omega^2}{(1+\omega^2)(1+4\omega^2)} - j\frac{3\omega}{(1+\omega^2)(1+4\omega^2)}$$

Now,
$$\lim_{\omega \to 0} G(j\omega) = 1 - 0j = 1\angle 0°$$

$$\lim_{\omega \to 0} G(j\omega) = -0 - 0j = 0\angle -180°$$

The plot of $G(j\omega)$ will cross the real axis where, Imaginary part of $G(j\omega)$ will be zero.

i.e.,
$$\omega = 0$$

It means, $G(j\omega)$ will not cross real axis. The polar plot of $G(j\omega)$ is drawn in figure.

(b)
$$G(s) = \frac{1}{s(1+s)(1+2s)}$$

$$G(j\omega) = \frac{1}{j\omega(1+j\omega)(1+2j\omega)}$$

$$= \frac{-j(1-2\omega^2)}{\omega[1+\omega^2][1+4\omega^2]} - \frac{3}{(1+\omega^2)(1+4\omega^2)}$$

$G(j\omega)$ will cross real axis, at the point where, imaginary part is zero.

i.e., $$\frac{1-2\omega^2}{\omega\left(1+\omega^2\right)\left(1+4\omega^2\right)} = 0$$

$$\Rightarrow \qquad \omega = \frac{1}{\sqrt{2}}$$

Now, $\lim\limits_{\omega\to 0} G(j\omega) = -3-\infty j=\infty\,\angle-90°$

$$\lim\limits_{\omega\to 0} G(j\omega) = -0-0j=0\,\angle-270°$$

The polar plot is drawn below.

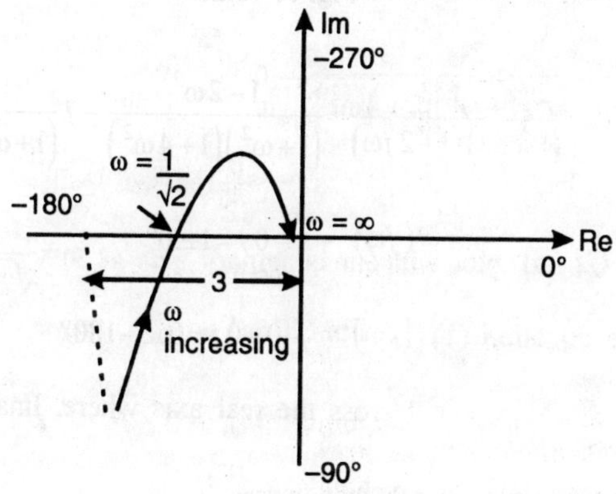

Fig. P. 10.20 (a)

Now, $|G(j\omega)|_{\omega=\frac{1}{\sqrt{2}}} = \left|\dfrac{1}{0.71(1+0.71j)(1+0.42j)}\right|$

$$= \frac{1}{(0.71)(1.23)(1.73)} = 0.67$$

(c) $\qquad G(s) = \dfrac{1}{s^2(1+s)(1+2s)}$

$$= \frac{1}{(j\omega)^2(1+j\omega)(1+2j\omega)} = -\frac{1-2\omega^2-3j\omega}{\omega^2\left(1+\omega^2\right)\left(1+4\omega^2\right)}$$

$$= \frac{2\omega^2 - 1}{\omega^2\left[1+\omega^2\right]\left[1+4\omega^2\right]} + j\frac{3\omega}{\omega^2\left(1+\omega^2\right)\left(1+4\omega^2\right)}$$

$$= \frac{2\omega^2 - 1}{\omega^2\left(1+\omega^2\right)\left(1+4\omega^2\right)} + j\frac{1}{\omega\left(1+\omega^2\right)\left(1+4\omega^2\right)} \tag{1}$$

$G(j\omega)$ plot will cross real axis where imaginary part being zero.

i.e.,

$$\frac{1}{\omega\left(1+\omega^2\right)\left(1+4\omega^2\right)} = 0$$

$$\Rightarrow \qquad\qquad\qquad \omega = \infty$$

If implies that polar plot will not cross real axis. Now equating real part to zero, we have,

$$\frac{2\omega^2 - 1}{\omega^2\left(1+\omega^2\right)\left(1+4\omega^2\right)} = 0$$

$$\Rightarrow \qquad\qquad\qquad \omega = \frac{1}{\sqrt{2}} \text{ rad/sec.}$$

i.e., $G(j\omega)$ plot will cut imaginary axis at $\omega = \dfrac{1}{\sqrt{2}}$ rad/sec.

From equation (1) $\lim\limits_{\omega \to 0} G(j\omega) = \infty\angle{-180°}$

or $\lim\limits_{\omega \to 0} G(j\omega) = 0\angle{0°}$

The polar plot is sketched below.

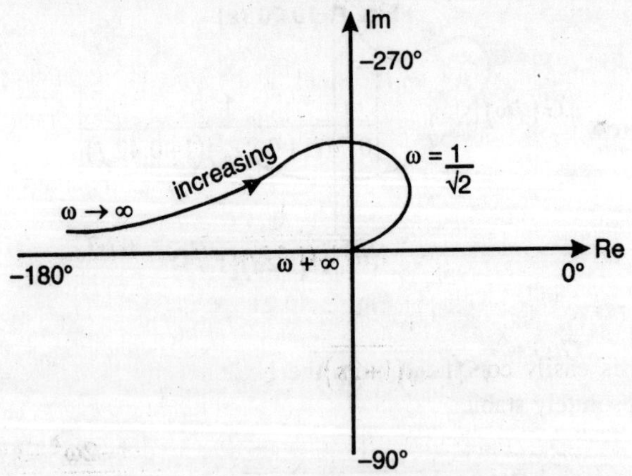

Fig. P. 10.20 (b)

Problem 10.21. Draw the Nyquist plot and determine therefrom the stability of the following open-loop transfer function of unity feedback control systems.

(a)
$$GH(s) = \frac{K(s+2)}{s^2(s+1)}$$

(b)
$$GH(s) = \frac{K}{s(s^2+s+4)}$$

If the system is conditionally stable, find the range of K for which the system is stable.

Solution:

(a)
$$GH(j\omega) = \frac{(K/2)(1+j\omega/2)}{(j\omega)^2(1+j\omega/4)}$$

The Nyquist plot is drawn in figure below.

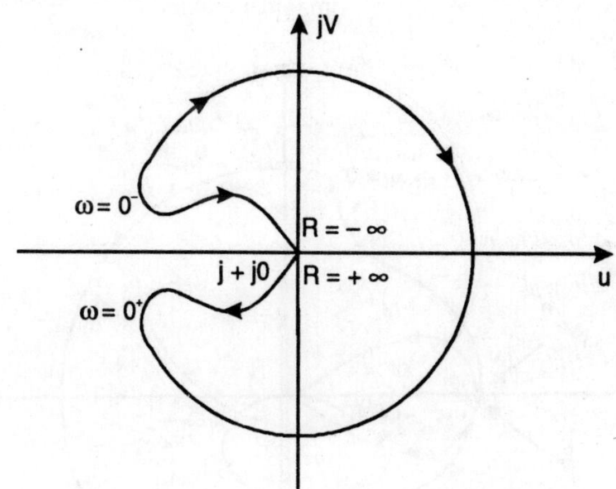

Fig. P. 10.21

It is easily concluded from the Nyquist plot that the system is absolutely stable.

(b)
$$GH(j\omega) = \frac{K}{j\omega(4-\omega^2+j\omega)}$$

Nyquist plot would cross the real axis at

$$GH(j\omega) = \frac{-K\left[\omega + j(4-\omega)^2\right]}{\omega\left[\omega^2 + \left(4-\omega^2\right)^2\right]}$$

Equating imaginary part to zero, we have

$$4 - \omega^2 = 0$$

or

$$\omega^2 = 4$$

$$\omega = \pm 2$$

$$\left|GH(j\omega)\right|_{\omega^2=4} = \frac{-K}{4}$$

So the Nyquist Plot crosses the real axis at $\omega = \pm 2$ with an

intercept of $\dfrac{-K}{4}$.

The plot is drawn below.

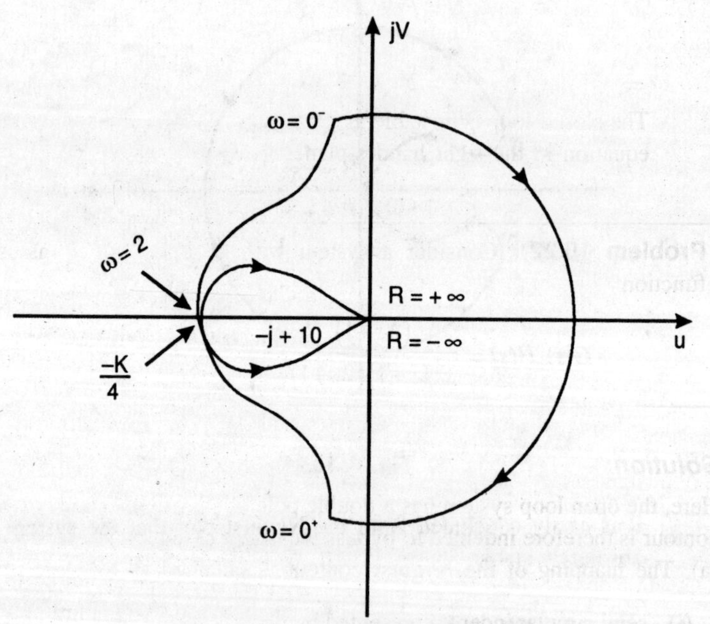

Fig. P. 10.21 (a)

From the Nyquist plot $\dfrac{K}{4} > 1$ or $K > 4$

From this value of K,

$$N = P - Z$$
$$-2 = 0 - Z$$

or

$$Z = 2$$

So, the system is unstable. It can be shown to be stable for $K < 4$

For $K > 1$, the Nyquist plot encircles $(1 + j0)$ one in anti-clockwise direction i.e., $N = 1$.

The open-loop function has $P = 1$

then

$$N = P - Z$$
$$1 = 1 - Z$$

$\Rightarrow$

$$Z = 0$$

Thus, the system is stable.

For $K < 1$, we can easily see that $(-1 + j\,0)$ will lie beyond $-K$. The crossing point of the plot.

$$N = -1,$$
$$P = 1$$
$$N = P - Z$$
$$-1 = 1 - Z$$

$\Rightarrow$

$$Z = 2$$

The closed-loop is unstable as it has two roots in its characteristic equation in the right hand s-plane.

Problem 10.22. Consider a system with an open-loop transfer function

$$G(s)\,H(s) = \frac{(4s+1)}{s^2\,(s+1)(2s+1)}$$

Solution:

Here, the open loop system has a double pole at the origin. The Nyquist contour is therefore indented to bypass the origin as shown in Fig. 10.22 (a). The mapping of the Nyquist contour is obtained as follows:

(i) semi circular indent represented by $s = \lim\limits_{\varepsilon \to 0} \varepsilon e^{j2\theta}$ where θ varies from $-90°$ through $(0° + 90°)$ is mapped into

$$\lim_{\varepsilon \to 0}\left[\frac{4\varepsilon e^{j\theta}+1}{\varepsilon^2 e^{j2\theta}\left(\varepsilon e^{j\theta}+1\right)\left(2\varepsilon e^{j\theta}+1\right)}\right] = \lim_{\varepsilon \to 0}\left(\frac{1}{\varepsilon^2 e^{j2\theta}}\right)$$

$$= \infty e^{-j2\theta}$$

$$= \infty \left(\angle 180° \to 0° \to \angle -180°\right)$$

This part of the Map is an infinite circle shown in fig. 10.22 (b)

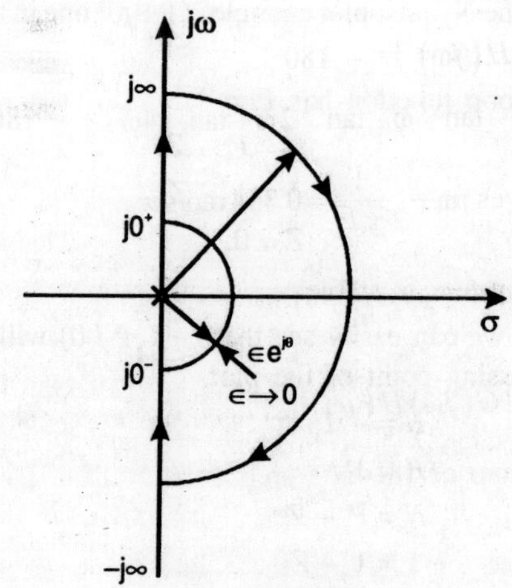

Fig. P. 10.22 (a)

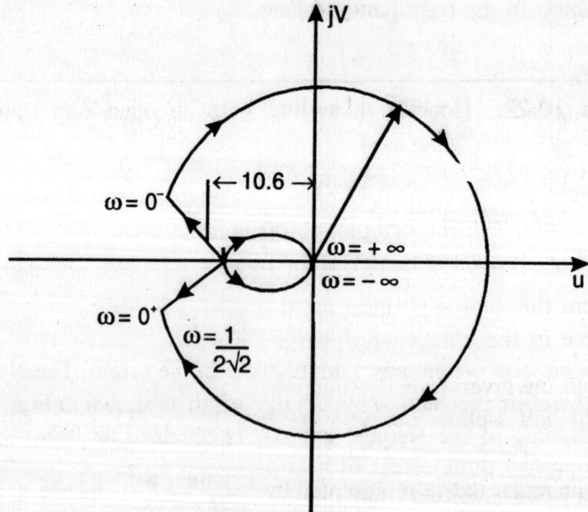

Fig. P. 10.22 (b)

(ii) Along the $j\omega$–axis

$$G(j\omega)H(j\omega) = \frac{(1+4\,j\omega)}{(j\omega)^2(1+j\omega)(1+j\,2\omega)}$$

For various values of ω, $G(j\omega)H(j\omega)$ is calculated and plotted as shown in figure 10.22 (b).

The $G(j\omega)H(j\omega)$ – locus intersect the real axis at a point where

$$\angle G(j\omega)H(j\omega) = -180°$$

or $-180° - \tan^{-1}\omega - \tan^{-1}2\omega + \tan^{-1}4\omega = -180°$

which gives $\omega = \dfrac{1}{2\sqrt{2}} = 0.354\,\text{rad/sec.}$

Therefore $|G(j\omega)H(j\omega)|_{\omega=\frac{1}{2\sqrt{2}}} = 10.6$

Further $|G(j\omega)H(j\omega)|_{\omega\to+\alpha} = 0\angle-270°$

i.e., the map of $j\omega$ -axis ends at $0\angle-270°$ $\omega\to+\infty$.

(iii) The infinite semicircle of the Nyquist contour represented by s
$= \lim\limits_{R\to\infty} \text{Re}^{5\phi}$ (ϕ varies from $+90°$ through $0°$ to $-90°$) is mapped into

$$\lim\limits_{R\to\infty} \frac{\left(1+4\,\text{Re}^{j\phi}\right)}{R^2\,e^{j2\phi}\left(1+\text{Re}^{j\phi}\right)\left(1+2\,\text{Re}^{j\phi}\right)} = 0e^{-J3\phi}$$

$$= 0\left(\angle-270°\to\angle0°\to\angle+270°\right)$$

The complete Mapped plot corresponding to the Nyquist contour of Fig. 10.22 (a) is shown in figure 10.22 (b).

From this plot it is observed that $(-1+j\,0)$ point is encircled twice in the clockwise direction. Therefore $N = -2$

From the given transfer function no pole to *G(s) H(s)* lies in the right half s-plane i.e., $P = 0$, Thus $-2 = 0 - Z$ or $Z = 2$

Hence two three errors of *q(s)* lie in the right half s-plane from which we conclude that the system is unstable.

Problem 10.23. The block diagram is shown as below in figure

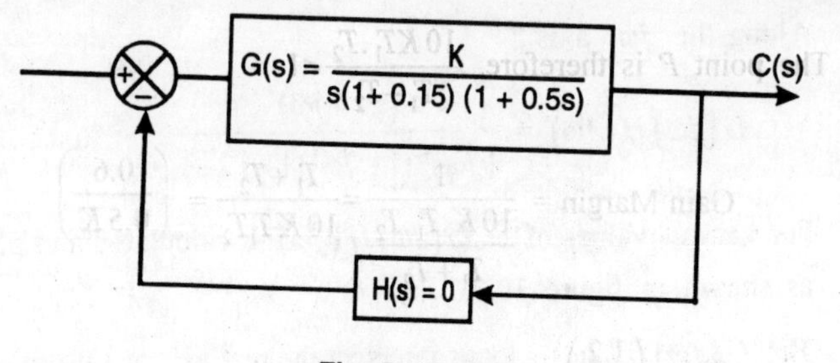

Fig. P. 10.23

(a) Draw the Nyquist plot for $0 < \omega < \infty$

(b) Find gain margin and phase cross over frequency

(c) Find phase margin of the system for K = 1.2

Solution:

Given

$$G(s) = \frac{K}{s(1+0.1s)(1+0.5s)}$$

and

$$H(s) = 10$$

Therefore

$$GH = \frac{10K}{s(1+0.1s)(1+0.5s)}$$

Here we can find T_1 as 0.1

and $T_2 = 0.5$

(a) The Nyquist plot is drawn as below:

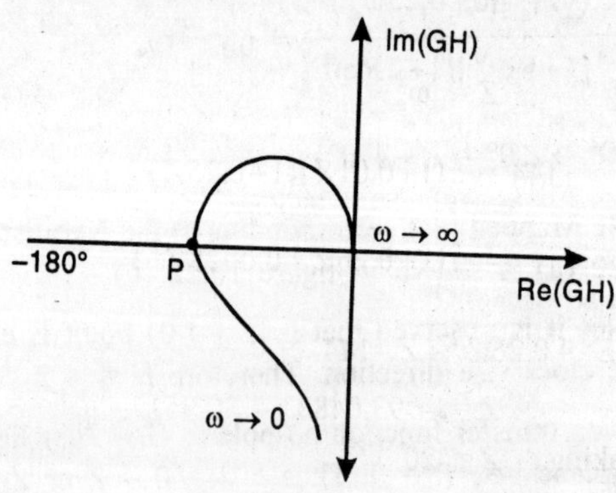

Fig. P. 10.23 (a)

(b)   The point P is therefore, $\dfrac{10\,KT_1.T_2}{T_1+T_2}<1$

$\therefore \qquad$ Gain Margin $= \dfrac{1}{\dfrac{10\,K.T_1.T_2}{T_1+T_2}} = \dfrac{T_1+T_2}{10\,KT_1T_2} = \left(\dfrac{0.6}{0.5\,K}\right)$

Then, $20\log\left(\dfrac{1.2}{K}\right) = 1.584 - 20\ \log\ K$

Now, the phase cross over frequency,

$$\omega_p = \dfrac{1}{\sqrt{T_1.T_2}} = \dfrac{1}{\sqrt{0.1\times0.5}} = 4.472\ \text{rad/sec.}$$

(c)   The gain cross over frequency ω_g is given by,

$$|GH(j\omega_g)| = 1$$

For $\qquad\qquad K = 1.2,$

$$GH(j\omega) = \dfrac{12}{j\omega(1+0.1\,j\omega)(1+0.5\,j\omega)}$$

$$|GH(j\omega_g)| = \dfrac{12}{\omega_g\sqrt{1+0.01\omega_g^2}\,.\sqrt{1+0.25\,\omega_g^2}} = 1$$

or $\quad \omega_g^2\left(1+0.01\omega_g^2\right)\left(1+0.25\,\omega_g^2\right) = 144$

Now putting $\qquad Z = \omega_g^2$

or $\qquad\qquad 144 = z(1+0.01Z)(1+0.25Z)$

$$= z\left(1+0.26\,Z+0.0025\,Z^2\right)$$

or $\quad 0.0025\,Z^3 + 0.26\,Z^2 + Z - 144 = 0$

Then, $\qquad\qquad Z = -\,93.0483,\ -\,30.9517,\ 20$

Therefore taking $\quad Z = 20$

and then, $\qquad \omega_p = \sqrt{Z} = 4.472$

$\therefore$ Phase margin, $P_m = \angle GH(j\omega_g) + 180°$

$$= 0 - 90° - \tan^{-1}(0.1\omega_g) - \tan^{-1}(0.5)\omega_g + 180°$$

$$= 90° - \tan^{-1}(0.4472) - \tan^{-1}(2.236)$$

$$= 90° - 24.1° - 65.9° = 0°$$

Problem 10.24. A unity feedback system characterised by the following open loop T.F.

$$G(s) = \frac{20}{s(s+1)(s+5)}$$

Apply the Nyquist cirterion for stability and indicate whether the above system is stable or not. Draw also the approx. Nyquist plot.

Solution:

Suppose the ω_1 be the phase cross-over frequency at which

$$\angle G(j\omega_1) = 180° = \angle \frac{20}{j\omega_1(j\omega_1+1)(j\omega_1+5)}$$

$$= -90° - \tan^{-1}\omega_1 - \tan^{-1}\left(\frac{\omega_1}{5}\right)$$

or $\qquad \tan^{-1}\omega_1 - \tan^{-1}\left(\frac{\omega_1}{5}\right) = 90°$

NNow $\qquad \tan^{-1}\left(\frac{\omega_1 + \omega_1/5}{1-\omega_1^2/5}\right) = 90°$

or, $\qquad \dfrac{\omega_1 + \omega_1/5}{1-\omega_1^2/5} = \infty$

This gives the result as; $1-\omega_1^2/5 = 0$

Therefore, $\qquad \omega_1 = \sqrt{5}$ rad/sec.

Then,   Magnitude,

$$|G(j\omega_1)| = \frac{20}{\omega_1\sqrt{1+\omega_1^2}\cdot\sqrt{5+\omega_1^2}} = \frac{20}{\sqrt{5}\cdot\sqrt{6}\cdot\sqrt{30}} = 0.667$$

Hence the system is stable.

The Nyquist plot is drawn as under,

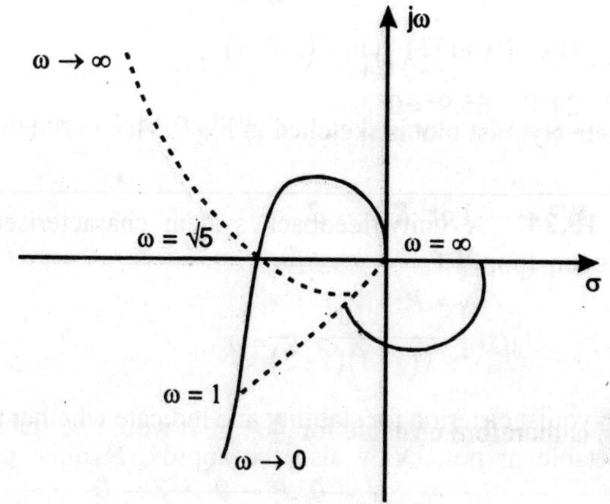

Fig. P. 10.24

Problem 10.25. Consider now a system with open-loop zero in right-half *s*-plane. Let

$$G(s)\,H(s) = \frac{K(s-2)}{(s+1)^2}$$

Investing in a Nyquist plot of the closed-loop system (unit feedback)

Solution:

This case corresponds to unidented Nyquist contour of Fig.P 10.25 Nyquist plot is obtained in the following steps:

$$j\omega \Rightarrow j0 \to \infty$$

$$G(s)H(s)\big|_{j\omega} = \frac{K(j\omega-2)}{(j\omega+1)^2}$$

If $j\omega = j0 = -2K$

If $j\omega = +j\infty = 0\angle -90°$

Crossing the axis of reals

$$G(j\omega)H(j\omega) = \frac{K\left[\left(4\omega^2-2\right)+j\omega\left(5-\omega^2\right)\right]}{\left(\omega^2+1\right)\left(\omega^2+1\right)^2}$$

At the crossing, imaginary part is zero i.e.,

$$s - \omega^2 = 0 \text{ or } \omega^2 = 5 \text{ or } \omega = \pm\sqrt{5}$$

At this value of ω $\left|G(j\omega)H(j\omega)\right|_{\omega^2=5} = K/2$

The complete Nyquist plot is sketched in Fig 9.14, examination of which reveals

For $\qquad K/2 > -1 \text{ or } K > -2$

$$N = -1, \ P = 0,$$
$$N = P - Z$$
$$-1 = 0 - Z \text{ or } Z = 1$$

The system is therefore unstable for $K > \dfrac{1}{2}$. It would be stable for $K < \dfrac{1}{2}$

when $\qquad\qquad\qquad N - 0, \ P - 0 > Z - 0$

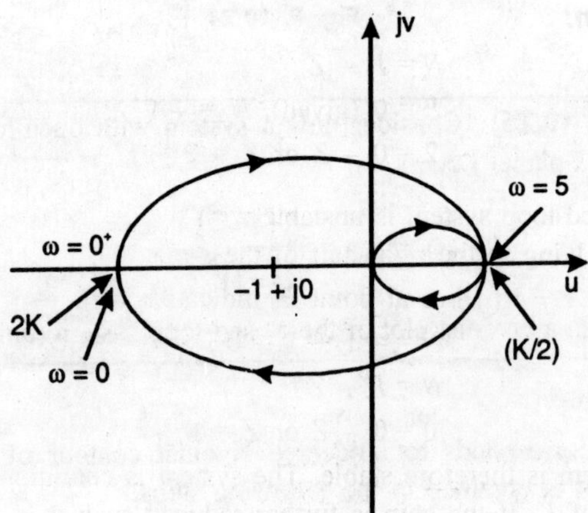

Fig. P. 10.25

Problem 10.26. The Nyquist plot of a system is shown in Fig. P 10.26. for a particualr value of gain of the open-loop system. The open-loop system has no right-hand s-plane poles. In the closed-loop (unity feedback) system stable? If the answer is no, determine of right-half s-plane zeros of the characteristic equation.

The systemgain is now adjusted such that the point (−1 + j0) lies at point P. Note that this would correspond to a new scale for the same figure. Determine one again the stability of the system. In the process has the gain been increased or decreased.

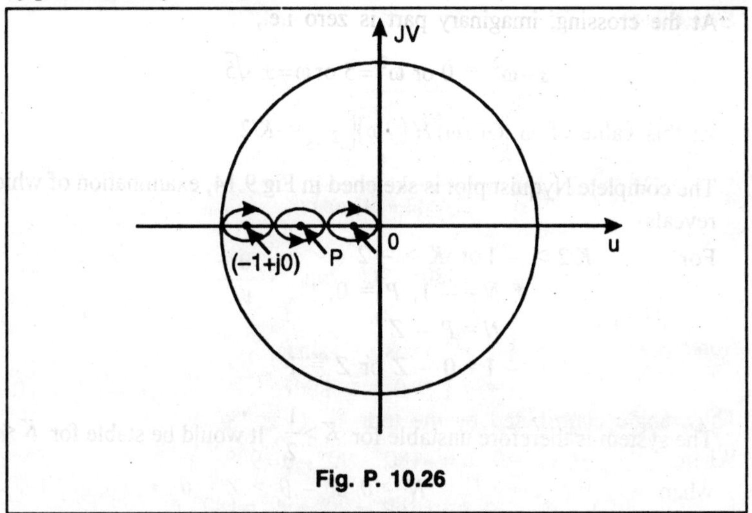

Fig. P. 10.26

Solution:

$$N = P - Z$$

$$P = 0 \text{ (given)}, N = -2, \text{ from the Nyquist plot}$$

Then $\qquad -2 = 0 - Z \text{ or } Z = 2$

The closed-loop system is unstable with two roots of the characteristic equation lying in the right half of the s-plane.

When $(-1 + j\,0)$ lies at point P, indicated, then

$$N = -1 + 1 = 0$$

$$N = P - Z$$

$$0 = 0 - Z \text{ or } Z = 0$$

The system is therefore stable. The system is conditionally stable for a range of K. If the gain is further reduced such that $(-1 + j\,0)$ lies inside the right most loop, it can easily be seen that the system would be stable.

$(-1 + j\,0)$ shifts to point P by reducing gain.

Problem 10.27. The open loop transfer function of a feedback control system is:

$$G(s)\ H(s) = \frac{-1}{2s(1 - 20s)}$$

Comment on the stability.

Solution:

$$G(s)\ H(s) = \frac{-1}{2\,s\,(1-20\,s)}$$

$$G(j\omega)\,H(j\omega) = \frac{-1}{2\,j\omega\,(1-20\,j\omega)}$$

$$\phi = +180° - 90° - \tan^{-1}\frac{-20\,\omega}{1}$$

Note:
$$- 1 = e^{j\pi} = \cos\pi + j\sin\pi$$
$$= -1 + 0 = -1$$

Thus, angle contributed by the term (-1) is $180°$

When
$$\omega = 0,\ \phi = +180° - 90° - 0 = 90°$$
$$\omega = \infty,\ \phi = +180° - 90° + 90° = 180°$$
$$\omega = 0.1,\ \phi = +180° - 90° + 63.43° = 153.43°$$
$$\omega = 0.0001,\ \phi = +180° - 90° + 0.115° = 90.115°$$

Thus, we see that when ω is very small, ϕ never becomes equal to $90°$. It is always slightly more than $90°$.

The Nyquist Plot is shown in the fig. below:

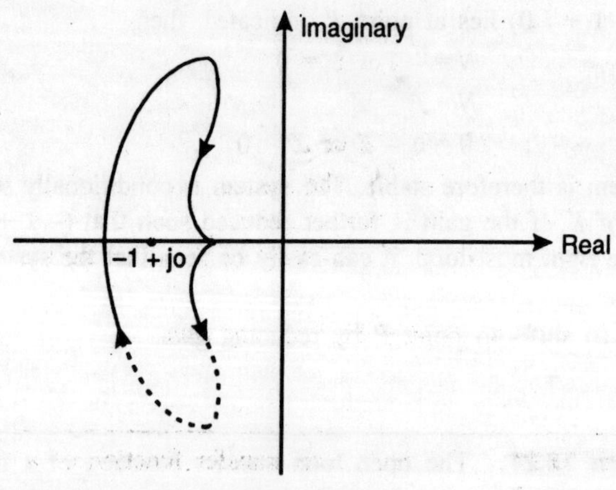

Fig. P. 10.27

The point $(-1 + j\,0)$ will always lie inside. There is one clockwise encirclement.

Hence, $N = -1$

Also, $P = 1$, because one open loop pole is lying on the RHS.

$$N = P - Z$$
$$-1 = 1 - Z$$
$$\therefore \qquad Z = 2$$

$\therefore$ Closed loop system is unstable.

Open loop system is also unstable as $P = 1$.

Reader's Notes

Reader's Notes